Recentering Tourism Geographies in the 'Asian Century'

This book considers what the transition into the Asian Century means for some of the most urgent issues in the world today, such as sustainable development, human rights, gender equality, and environmental change. The book critiques Anglo-Western centrism in tourism theory and calls on tourism scholars to make radical shifts toward more inclusive epistemology and praxis.

From the British Century of the 1800s to the American Century of the 1900s to the contemporary Asian Century, tourism geographies are deeply entangled in broader shifts in geopolitical power. In the shadow of the COVID-19 pandemic, the significance of shifts in tourism geographies and the themes addressed in this volume are more urgent than ever. That the world faces increasing turmoil is abundantly clear. Yet, amidst the disruption to the everyday, it is hope and compassion, but also political-economic restructuring that is needed to reset the tourism industry in more sustainable, equitable, and ethical directions. In no uncertain terms, the pandemic has forever changed the tourism industry as the world once knew it. This book, therefore, sets out to collectively build on the momentum of the inclusive scholarship that Critical Tourism Studies-Asia Pacific is renowned for, while also asking readers to pause and reflect on the possibilities and challenges of tourism in a post-pandemic Asian Century.

The chapters in this book were originally published as a special issue of the journal, *Tourism Geographies*.

Harng Luh Sin is former Associate Professor at Sun Yat-Sen University and a Visiting Fellow at Singapore Management University. Her work looks at volunteer and responsible tourism, sustainable development, and the critical tourism in Southeast Asia and China. She is the co-founder of the Critical Tourism Studies Asia-Pacific network.

Mary Mostafanezhad is Associate Professor in the Department of Geography and Environment at the University of Hawai'i at Mānoa. Her scholarship is focused on tourism, development, and socio-environmental change. She is Co-Editor-in-Chief of *Tourism Geographies* and the Critical Green Engagements Series of the University of Arizona Press.

Joseph M. Cheer is Co-Editor-in-Chief of *Tourism Geographies*. He is Professor at Center for Tourism Research, Wakayama University, Japan, and holds adjunct appointments at AUT, New Zealand, UCSI University, Malaysia, and Monash University, Australia. Joseph is also a board member of Pacific Asia Travel Association (PATA), the region's main industry body.

Recentering Tourism Geographies in the 'Asian Century'

Edited by

Harng Luh Sin, Mary Mostafanezhad and Joseph M. Cheer

Routledge
Taylor & Francis Group

LONDON AND NEW YORK

First published 2022
by Routledge
4 Park Square, Milton Park, Abingdon, Oxon OX14 4RN

and by Routledge
605 Third Avenue, New York, NY 10158

Routledge is an imprint of the Taylor & Francis Group, an informa business

British Library Cataloguing in Publication Data
A catalogue record for this book is available from the British Library

ISBN: 978-1-032-20828-2 (hbk)
ISBN: 978-1-032-20829-9 (pbk)
ISBN: 978-1-003-26542-9 (ebk)

DOI: 10.4324/9781003265429

Typeset in Myriad Pro
by Newgen Publishing UK

Publisher's Note
The publisher accepts responsibility for any inconsistencies that may have arisen during the conversion of this book from journal articles to book chapters, namely the inclusion of journal terminology.

Disclaimer
Every effort has been made to contact copyright holders for their permission to reprint material in this book. The publishers would be grateful to hear from any copyright holder who is not here acknowledged and will undertake to rectify any errors or omissions in future editions of this book.

Contents

Citation Information

The chapters in this book were originally published in the journal *Tourism Geographies*, volume 23, issue 4 (2021). When citing this material, please use the original page numbering for each article, as follows:

Afterword

Afterword: a critical reckoning with the 'Asian Century' in the shadow of the anthropocene
Tim Oakes
Tourism Geographies, volume 23, issue 4 (2021), pp. 937–943

For any permission-related enquiries please visit:
www.tandfonline.com/page/help/permissions

Notes on Contributors

Kathleen M. Adams is a professorial research associate in the Department of Anthropology and Sociology at SOAS University of London and Professor Emerita in the Department of Anthropology at Loyola University Chicago. Her research addresses tourism, heritage, arts, museums, and identity politics in island Southeast Asia.

T. C. Chang is Associate Professor in the Department of Geography at the National University of Singapore. His research interests include Asian tourism, urban tourism, arts/culture/heritage. He has co-edited two books on Southeast Asian and Asian tourism.

Joseph M. Cheer is Co-Editor-in-Chief of *Tourism Geographies*. He is Professor at Center for Tourism Research, Wakayama University, Japan, and holds adjunct appointments at AUT, New Zealand, UCSI University, Malaysia, and Monash University, Australia. Joseph is also a board member of Pacific Asia Travel Association (PATA), the region's main industry body.

Yinn Shan Cheong is a Masters student in the Department of Geography at the National University of Singapore. Her research is interested in unpacking the place of social relationships in tourism experiences, particularly that of family, friendship, and class relations within the context of Singapore.

Jaeyeon Choe works at the School of Management, Swansea University, UK, and is Visiting Fellow at the School of Hospitality and Tourism, Hue University, Vietnam. Her research revolves around spiritual tourism and sustainable community development in Southeast Asia.

Chris Gibson is Professor of Human Geography and Executive Director of the inter-disciplinary research program, Global Challenges, at the University of Wollongong, Australia. He has published widely on the topics of cultural economy, festivals, and critical tourism research. He is currently Editor-in-Chief of the academic journal, *Australian Geographer*.

Stuart Hayes is a PhD student in the Department of Tourism, University of Otago. His doctoral research focuses on investigating contemporary postgraduate tourism education.

Kumi Kato is Professor at the Faculty of Tourism, Wakayama University, Japan. Her research area includes sustainability, community resilience, and ethics related to tourism, focusing on disaster recovery, slow tourism, and dark tourism.

Gabriel Laeis is a lecturer at IUBH International University, Germany, where he teaches undergraduate and postgraduate courses on hospitality and tourism management. His teaching is largely based on practical experience in the hospitality industry as a chef, waiter, and consultant. He has conducted research projects on agriculture–tourism linkages in South Africa and Fiji.

Claudio Minca is Professor in the Department of History and Cultures at Bologna University, Italy. His research centers on three major themes: the spatialization of (bio)politics; tourism and travel theories of modernity; and the relationship between modern knowledge, space, and landscape in postcolonial geography.

Mary Mostafanezhad is Associate Professor in the Department of Geography and Environment at the University of Hawai'i at Mānoa. Her scholarship is focused on tourism, development, and socio-environmental change. She is the Co-Editor-in-Chief of Tourism Geographies and the Critical Green Engagements Series of the University of Arizona Press.

Tim Oakes is Professor of Geography and Director of the Center for Asian Studies at the University of Colorado Boulder. He has been a visiting professor at the University of Technology Sydney, Guizhou Minzu University, Wageningen University, National University of Singapore, and the University of Hong Kong.

Can Seng Ooi is a sociologist and professor of Cultural and Heritage Tourism at the University of Tasmania. His research in tourism includes policy development, destination branding, sustainability, cultural development, and the distribution of tourism benefits to the community. His works often draw comparative lessons from Denmark, Singapore, China, and, more recently, Australia.

Giang Thi Phi is Assistant Professor at the College of Business & Management, VinUniversity. Her research interest focuses on tourism/event management, design thinking, innovation, entrepreneurship and sustainable development.

Ricardo Nicolas Progano is Lecturer at the Center for Tourism Research of Wakayama University, Japan. His research interests include pilgrimage tourism, heritage management, and cross-cultural studies. He has carried his fieldwork on the recent tourism development of Japanese pilgrimage sites, especially Kumano Kodo.

Maartje Roelofsen is Postdoctoral Researcher in the Faculty of Economics and Business at Universitat Oberta de Catalunya, Barcelona, Spain. Maartje's current research focuses on the digitization of tourism, the datafication of work and life in tourism's platform economies, and algorithmic placemaking in tourism.

Ian Rowen is Assistant Professor of Sociology, Geography and Urban Planning at Nanyang Technological University, Singapore. A Fulbright Scholar (2013–2014), he has written about regional politics, social movements, and tourism. Prior to earning a PhD in Geography from the University of Colorado Boulder, he worked as a tour guide, translator, and journalist in China, Taiwan, and elsewhere.

Harng Luh Sin is former Associate Professor at Sun Yat-Sen University and Visiting Fellow at Singapore Management University. Her work looks at volunteer and responsible tourism, sustainable development, and the critical tourism in Southeast Asia and China. She is the co-founder of the Critical Tourism Studies Asia-Pacific network.

Regina Scheyvens is Professor of Development Studies at Massey University, New Zealand, where she combines a passion for teaching about international development with research on tourism and sustainable development. She has written two books, and articles on themes such as inclusive tourism, backpacker tourism, ecotourism and sustainable tourism. Regina's field research is mainly in the South Pacific.

Nicole Tarulevicz is a senior lecturer in History in the School of Humanities at the University of Tasmania, Australia. She is the author of *Eating Her Curries and Kway: A Cultural History of Food in Singapore* (2013) and a recipient of the ASFS Award for Food Studies Pedagogy (2013). She is the recipient of an Australian Research Council Discovery Grant (2019–2021).

Hazel Tucker is Professor in the Department of Tourism at the University of Otago, New Zealand. Her research focuses on tourism and postcolonialism, tourism encounters, emotion, heritage, and gender.

Carol Xiaoyue Zhang is Assistant Professor in Tourism Marketing at Nottingham University Business School. Carol has research interests predicated on a philosophical perspective investigating the sociopolitical aspects of tourism. Taking a postmodern perspective, she has conducted different identity research in tourism.

Jundan Jasmine Zhang is a postdoctoral researcher in the Department of Ecology at the Swedish University of Agricultural Sciences, Sweden. In the past, she has conducted long-term research in Shangri-La County, Southwest China, adopting a poststructuralist political ecology approach.

INTRODUCTION

Tourism geographies in the 'Asian Century'

Harng Luh Sin (iD), Mary Mostafanezhad and Joseph M. Cheer

ABSTRACT

From the British Century of the 1800s to the American Century of the 1900s to the contemporary Asian Century, tourism geographies are deeply entangled in broader shifts in geopolitical power (Luce, 1999; Scott, 2008; Shenkar, 2006). This paper considers what the transition into the Asian Century means for some of the most urgent issues of our time such as sustainable development, human rights, gender equality, and environmental change. We critique Anglo-Western centrism in tourism theory and call on tourism scholars to make radical shifts toward more inclusive epistemology and praxis. In the shadow of the COVID-19 pandemic, the significance of the themes addressed are more urgent than ever. The pandemic has hastened claims that the Asian century has further accelerated given the contrasting successes of many Asia-Pacific countries, especially as compared to their Euro-American counterparts (Park, 2020). As critical tourism scholars, we are faced with an unprecedented situation, even as the pandemic looks set to become globally endemic and the true extent of its fullest impacts are only beginning to emerge, with more to surface in the years ahead. That the world faces increasing turmoil is abundantly clear. Yet, amidst the disruption to the everyday, it is hope and compassion, but also political-economic restructuring that is needed to reset the tourism industry in more sustainable, equitable, and ethical directions (Cheer, 2020; Lew, Cheer, Haywood, Brouder, Salazar, 2020; Mostafanezhad, 2020). While in no uncertain terms, the pandemic has forever changed the tourism industry as we once knew it, it is our hope that we can collectively build on the momentum of the inclusive scholarship that Critical Tourism Studies-Asia Pacific is renowned for (Edelheim, 2020; Pernecky, 2020) as we pause to reflect on the possibilities and challenges of tourism in a post-pandemic Asian Century.

The 19th century belonged to the United Kingdom, the 20th century to the United States. Many market experts and analysts now speculate that the 21st century will be remembered as the 'Asian Century', dominated by rising superpowers such as Indonesia, India and China (Holmes, 2017).

Coined in 1988 by Deng Xiaoping during his meeting with India's then Prime Minister Rajiv Gandhi, the term 'Asian Century' has become a compelling characterization of the region's prospects in the 21st century. In the recent podcast, 'The Asian Century has Arrived', Oliver Tonby (2019) proclaimed, 'it's not about if and when Asia will rise, but how Asia is going to lead'. Today, the term refers to the economic and political ascendency of Asia as well as to the so-called 'Chinese Century' as a key pillar alongside tremendous growth in the economies of India, South Korea, and the tiger economies of Southeast Asia. Apropos to a wider Asia-directed pivot, global coloniality and the Asian century are intimately intertwined; Asia's rise does not mean coloniality has ended, but rather that neocolonial power is increasingly shared with non-western states (Lee et al. 2015, p. 187). Asia's ascendancy marks a geopolitical reckoning that has intensified since Barrack Obama's 2011 nod to Asia urging more productive relations drawing on common interests (Silove, 2016). For many analysts, the 2016 Trans-Pacific Partnership Agreement (TPP) and the American withdrawal by Donald Trump reflects the declining hegemony of Western states (Beeson, 2020). Today, China's shadow over the region continues to widen the scope of its geopolitical and foreign policy deliberations and the prospects of Asia are increasingly and inextricably entwined with China.

As Jonathan Woetzel argues, Asia's growing dominance in world affairs is irrefutable – 'the Asian century is an indisputable reality' (McKinsey, 2019a, n.p.). Parag Khanna (2019) similarly notes how, 'Asians once again see themselves as the center of the world – and its future'. The drift to the East is more or less complete with the recenterings at play not only about realignment of wide-ranging hegemonies, historical legacies, economic and geopolitical power shifts, but it also encompasses cultural repositioning away from dominant Western-centric invocations, to others that increasingly take in diverse influences and most notably Asia-centric ones. Indeed, the current COVID-19 pandemic and resultant tensions emerging in increasingly divided worldviews reflect these broader geopolitical anxieties (Mostafanezhad et al., 2020). Tharoor (2020) notes that 'Since the late 20th century – dubbed by Time Magazine's founder as the "American Century" – there has been talk of the 21st century as the Asian one. It took several geopolitical shocks for the vision of the "American" century to take hold … The pandemic may be another epochal jolt to the system, reshaping how we think about the course of global affairs'. Facing unprecedented challenges, the global tourism landscape where the Asian footprints loom large is now on hold as travel restrictions and quarantine orders worldwide have curtailed international travel. Yet, emerging evidence suggests that tourism recovery will first center around domestic markets and progressively among neighboring countries in Asia (McKinsey, 2019b), and as for tourism recovery in China, how this unfolds will be a harbinger of what is to come for the global industry (Zhang, 2020).

From the British Century of the 1800s to the American Century of the 1900s to the contemporary Asian Century, tourism geographies are deeply entangled in broader shifts in geopolitical power (Luce, 1999; Scott, 2008; Shenkar, 2006). While Gillen (2016) and others (e.g. Mascitelli & O'Mahony, 2014) have critiqued the potential of the Asian Century to clarify more than it obscures, what is clear is that the cultural and political-economic climate in which tourism operates has transitioned along with broader power shifts. However, as Gillen notes, the area we call 'Asia' is extremely diverse, constitutes 75 percent of the global population, and has notably nebulous borders. Moreover, the geographical imaginary of Asia itself is an orientalist fantasy (2016). As such, the dynamism of Asia demands ongoing reappraisal (Connors et al., 2017), especially regarding the ways in which it is affected by and affects global flows of culture, finance, and politics.

In this collection, the papers demonstrate how tourism geographies reflect these flows of socio-cultural and political-economic flows in a particularly visible ways, where tourists and the tourism industry are dramatically swept up in the swirl of geopolitical repositionings. Given the economic and political prominence of the industry in the region, the authors address a range of shifts characterized by the 'Asian Century' through critically engaged and empirically rich accounts of tourism in Asia and the Pacific. The papers consider what the transition into the Asian Century means for some of the most urgent issues of our time such as sustainable development, human rights, gender equality, and environmental change. They also critique Anglo-Western centrism in tourism theory and call on tourism scholars to make radical shifts toward more inclusive epistemology and praxis. Building on these trends, in this introduction to the Special Issue, *Tourism Geographies in the Asian Century*, we highlight how the unfolding of the Asian Century has been underway over the past several decades. Further, the post-pandemic tourism landscape, we argue, will require the tourism practitioners and scholars to account for the radical recentering of geopolitical and political-economic power and the broader Asia pivot.

For tourism scholars, this is a clarion call to consider how tourism geographies are being reshaped by these wide-ranging shifts. Authors in this collection consider tourism as a lens through which to examine some of the most urgent political-economic, environmental, and geopolitical implications of this recentering of tourism geographies in the Asian Century. Tourism research has been notably affected by the continuing rise in research knowledge production out of Asia based universities (Qian et al., 2019). This contrasts with Pearce's (2004) sentiments nearly two decades earlier where he cited limitations evident in theory development from Asia-Pacific based research and researchers. Pearce (2004, p. 66) was sanguine about regionally-based solutions to address the inequities suggesting that 'the promise of a special kind of contribution to theoretical tourism innovation in research from the Asia-Pacific region lies in the cultural traditions of the researchers' own countries'. Mura and Pahlevan Sharif (2015, p. 833) similarly encounter the binaries at the positivist-qualitative tourism research intersection in Southeast Asia citing that the status quo 'needs to be understood in relation to complex historical colonialist and postcolonialist influences as well as global structures of power'.

Collectively, the papers in this issue advance scholarship on the multiple modalities and recenterings of critical tourism studies and tourism geographies in the Asian Century. The papers consider empirical and theoretical shifts that reflect this recentering in extant Asia-Pacific scholarship as well as the broader implications of the supercharging of global tourism out of the region. Using China as an exemplar, the rise of tourism in the everyday, as well as the ubiquity and predominance of Chinese tourists around the globe is not reflected in a commensurate production of relevant research, nor does it acknowledge the vast production of knowledge on tourism outside the English-language academic circuit, especially those published in Chinese (Xu et al., 2014). Recentering is akin to a reorienting of dominant discourses from the margins back to the centre. We envision this collection to be less about the particulars of tourism in the Asia-Pacific or the tourists from the Asia-Pacific, than about the political-economy and cultural politics of historical and contemporary research in the region. To recenter critical tourism scholarship requires maintaining an equilibrium between a strict anti-essentialism and an openness to accommodate diverse ways of knowing. This act of conciliation and transformation is beset with hegemonic and counter-hegemonic engagements. While scholars have long called for the incorporation of a range of culturally diverse conceptualizations, tourism research continues to be dominated by English language voices. Yet, as new technologies and modes of cross-cultural translation and exchange take hold, and the momentum shifts away from the Anglo-centric axis toward more inclusive and diverse standpoints, tourism scholarship is destined to take shape in unprecedented ways.

To be sure, in providing an entrée to this curated collection, it has been an oddly mournful yet reassuring exercise. Borne out of the 2018 Critical Tourism Studies Asia-Pacific (CTS-AP) Inaugural Conference at Universitas Gadjah Mada in Indonesia, and now in the shadow of the COVID-19 pandemic, the significance of the themes addressed in this issue are more urgent than ever. The pandemic has hastened claims that the Asian century has further accelerated given the contrasting successes of many Asia-Pacific countries, especially as compared to their Euro-American counterparts (Park, 2020). The Asia-Pacific chapter of Critical Tourism Studies was first mooted with the desire to create the space for recognizing and engaging the plurality of tourism scholarship outside universalized Western discourses. There has long been a call in CTS to decenter 'tourism's intellectual universe' with the 'willing[ness] to learn from every knowledge tradition, from Africa, Asia and from indigenous peoples around the world' (Pritchard & Morgan, 2007, p. 25). The Asian Century offers new opportunities to rethink and reconfigure contemporary tourism discourses. In this vein, Lee, Hongling and Mignolo argue that the project 'is both challenging and necessary, as it invites us to re-conceptualize the past (i.e. existing and hegemonic narratives of the past) in the present' (2015, p. 189), even as we recognise the critical need to avoid perpetuating its own insular thought in what Chang (forthcoming) describes as Asian-centrism.

In founding the broader Asia-Pacific chapter of Critical Tourism Studies in this time of crisis, we realise that our premises remain – our past lives and work as pre-pandemic scholars were inspired by the critical turn in tourism studies (Ateljevic et al., 2007, 2012) and responds to the call to 'dispute hegemonic neoliberal ways of producing and disseminating tourism knowledge' (Ayikoru et al., 2009). Many issues familiar to critical tourism scholars – race, class, gender, sexuality, and disability – have become increasingly visible through social movements, globally. The Black Lives Matter (BLM), Stop Asian Hate, Occupy Wallstreet, and Me Too movements in recent years brought about increasing awareness of identity politics and became triggers for many Asian societies to begin unpicking similar issues in their public domains. The papers in this collection similarly tackle this broad spectrum of issues, many of which have become increasingly visible in recent months. Critical tourism scholarship is characterized in part by a commitment to fundamentally question hegemonic knowledge making practices in the broader service of social justice and sustainable development agendas (Mura & Wijesinghe, 2021).

The issue begins by addressing the philosophical drivers and theoretical implications of recentering tourism research in the Asian Century. In *Critical tourism studies: new directions for volatile times*, Gibson (2019) outlines key themes and future research agendas for tourism in the Anthropocene. As human consumption exceeds planetary thresholds, he writes, three 'unfurling forces' are unleashed: 'excess capital and its territorial fixes; excessive mobilities and accompanying sociomaterial struggles; and biopolitical limits and excesses' (this issue). Despite the ongoing challenges of tourism encounters, Gibson also highlights the potential for tourism to become a space of empathy and learning from and with each other. Gibson's work challenges us to think about the role of tourism in the economic and environmental crisis. In, *What western tourism concepts obscure: intersections of migration and tourism in Indonesia*, (Adams 2020; Adams et al., 2021) addresses how western concepts fail to account for the nuance of nonwestern experiences of migration and tourism in Indonesia. Based on long-term ethnographic fieldwork and online research, she focuses on historically rooted Toraja travel for purposes of experiential/financial enrichment (*merantau*). By describing how Western identity categories of 'tourist' and 'migrant' do not fully capture the experiences of Indonesian global labor and education migrants as well people's return to their homelands, her work breaks down these Western binaries and opens space for new, more culturally relevant tourism knowledges and practices. Tucker and Hayes (2019) further deconstruct western tourism binaries, in *Decentering scholarship through learning with/from each 'other'*. They consider how we might cultivate a decentring of Anglo-Western-centrism in tourism scholarship. Based on a field school program in Pai, Thailand, they examine how Western and non-Western researchers engage with tourism dynamics. This practice facilitated opportunities to embody the notion of 'being Other' in the formation of tourism knowledge production. Their work not only highlights the hegemony of Western knowledge practices, but also the potential role of the 'international classroom' in creating counter-hegemonic scholarship. Finally, Chang (2019) examines the relationship between CTS and Asian tourism research in, *'Asianizing the field': questioning Critical Tourism Studies in Asia*. His work identifies the role of CTS in challenging conventional concepts and theories in tourism studies. Focused on six critical points of entry, Chang addresses issues related to principle, language, authorship, concepts, emancipation, and pedagogy and pushes the boundaries of CTS by calling for a Critical Asian Tourism Studies that celebrates the relationship between CTS and Asian tourism scholarship.

The second tranche of papers focuses on the lived experience of tourism for communities whose lifestyles and well-being have become increasingly bound to tourism and where the ramifications of tourism expansion are felt firsthand. Minca and Roelofsen (2019) enquire into the biopolitical implications of Airbnb on stakeholder communities. While benignly couched in terms of the sharing economy, they unpack the socio-economic implications of the rating system as a mechanism that unevenly affects hosts and guests. The conceptual registers of home are further questioned in Cheong and Sin (2019) incisive piece on family tourism. They argue that critical inquiry into family tourism allows us to understand the role tourism plays in the reproduction of normative social ideas such as that of a family. Their piece captures the role of family tourism in a time when the practice is very much curtailed due to the current pandemic, thus offering the opportunity for future studies to consider what happens in its absence. Scheyvens and Laeis (2019) are similarly focused on the concerns of stakeholders. They draw attention to the possibilities of optimal outcomes of tourism development through closer more productive linkages between agricultural production. This is especially important in contexts where economic leakages hold considerable implications for stakeholder communities. Tarulevicz and Ooi (2019) describe the stubborn barriers of tourists' food preferences. They address how food is appropriated in place-making practices in ways that are often hierarchical and perpetuate historically rooted racisms. Tourism expansion highlights and intensifies the contradictions at play where the 'authenticity' of food can become sanitized and regulated in the interests of tourist safety. In rural Japan, Progano et al. (2020) highlight how visitors shape the pilgrimage experience and the implications this has for host communities. This can have enormous implications for destination managers and communities in peripheral places who often lack the wherewithall and capacity to adapt quickly to the changing demands of tourists. Finally, Mostafanezhad addresses tourism during the annual 'smoky season' in northern Thailand. She describes how environmental narratives about the causes and effects of seasonal air pollution are reshaping rural and urban relations, especially between farmers and tourism practitioners whose livelihoods are threatened by the circulation of particular matter. Her article brings emerging work at the intersection of urban political ecology and new materialism to bear on tourism to reveal the more-than human sociality of the tourism industry.

The special issue dovetails with two papers that address the conference as a space of knowledge production and politics. Apropos to the CTS-AP Inaugural Conference that itself aimed to decentre knowledge production and build a more inclusive scholarship in tourism spaces, these papers push the boundaries of traditional 'fieldsites' in tourism geographies. Zhang and Zhang (2020) highlight how Chinese scholars are increasingly pressed to join the fray in established 'Western' conferencing circuits while negotiating their own research within the context of their cultural traditions and identities. Rowen (2019) goes one step further to suggest that geopolitical shifts between state actors can themselves have a direct influence the social production of academic conferences. Both papers home in on the increasingly significant role Chinese academics play in the international academic community (whether based in China or of Chinese national or ethnic relations). Each paper ethnographically highlights the politics of academic knowledge production, publications, and conferences. They also remind us of the often-nuanced power dynamics of our field sites, collaborations, and publishing practices. Efforts to develop a more inclusive and pluralised tourism studies requires that we ask ourselves difficult questions of how and where we position CTS-AP as an intellectual community.

The special issue concludes with two pieces considering our work against the current COVID-19 pandemic – the first highlighting the varied outcomes of the pandemic in Southeast Asia, where Adams et al (2020) outline four trends that have proliferated throughout the region in the wake of the COVID-19 pandemic including livelihood diversification, ecosystem regeneration, cultural revitalization, and domestic tourism development. These trends highlight the shifting tourism landscape and offer insight into what Southeast Asia's post-pandemic tourism landscape may hold. Finally, an Afterword by established Asianist Tim Oakes (2020) then critically questions the heightened relevance of this project against two interrelated issues – the current geo-political tensions between USA and China, and the ongoing COVID-19 pandemic. He argues that it is now 'crucial to recognize how attention to the politics of knowledge, in turn, reveals the need for a reset on how we conceive criticality itself'. These four concluding papers highlight the competing tensions in the current structure of academia. As CTS scholars endeavour to create hopeful and inclusive scholarship (Ateljevic et al., 2007, 2012; Pernecky, 2020), in the act of naming, we also – albeit inadvertently – construct new boundaries around what constitutes 'critical' in tourism studies. The practical impossibility of accounting for the full range of knowledge making practices will continue to challenge the CTS collective. Indeed, how to account for this diversity is at the core of these papers where authors collectively ask, whose knowledge counts and to what end?

As scores of Asian tourists challenged the Western normative assumptions, they are now conspicuously absent in once overtouristed sites. The family vacations that Cheong and Sin (2019) posit as fundamental to the imagination of a 'happy family' in Singapore where they have limited time together at home has all but flipped around – during the pandemic, home was one of the only places Singapore residents were bound to. Conferences around the globe that Zhang and Zhang (2020), and Rowen (2019) discuss have fallen like dominoes in their cancelations and postponements, making the February 2020 second biennial CTS-AP Conference at Wakayama University in Japan one of the last international tourism research conferences of that year. As critical tourism scholars, we are faced with an unprecedented situation, even as the pandemic looks set to become globally endemic and the true extent of its fullest impacts are only beginning to emerge, with more to surface in the years ahead.

At Wakayama in February 2020, alongside anxieties and apprehensions of what turned out to be just the beginning of an outbreak of what was then an unnamed pathogen, diligent hand-washing and sanitising ensued religiously amidst the uncertainty and fear. For many, the conference commenced what would eventually become an epoch moment in our personal and professional lives. Yet, there was palpable camaraderie, community, and a sense of togetherness despite the looming threat to our very understanding of tourism as an industry, practice, and academic enterprise. Casting a gaze forward, the third CTS-AP conference is scheduled to be hosted at Vietnam National University, Hanoi in December of 2022. Whether academic research and conferences will have resumed by then remains in limbo. That the world faces increasing turmoil is abundantly clear. Yet, amidst the disruption to the everyday, it is hope and compassion, but also political-economic restructuring that is needed to reset the tourism industry in more sustainable, equitable, and ethical directions (Brouder, 2020; Cheer, 2020; Crossley, 2020; Lapointe, 2020; Lew et al., 2020; Mostafanezhad, 2020a, 2020b; Tomassini & Cavagnaro, 2020). While in no uncertain terms, the pandemic has forever changed the tourism industry as we once knew it, it is our hope that we can collectively build on the momentum of the inclusive scholarship that CTSAP is renowned for (Edelheim, 2020; Pernecky, 2020) as we pause to reflect on the possibilities and challenges of tourism in a post-pandemic Asian Century.

Disclosure statement

No potential conflict of interest was reported by the author(s).

ORCID

Harng Luh Sin (ID) http://orcid.org/0000-0001-7850-993X

References

Ateljevic, I., Morgan, N., & Pritchard, A. (2012). *The critical turn in tourism studies: Creating an academy of hope.* Elsevier.

Ateljevic, I., Pritchard, A., & Morgan, N. (2007). *The critical turn in tourism studies: Innovative research methodologies.* Routledge.

Ayikoru, M., Tribe, J., & Airey, D. (2009). Reading tourism education: Neoliberalism unveiled. *Annals of Tourism Research, 36*(2), 191–221. https://doi.org/10.1016/j.annals.2008.11.001

Beeson, M. (2020). Donald Trump and post-pivot Asia: The implications of a "transactional" approach to foreign policy. *Asian Studies Review, 44*(1), 10–27. https://doi.org/10.1080/10357823.2019.1680604

Brouder, P. (2020). Reset redux: Possible evolutionary pathways towards the transformation of tourism in a COVID-19 world. *Tourism Geographies, 22*(3), 484–490. https://doi.org/10.1080/14616688.2020.1760928

Chang, T. C. (2019). 'Asianizing the field': Questioning critical tourism studies in Asia. *Tourism Geographies,* 1–18. https://doi.org/10.1080/14616688.2019.1674370

Chang, T. C. (forthcoming), 'Insularities: Anglo- and Asia-centrism in tourism', *Annals of Tourism Research.*

Cheer, J. M. (2020). Human flourishing, tourism transformation and COVID-19: A conceptual touch-stone. *Tourism Geographies, 22*(3), 514–524. https://doi.org/10.1080/14616688.2020.1765016

Cheong, Y. S., & Sin, H. L. (2019). Going on holiday only to come home: Making happy families in Singapore. *Tourism Geographies,* 1–22. https://doi.org/10.1080/14616688.2019.1669069

Connors, M. K., Davison, R., & Dosch, J. (2017). *The new global politics of the Asia-Pacific: Conflict and cooperation in the Asian Century.* Routledge.

Crossley, É. (2020). Ecological grief generates desire for environmental healing in tourism after COVID-19. *Tourism Geographies, 22*(3), 536–546. https://doi.org/10.1080/14616688.2020.1759133

Edelheim, J. (2020). How should tourism education values be transformed after 2020? *Tourism Geographies, 22*(3), 547–554. https://doi.org/10.1080/14616688.2020.1760927

Gibson, C. (2019). Critical tourism studies: New directions for volatile times. *Tourism Geographies,* 1–19. https://doi.org/10.1080/14616688.2019.1647453

Gillen, J. (2016). Some problems with "the Asian Century". *Political Geography, 50,* 74–75. https://doi.org/10.1016/j.polgeo.2014.08.006

Holmes, F. (2017). One easy way to invest in the 'Asian Century'. *Forbes,* June 15. https://www.forbes.com/sites/greatspeculations/2017/06/15/one-easy-way-to-invest-in-the-asian-century/-3c251f4e364b

Khanna, P. (2019). *The future is Asian.* Simon and Schuster.

Lapointe, D. (2020). Reconnecting tourism after COVID-19: The paradox of alterity in tourism areas. *Tourism Geographies, 22*(3), 633–638. https://doi.org/10.1080/14616688.2020.1762115

Lee, V. P., Hongling, L., & Mignolo, W. D. (2015). Global coloniality and the Asian century. https://doi.org/10.1080/14616688.2020.1762115

Lew, A. A., Cheer, J. M., Haywood, M., Brouder, P., & Salazar, N. B. (2020). Visions of travel and tourism after the global COVID-19 transformation of 2020. *Tourism Geographies*, *22*(3), 455–412. https://doi.org/10.1080/14616688.2020.1770326

Luce, H. R. (1999). The American Century. *Diplomatic History*, *23*(2), 159–171. https://doi.org/10.1111/1467-7709.00161

Mascitelli, B., & O'Mahony, G. (2014). Australia in the Asian century-A critique of the white paper. *Australasian Journal of Regional Studies, The*, *20*(3), 540.

McKinsey, C. (2019a). A conversation about the future of Asia. Retrieved August 9, 2020, from https://www.mckinsey.com/featured-insights/asia-pacific/a-conversation-about-the-future-of-asia

McKinsey, C. (2019b). The Asian Century has arrived. Retrieved August 11, 2020, from https://www.mckinsey.com/featured-insights/asia-pacific/the-asian-century-has-arrived

Minca, C., & Roelofsen, M. (2019). Becoming Airbn*beings*: On datafication and the quantified self in tourism. *Tourism Geographies*. https://doi.org/10.1080/14616688.2019.1686767

Mostafanezhad, M. (2020a). The materiality of air pollution: Urban political ecologies of tourism in Thailand. *Tourism Geographies*. https://doi.org/10.1080/14616688.2020.1801826

Mostafanezhad, M. (2020b). Covid-19 is an unnatural disaster: Hope in revelatory moments of crisis. *Tourism Geographies*, *22*(3), 639–645. https://doi.org/10.1080/14616688.2020.1763446

Mostafanezhad, M., Cheer, J. M., & Sin, H. L. (2020). Geopolitical anxieties of tourism: (Im)mobilities of the COVID-19 pandemic. *Dialogues in Human Geography*, *10*(2), 182–186. https://doi.org/10.1177/2043820620934206

Mura, P., & Pahlevan Sharif, S. (2015). The crisis of the 'crisis of representation'–Mapping qualitative tourism research in Southeast Asia. *Current Issues in Tourism*, *18*(9), 828–844. https://doi.org/10.1080/13683500.2015.1045459

Mura, P., & Wijesinghe, S. N. (2021). Critical theories in tourism–A systematic literature review. *Tourism Geographies*, 1–21. https://doi.org/10.1080/14616688.2021.1925733

Oakes, T. (2020). Afterword: A critical reckoning with the 'Asian Century' in the shadow of the Anthropocene. *Tourism Geographies*.

Park, J. (2020). COVID-19 accelerating arrival of Asian Century. Retrieved August 9, 2020, from https://www.koreatimes.co.kr/www/biz/2020/05/488_290251.html?fbclid=IwAR0H2hbXxLnTKvyc1yEk_3uiKBlRvciXQJ37YCatBUVjUDr3eBVEK1UPevE

Pearce, P. L. (2004). Theoretical innovation in Asia Pacific tourism research. *Asia Pacific Journal of Tourism Research*, *9*(1), 57–70. https://doi.org/10.1080/1094166042000199639

Pernecky, T. (2020). Critical tourism scholars: Brokers of hope. *Tourism Geographies*, *22*(3), 657–666. https://doi.org/10.1080/14616688.2020.1760925

Pritchard, A., & Morgan, N. (2007) De-centring tourism's intellectual universe, or transversing the dialogue between change and tradition, in Irena Ateljevic, Annette Pritchard, Nigel Morgan (eds), *The Critical Turn in Tourism Studies*, 11–28.

Progano, R. N., Kato, K., & Cheer, J. M. (2020). Visitor diversification in pilgrimage destinations: Comparing national and international visitors through means-end. *Tourism Geographies*. https://doi.org/10.1080/14616688.2020.1765013

Qian, J., Law, R., Wei, J., & Wu, Y. (2019). Trends in global tourism studies: A content analysis of the publications in tourism management. *Journal of Quality Assurance in Hospitality & Tourism*, *20*(6), 753–768. https://doi.org/10.1080/1528008X.2019.1658149

Rowen, I. (2019). Tourism studies is a geopolitical instrument: Conferences, Confucius Institutes, and 'the Chinese Dream'. *Tourism Geographies*, 1–20. https://doi.org/10.1080/14616688.2019.1666912

Scheyvens, R., & Laeis, G. (2019). Linkages between tourist resorts, local food production and the sustainable development goals. *Tourism Geographies*, 1–23. https://doi.org/10.1080/14616688.2019.1674369

Scott, D. (2008). *The Chinese Century?* Springer.

Shenkar, O. (2006). *The Chinese century: The rising Chinese economy and its impact on the global economy, the balance of power, and your job*. Pearson Education.

Silove, N. (2016). The pivot before the pivot: US strategy to preserve the power balance in Asia. *International Security*, *40*(4), 45–88. https://doi.org/10.1162/ISEC_a_00238

Tarulevicz, N., & Ooi, C. S. (2019). Food safety and tourism in Singapore: Between microbial Russian roulette and Michelin stars. *Tourism Geographies*, 1–23. https://doi.org/10.1080/14616688.2019.1654540

Tharoor, I. (2020). The pandemic and the dawn of an 'Asian Century'. *The Washington Post*, July 10. https://www.washingtonpost.com/world/2020/07/10/pandemic-dawn-an-asian-century/

Tomassini, L., & Cavagnaro, E. (2020). The novel spaces and power-geometries in tourism and hospitality after 2020 will belong to the 'local". *Tourism Geographies*, *22*(3), 713–719. https://doi.org/10.1080/14616688.2020.1757747

Tonby, Oliver (2019). The Asian Century has arrived, https://www.mckinsey.com/featured-insights/asia-pacific/theasian-century-has-arrived.

Tucker, H., & Hayes, S. (2019). Decentring scholarship through learning with/from each 'other'. *Tourism Geographies*, 1–21. https://doi.org/10.1080/14616688.2019.1625070

Xu, H., Zhang, C., & Lew, A. A. (2014). Tourism geography research in China: Institutional perspectives on community tourism development. *Tourism Geographies*, *16*(5), 711–716. https://doi.org/10.1080/14616688.2014.963663

Zhang, J. J., & Zhang, C. X. (2020). Ontological mingling and mapping: Chinese tourism researchers' experiences at international conferences. *Tourism Geographies*. https://doi.org/10.1080/14616688.2020.1757745

Zhang, M. (2020). Where does tourism in China now stand? *Caixin Global*, August 8. https://www.caixinglobal.com/2020-08-08/zhang-mei-where-does-tourism-in-china-now-stand-101590172.html

Critical tourism studies: new directions for volatile times

Chris Gibson (iD)

ABSTRACT

Key themes in critical tourism geographies are reviewed and future research agendas suggested in light of a growing global sense of ecological and political-economic instability. Three cross-cutting threads in critical tourism studies are proposed as frames to recalibrate existing knowledges. Following theorisation of the Anthropocene as humans exceeding planetary thresholds, each frame encircles a key excess unleashed in the context of tourism that looms large amidst unfurling forces: excess capital and its territorial fixes; excessive mobilities and accompanying socio-material struggles; and biopolitical limits and excesses. Concluding thoughts focus on responsibility and an interlocutory optics required for tourism studies in the Anthropocene: with an eye squarely focussed on a suite of tenacious and oppressive forces – dispossession, displacement, commodification, exclusion, extinction – and another drawn to resistances as well as tourism's quieter, but perhaps no less significant, possibilities to learn, to address wrongs and to extend empathy that emerge from every-day moments of encounter.

摘要

鉴于生态和政治经济不稳定的全球意识日益增强, 对批判旅游地理研究的主要议题进行了回顾, 并提出了未来的研究议程。本文提出了批判旅游研究中三个相互交叉的线索, 作为框架来重新校准现有的知识。根据人类世理论, 人类超越了地球的极限, 每一个框架都围绕着旅游背景下展现出来的一个关键过剩展开:资本过剩及其区域修复;过度的流动及伴随的社会物质斗争;以及生物政治的限制及过度使用。结论部分的思考强调责任和人类世旅游研究中一个对话性的聚焦:一个方面直接关注一套顽强的、压制性的力量——剥夺,取代,商品化,排斥、灭绝;另一个方面是关注抵抗及在旅游业中存在的、比较平静但或许不太重要的各种可能性,用来了解错误,解决错误和拓展日常邂逅时刻表现出来的同理心。

Introduction

My family had the privilege to spend this summer on the road in the Canadian Rockies. As well as hiking in forests anticipating encounters with grizzlies and elk, we sought to witness the glaciers of the Columbia Icefield, before they are gone. En route, we found that millions of trees in the Rockies have succumbed to insect eruptions, as changed rainfall patterns stress trees, and warmer winters fuel unhindered growth in beetle populations and spread fungal infections (Goodsman et al., 2018). From a distance the mountain slopes look as if in full fall colour, except the trees are evergreen conifers, and they're dead. Later, in the evening comfort of our lodge, we watched newsfeed images of, among other things: catastrophic fires in California; asylum seekers at sea in search of Mediterranean shores; Trump removing the US from the Paris climate agreement and escalating a trade war with China; protestors in Barcelona resisting over-tourism; coral bleaching on the Great Barrier Reef; and holidaymakers in Greece fleeing wildfires by running into the water on beaches, escaping the flames. A sense of perversity was overwhelming. International tourism has enjoyed unparalleled growth (Saarinen, 2018a) while our climate warms to perilous levels (Steffen et al., 2011). Unprecedented acts of state violence have sought to secure borders, while space for mass holidaymaking simultaneously expands (Büscher & Fletcher, 2017). My family's own personal privilege to travel seemed paltry set against the suffering of others, as the ranks of the homeless, stateless, and property insecure grow.

Here, I take this sense of perversity as a trigger to review ideas in critical tourism studies regarding our planet's future, and in anticipation of the more profound upheavals in existing ways of doing things that now appear inevitable. Inspired by theorisation of the Anthropocene as humans exceeding planetary thresholds (Steffen et al., 2011), I assemble and review below three cross-cutting threads of critical tourism scholarship. Each is structured as a subsection, and framed around a key excess unleashed in the context of tourism, that looms large amidst unfurling forces. They are: (1) excess capital and its territorial fixes; (2) excessive mobilities and accompanying material struggles, and (3) biopolitical limits and excesses.

The timing seems right for an integrative update on critical tourism studies. A decade ago, I was asked to review progress in tourism studies in three broad essays (Gibson, 2008, 2009, 2010). Among the pressing concerns in an emergent critical thread of tourism studies in the 1990s and 2000s were (post)colonial power relations, uneven development and local livelihoods, and the rise of ethical tourism. Much has transpired since then. A global financial crisis (followed by austerity governance); the rapid rise of new social media; generationally-catastrophic storms and floods in China, Nepal, Puerto Rico, Brisbane, and Houston; populist movements exemplified by Brexit and Trump but prevalent across Eastern Europe, Turkey, and the Philippines; and arguably the most profound of all: the 2016 ratification of the Anthropocene as a new geological epoch in which the human imprint is permanently observable in the earth's very strata.

Accompanying these events, tourism geographers' engagements with critical theory have broadened and deepened. Such themes as nature-cultures, violence, and everyday embodied relations in material and cultural geographic spaces, are now more common (Büscher & Fletcher, 2017; de Jong, 2017; Devine & Ojeda, 2017). A growing interface now exists between tourism research and disaster studies, moving beyond normative assessments of preparedness and recovery towards more nuanced analyses of asymmetries in power relations and micro-scale complexities (Schmude, Zavareh, Schwaiger, & Karl, 2018). New critical scholarly concepts have proliferated, while critical analysis of the systemic, big picture issues (capitalism, colonialism, militarism, and gender) has sharpened (Gonzalez, 2013; Mostafenezhad, 2018). Critical tourism studies have coalesced into a discrete community of scholarship that more confidently stands apart from the growth-management paradigm: overtly engaging with critical theory, and keen to document and diagnose systems of domination, injustices and oppressions. Critical tourism researchers have adopted a more militant tone – fitting, I believe, for the times.

For the world *has* grown more unstable, haunted by the spectre of catastrophic ecological futures, while grappling with unparalleled mobilities, new disruptive technologies, perverse combinations of financialization and austerity, and economic and geopolitical uncertainty. While significant debate surrounds the idea of the Anthropocene, its culprits and alternative monikers (Haraway, 2015; Moore, 2016), what remains clear is the exceeding of planetary boundaries by humans across a range of spheres (Steffen et al., 2011). Ours is an age of contradiction in such excess – in which tourism growth and related carbon emissions have increased faster than other sectors than the economy, despite global compacts to reduce carbon emissions (Lenzen et al., 2018). The heralding of the Anthropocene marks a point of no return, of profound rupture (Hamilton, 2016; Ooi, Duke, & O'Leary, 2018) as global thresholds on climate change and habitat loss are irretrievably exceeded, and air, ocean, and chemical cycles disrupted from their Holocene equilibria (Castree, 2016; Steffen et al., 2011). Volatility and excess abound.

Tourism – as an industry, a pursuit, and a scholarly concern – both reflects and amplifies such upheavals. Literatures on tourism and geopolitics, racialized biopolitics, and political ecologies of disaster, vulnerability and resilience, have emerged and diversified (Alderman, 2018; Jedd et al., 2018; Mostafenezhad, 2018; Neef & Grayman, 2018). Tourism scholars were among the earliest talking about climate change and adaptation – a now considerable corpus of work captured in successive global reviews (Hoogendoorn & Fitchett, 2018; Kaján & Saarinen, 2013; Pang, McKercher, & Prideaux, 2013). Only recently has tourism studies engaged with conceptual debates surrounding the Anthropocene to the extent that has occurred elsewhere in geography (Gren & Huijbens, 2016; Moore, 2018a). Conceptually, the Anthropocene invokes more fundamental existential and ontological challenges: to capitalist commodifications of nature, to prevailing tenets of western philosophy, to the very notion of 'the human' and the assumption of Homo sapiens' exceptional subjectivity (Head, 2016; Ruddick, 2017). While the growing literatures on tourism and climate change adaptation for example are much needed, it is no longer conceptually valid to talk about 'natural' environments 'impacted upon' by human tourism activities (Hall, 2016), to describe problems such as climate change as merely 'environmental' (to which humans 'respond' in managed causal fashion) (Head, 2017; Lejano, 2017); or to think of climate as a singular phenomenon requiring responses detached from other processes. As Head (2010, p. 234) has argued, 'the stimuli to which we are adapting are complex assemblages comprising more-than-climate'. Coming to grips with present and imminent volatility means contemplating the interconnectedness of things, and questioning the very categories into which our analyses are corralled. Notwithstanding urgent calls for tourism studies to come to grips with the Anthropocene (Gren & Huijbens, 2016) much tourism scholarship proceeds within stable frames, deaf to such conceptual challenges.

Here, I seek to respond to this by taking stock of ideas in tourism geography that assist in making sense of growing volatility, while pointing towards new directions. In so doing, I seek to move beyond 'environmental', 'geopolitical', 'cultural', or 'economic' themes as limiting categories. Instead, the narrative follows three cross-cutting threads as attempts to recalibrate existing knowledges in light of Anthropocenic critique. Axiomatic concerns that have dominated debate such as gender, capitalism, labour, class, and colonialism continue to reverberate everywhere (Figueroa-Domecq, Pritchard, Segovia-Péreza, Morgan, & Villacé-Molinero, 2015; Ioannides & Zampoukos, 2018). In what follows, such concerns appear interwoven throughout rather than as separate domains: intersecting axes rather than compartmentalised perspectives. I conclude with thoughts on responsibility and an interlocutory optics attuned to the Anthropocene. While focussing empirical and conceptual attention on oppressive forces – dispossession, displacement, corruption, commodification, exclusion, extinction – critical scholars, I argue, must also account for tourism's quieter, but perhaps no less significant, possibilities to learn, to address wrongs and to extend empathy that emerge from everyday moments of encounter.

Excess Capital and its territorial fixes

At the outset, critical tourism scholars have been keen to stand apart from tourism economists and management researchers that feed and depend upon the tourism industry in line with a 'growth is good' mentality (Fletcher, Blanco-Romero, Blázquez-Salom, & Murray, 2018). Antecedent was Britton's (1991) analysis of the political economy of tourism. For Britton (1991, p. 451), tourism had become 'a major internationalized component of Western capitalist economies; it is one of the quintessential features of mass consumer culture and modern life'. The rise of tourism correlates with the generally agreed timeframe in which Anthropocenic activities escalated.

In the past two decades, with consolidation of corporate power and promulgation of neoliberal policies we have witnessed what Bianchi (2009, p. 487) describes as 'the most aggressive restructuring of class power and privilege'. Earlier phases of corporate concentration and expansion have been followed by further privatisation, concentration and sophistication in market competition and control. National airlines have been merged and/or privatised while new low-cost carriers have proliferated, hotel chains incorporated into conglomerates and global financial portfolios, and ever-more elaborate systems evolved for loyalty rewards, flight service bundling, and travel and accommodation booking (Duval, 2013; Yrigoy, 2018a). Cruise ships have boomed in popularity as distinctive 'spaces of containment and revenue capture' (Weaver, 2005, p. 165). In this phase, powerful interests extract maximising value from previously latent opportunities, all the while gathering large amounts of data on customer preferences, habits and purchasing decisions. Corporate interests develop novel means to secure monopolies, contain tourists and capture profits.

While concentration has intensified, tourism's thirst for 'the new', reliance on local knowledge and low entry barriers have triggered further fragmentation, specialization, and diversification (Hampton & Hamzah, 2016). Western baby boomers reaching retirement age seek more elite, luxury experiences, carefully curated small group tours, specialist cruises, arts trails, walking, and architecture tours, with cultural, historical or ecological themes. Even the advent of the Anthropocene itself has created opportunities to repackage and represent sustainable tourism destinations (Moore, 2018a). Underlying contradictions – consolidations of corporate power and simultaneous market fragmentations; geographic concentrations and expansions; mass travel and Anthropocenic tourism – point to capitalism's 'symbolic and material creativity' (Moore, 2018a, p. 1), but also the limits of singular interpretations of tourism's political economy (cf. Debbage, 2018). Britton (1991, p. 455) originally theorised a 'tourism production system' – less a single industry than an amalgam of sectors, each with their own geographies, divisions of labour and competitive dynamics. Beyond the boundaries of a formal industry, and state regulation, are both undergirding capitalist valuation processes (Young & Markham, 2019), and loose assemblages of economic actors that produce tourism places and circuits.

As the edges of tourism-based capitalism become more porous, other actors enter markets, disrupt establish practices, and fuel further cycles of destination production, acquisitions, mergers, and consolidations of market power. In the past decade alone, tourism has been disrupted and transformed by Airbnb (which now boasts 200 million members globally), Instagram (which has grown to 800 million users since its launch in late 2010), Trip Advisor and Google Translate – all now dominant players bridging ecommerce, property, social media, and big data services. Such trends reinforce the point now repeated almost to the point of cliché, that tourism be taken more seriously by critical geographers of all persuasions, and more thoroughly integrated into social theory generally (Büscher & Fletcher, 2017; Young & Markham, 2019). Far from a standalone sector merely concerned with 'leisure' (as opposed to parts of the economy deemed more 'productive'), tourism is utterly imbricated in the spatio-temporal dynamics of contemporary capitalism, *writ large*.

The nexus between financialization and geographic space is just one example. Voluminous research has documented and debated the manner in which institutional and private investors, flushed with excess surplus capital, seek spaces in which to reinvest; they require 'a spatio-temporal fix, or quasi-resolution of the crisis tendencies of contemporary capitalism' (French, Leyshon, & Wainwright, 2011, p. 798). While planetary affairs become increasingly unstable, enormous flows of investment capital on global markets seek opportunities in specific places and times to deploy surplus capital to deliver returns (Harvey, 2011). Rarely acknowledged in economic and urban geography is, however, the contribution of tourism to such spatial fixes, and the touristic mobility routes that make capital flows into them possible (for important exceptions, see the work of Fletcher (2018) and Yrigoy (2014, 2018a)). In global markets dominated by institutional investors and foreign landlords, tourist places associated with lifestyle and culture provide profitable, relatively lower-risk options (Smet, 2016). Tourism propels forces of inward investment from global real estate capital across distinctive geographies. Hotels in Spain, Greece, and Portugal are increasingly woven into global real estate investment flows (Yrigoy, 2018a). High-amenity/lifestyle cities such as Vancouver, Sydney, Barcelona, and New York and Lisbon are meanwhile elevated into a different category of underlying land rent valuation (Montezuma & McGarrigle, 2018). In an era of excess capital seeking safe returns in volatile times, tourism is deeply interwoven in diversified strategies among investment institutions to extract value, leverage debt, and access and transform space for profit (Büscher & Fletcher, 2017).

Further research is required, conceptually and empirically, to more fully sketch the contours of this, especially amidst global ecological and political-economic volatility. The work of Ismael Yrigoy provides an instructive example: hotels, for example, have become devalued assets in the post-crash era, and yet are also highly profitable – fuelling a shift away from securitisation and increased ownership by foreign real estate investment trusts and private equity funds (Yrigoy, 2016, 2018a). Hotel corporations have in turn sought to discipline labour, not just to cut costs, but to 'smooth the process of revaluing hotel assets' (Yrigoy & Cañada, 2019, p. 180) and thus improve their image to potential investors. Mega-resorts in small island states meanwhile lock-in new dependencies to global financial markets (Lee, Hampton, & Jeyacheya, 2015), so much so that recently Hampton and Bianchi (2018) have argued that a paradigmatic form and phase of financialized real estate-tourism development has emerged, with Southeast Asia its prime spatial target, and Singapore and Hong Kong as its investment fulcrums: it revolves around spectacular developments, marketed with much fanfare, mixing large scale accommodation, residential property towers, shopping malls, and casinos. Such developments exemplify what Büscher and Fletcher (2017, p. 651) describe as 'destructive creation': damaging environmentally – especially in delicate tropical coastal locations – as well as disruptive socially, culturally and economically, relying on land dispossessions and/or evictions, with scant participation or benefit sharing among local people. While urban mega-developments and coastal resorts aren't new, there is a new form and scale of spatial fix propelled by flows from international investment institutions and mega-rich individuals seeking safe harbours for otherwise unmoored surplus capital, amidst persistent volatility.

Tourism is thus more deeply entwined with excess capital and its desires for spatial fixes, in increasingly diverse ways. In Miami, condominiums and other mega-developments in glamorous tourist districts have become target investments for money laundering (McPherson, 2017). In the Bahamas, 'Anthropocene imaginaries participate in the recreation, redesign, and rebranding of specific spaces as emergent "tourism products"' (Moore, 2018a, p. 1) that include 'sustainable' second-home destinations inspired by Anthropocene ideas. Ecotourism, itself framed as a solution to the problems wrought by Anthropocene processes, provides another form of capitalist fix (Fletcher, 2018).

Meanwhile, Airbnb and other home-sharing platforms have moved beyond an initial phase inviting homeowners to rent out spare bedrooms, to a consolidated phase where significant investment in housing, including in the style of new major urban redevelopments discussed above, is underpinned by anticipated short-term rental trade. In consequence, critics have pointed to a range of social and economic impacts including housing market inflation, evictions of low-income tenants, conflicts with neighbours over noise, and belated regulatory responses among cities that struggle to adapt (Gurran, 2018; Gurran, Searle, & Phibbs, 2018; Yrigoy 2018b). In Paris, intense touristification has spread beyond traditional sight-seeing hotspots into widely gentrified residential neighbourhoods, in large part the result of Airbnb, but also linked to pedestrian and cycling mobilities (in turn flourishing as a consequence of Instagram, Google Maps, etc.) (Freytag & Bauder, 2018). Novy (2018) describes a 'pentagon of mobility and place consumption' in which neighbourhoods are gentrified and made more exclusive through intertwining flows of home-sharing tourists, real estate capital, social media discourses, and local urban aesthetes.

While many (or even most) tourists may not be themselves cashed-up entrepreneurs, potential investors, overseas owners, and/or prospective business migrants do increasingly visit places prior to investing, blurring boundaries between leisure, residential, and financial subjectivities (Montezuma & McGarrigle, 2018; Mostafanezhad & Norum, 2018). Global real estate has become an asset class for foreign investors who harbour both wealth and mobility motivations (Rogers & Koh, 2017). The mega-rich, meanwhile, seek private luxury retreats in paradisiacal places, 'moving away from the anchor resort model of enclave mass tourism' (Moore, 2015, p. 513). National population policies are recalibrated to attract 'investor immigrants' – a wealthy cohort of mobile individuals, often heavily debt-leveraged, to whom luxury projects are typically marketed (Ley, 2010). While quality of life, amenity and climate are key attractors, 'economic motivations are the most significant, even if diverse, among the residential investor typologies with some seeking a safe haven and others geoarbritage or income optimization' (Montezuma & McGarrigle, 2018, p. 1). Tourism growth and financialized urban transformations in service of excess capital are, in short, mutually constitutive.

Tourism assemblages (encompassing tourists, destinations, investors, mobilities, built environments) are thus not merely a distinctive form of capitalist phenomena, but central to capitalism's compulsion towards finding territorial fixes to resolve its own excesses (Fletcher, 2018; Fletcher & Neves, 2012). Diverse places are transformed into capitalist value through differentiated strategies of enclosure and value capture (Young & Markham, 2019). In an Anthropocene context characterised by volatility, this is all the more amplified. As the global economy reels from the aftermath of the Global Financial Crisis, and belatedly divests from fossil fuels, surplus capital seeks secure spatial fixes in diverse opportunities. The capitalist production of tourism circuits, flows and spaces by definition feeds such diversity.

This line of reasoning opens up connections to other critical tourism research on the state-capitalist and colonial violence required to secure space for tourists, and land for tourism investments. The 'sweet vision of tourism' has meant that 'social conflict, environmental destruction and colonialism are usually its hidden face' (Devine & Ojeda, 2017, p. 607). Tourism landscapes are 'carved out of practices of material and intangible dispossession, as well as practices of state territorialization' (Devine, 2017, p. 634). In such contexts where 'paradise' is 'produced' (from Hawai'i to Latin America, Southern Africa to Indonesia), tourism requires 'constitutive violence' that takes various forms, from loss of land, community, and language, to extraction, evictions, and enclosures, erasures, and (neo)colonialism (Büscher & Fletcher, 2017; Devine & Ojeda, 2017, p. 605). The Serengeti for example is 'a landscape that has been produced in the image of an idea of African nature and what it should be' (Gardner, 2016, p. ix), yet with foreign investors acquiring land, Maasai villagers have been dispossessed of grazing land they had used for over a century. In Mumbai, slum tourism depoliticises poverty – erasing problems of poor sanitation and overcrowding in favour of dis-courses of resourcefulness and industry that 'enable wealthy middle-class westerners to feel "inspired"' (Nisbett, 2017, p. 37). Tourism generates exploitative conditions but also masks relations of power 'through the seductive romance of uplift and develop-ment' (Gonzalez, 2013, p. 6)

Tourism development is thus 'linked to the consolidation of state territorialization, practices of nation-making and military and colonial projects' (Devine & Ojeda, 2017, p. 606). As Gonzalez (2013, p. 4) illustrates in Hawaii and the Philippines, tourism and militarism are mutually constituted and interdependent; two of 'the most dominant apparatuses by which the United States extends its reach'. Tourism and militarism do mutual work in securing space to set the stage for foreign dominance. In analogous vein, Rowen's (2014, p. 62) work demonstrates how Chinese tourism operates as a pol-itical technology of state territorialisation, a 'mode of social and spatial ordering that produces tourists and state territory as effects of power'. Amidst geopolitical instabil-ity, tourism is for China a 'creative territorialization strategy' linked to its growing geo-political and trade ambitions; hence through patterns of investment in ports, hotels, and airports as well as 'soft power', 'deployment of outbound tourism achieves polit-ical objectives' (Rowen, 2018, p. 61).

Meanwhile, in the context of growing frequency and severity of disasters, excess capital both shapes pre-disaster inequalities and post-disaster state-territorial actions (Blakie, 2016, p. xii). Post-disaster recovery resources often flow faster to touristed areas over those less familiar to international donors; hotel chains backed by multina-tionals rebuild faster and thus consolidate market share; while local elites and well-connected tourism operators 'exploit distorted recovery governance mechanisms and take advantage of the legal and institutional uncertainties triggered by disasters' (Neef & Grayman, 2018, p. xi). China has been criticised, for example, for its 'debt-trap diplomacy' (Fox, 2018, p. 1): making strategic loans to Pacific small island states to rebuild tourism infrastructure after devastating cyclones. The effect is to open up new markets for Chinese outbound tourism (as soft power), and deepening strategic geo-political influence, yet saddling smaller vulnerable nations with debt. In an era defined by excess capital and growing volatility (in markets, geopolitical relations, and climate), global flows of finance, vested interests, and tourists-as-investors-and-mobile-geopolit-ical-agents are now inextricably entwined.

Excessive mobilities and their material struggles

Another excess of the Anthropocene relates to mobile human bodies. In the context of globalisation and transnationalism, distinctions between forms of mobility have blurred, along with diverging degrees of political contestation and scrutiny. Multicultural societies, global mobile labour markets, and diasporic networks all fuel discrete modalities of mobility, from digital nomadism to return migrants visiting fam-ily (see Adams, this issue). Although once the preserve of the western elite, tourism is now much more common across Asia, Africa, and South America, with vacations increasingly seen as an entitlement, a regular excursion in the seasonal rhythms of everyday life. Yet at the same time that millions are boarding planes or trains on gap years and exchanges, working holidays, cruises, conference trips, and family reunions, millions more work long hours away from home on cruise ships and at resorts, or are fleeing armed conflict and environmental catastrophes in search of safety and hopeful futures (Dickson, 2015). Nowadays borders both matter more, and less; mobilities are both increasingly surveilled, and made easier by cheap flights, tablets, and apps. Although vast numbers of the world's population still cannot afford travel for sheer leisure (or indeed, cannot move freely because of travel restrictions), there is hardly a location on Earth – even in the poorest or most war-torn regions – not already touched by the tentacles of the tourism industry.

With increased mobility and geographic extent are a suite of related issues and conflicts associated with the sheer numbers of tourists, as travelling becomes more accessible, and as fads and fashions emerge. Perhaps a distinctive marker of this phase of the Anthropocene will prove to be the politics of excessive human movements – what Gonzalez (2013, p. 6) calls 'unfettered mobilities' beyond a threshold within which host communities and governments can cope.

Such excesses are accompanied by interrelated ecological, economic and cultural disruptions that appear unresolvable. Of immediate concern are worsening carbon emissions and ecological damage (Lenzen et al., 2018). More off-the-beaten-path travel means capitalist valuation and enclosure extend further (cf. Young & Markham, 2019), unleashing more widespread 'destructive creation' (cf. Büscher & Fletcher, 2017). The 'dual violent practices' (Devine, 2017) of commodifying place and state territorialisation are further extended geographically. Accompanying ecological problems include air, water, and soil pollution, habitat destruction and fragmentation, and biodiversity loss (Hall, 2016). It is an old truism that tourism is paradoxically dependent on natural resources and environmental amenity, even though it can produce enormous environmental problems. But this seems especially intense with the explosion of tourist numbers, and their greater geographic reach.

Social media has further amplified and refracted what we once understood as the tourist gaze: fuelling the search for the 'perfect Instagram photo' (cf. Mostafanezhad & Norum, 2018). In search of 'viral' photos (which in turn leads to followers, invitations for sponsored posts and free travel), 'Trojan horse' Instagrammers are pushing further beyond anticipated places. Concerns grow about unregulated mobilities – literally off the beaten path – as swelling numbers of visitors seek to recreate stunning photos in ecologically fragile places, spurring calls for 'responsible geo-tagging' to limit Instagrammers' damaging effects (Salmon, 2018).

Less recognised in both the tourism and ecological literatures are the more nuanced entanglements unleashed by excessive mobilities – for example, the role of international tourism and mobility in the spread of invasive species and diseases (in turn prompting challenges to existing biosecurity paradigms) (Hall, 2015). The Arctic – for which polar 'wilderness' discourses are especially powerful (and saleable) – is increasingly under threat from invasive species as a consequence of cruise ship and marine expedition markets (Hall, James, & Wilson, 2010).

Meanwhile, excessive mobilities have themselves become more volatile, numerically and politically. Fashions have become fickler, with surges in visitor numbers in some places bewildering even pro-growth tourism advocates. New social media, globalisation of popular culture (e.g. Netflix), and the advent of cheap flights have combined to drive huge annual increases in visitation to certain destinations in only a couple of seasons. The enormous popularity of the TV show Game of Thrones led Iceland's tourism visitor numbers to triple between 2011 and 2016 (Groundwater, 2017). Amsterdam, with a residential population of 850,000, in 2016 had an estimated 18 million visitors – double the number that visited the entire continent of Australia (itself experiencing an unprecedented upsurge in inbound Chinese tourist numbers) (Connell & McManus, 2019; Millar, 2017). Cruise ship arrivals in Venice have increased fivefold in the past 15 years (Nadeau, 2012). With a residential population of only 50,000, it now receives 20 million visitors annually.

A host of logistical and political problems accompanies excess mobilities, impelling questions of materiality, and associated social and political struggles over space. Popularity overwhelms cities with insufficient infrastructure or populations to provide a tax base to cover necessary services. In Venice, the proliferation of cruise ship visitors has increased visitor numbers, and the pressure on the city and its infrastructure, while paradoxically exacerbating decreases in overnight hotel stays in the city, reducing hotel income and city tax revenue (Nadeau, 2009). The advent of outbound Chinese 'zero-dollar' mass tourism (where all revenues are captured by mainland tour agents rather than in situ at the destination) has fuelled dissent and attempts at regulation in Thailand and Bali (Connell & McManus, 2019).

Sheer overcrowding from rapid shifts in market dynamics angers local residents (especially where cultural mores are viewed as quite different from those of visitors, as in Japan), and deteriorates the tourist experience (Ryall, 2018), while amplifying vulnerabilities and risks in more-frequent heatwave conditions. Officials in Amsterdam have banned new souvenir shops and have taken to live streaming crowds to deter mass tourism (Millar, 2017). In Barcelona and Berlin, growing tensions between residents and short-term rentals, noise and nightlife is an 'increasing source of dispute and residents' contestation' (Nofre, Giordano, Eldridge, Martins, & Sequera, 2018, p. 377). In such places, there may no longer be a 'high' vs 'low' season, with a chance for local residents to recover and lead more anonymous lives in the winter. Private residences are converted to more lucrative Airbnb lettings, driving up housing costs, fuelling gentrification, displacing residents and in the cases of Dubrovnik and Venice, resulting in overall permanent population decline (Minoia, 2016). Some 1,157 people live in Dubrovnik's Old Town, down from 5,000 in 1991. In the 1980s, more than 120,000 people lived in Venice. By 2030, some demographers predict, Venice will have no more full-time residents (Nadeau, 2009).

Inevitably, such pressures lead to protests and various forms of resistance, from graffiti in tourist areas, to street marches, verbal abuse of tourists in crowded market squares, and more organised activisms, such as Venice's 'No Grandi Navi' (no cruise ship) campaign. Nuances of such campaigns extend beyond anti-tourist sentiment, to struggles to define the identity of material spaces, contested claims over neighbourhoods, and for physical access to them (Novy, 2018). While such conflicts raise pragmatic questions of the role of the state, regulators and planning agencies in mitigating impacts (Gurran et al., 2018), the underlying Anthropocenic dilemma is much more difficult to resolve: that for confined, popular places, there simply may be too many tourists on the move.

Biopolitical limits and excesses

My third and final proposed framing of critical tourism studies for the Anthropocene builds on the previous two, asking what kinds of ethical and political consequences arise 'from the everyday bodily politics of moving through and occupying space and make place' (Alderman, 2018, p. 3). A decade ago, embodiment seemed a liminal concept limited to gender studies and cultural geography. Nowadays to suggest that tourists 'eat place' through gastronomic experiences is almost mundane, very much at the centre of tourism marketing, cooking and travel shows (de Jong & Varley, 2017). No longer is the prospect of lying on a beach or gazing upon scenic landscapes enough to attract tourists; the 'experience economy' involves more active corporeal participation: eating, drinking, cycling, and soaking up atmospheres.

It is important, I believe, to not lose sight of the critical purchase of the concept of embodiment (and related concepts such as encounter), as Anthropocenic uncertainty escalates. At more intimate scales, tourism mobilities unleash distinctive biopolitics. Alongside processes of commodification and state territorialisation discussed above, tourism entails discourses, technologies and practices that govern bodies, lives, and affects (Alderman, 2018). Such discourses, technologies and practices influence what it means to be 'good' tourists, consumers, hosts, or workers – to be hospitable, 'authentic' or to have good 'taste' (de Jong & Varley, 2017). Amidst Anthropocenic volatility – its disasters, unanticipated booms, and collapses, its crush of bodies, wanted and unwanted, in lived material spaces – there is much value in an ability to locate precisely the agents, moments and techniques of the exercise of power. Research in this vein attends to 'the marginalising and silencing effects tourism policy exerts when the power values and interests involved in its formation are not critically appraised' (de Jong & Varley, 2017, p. 212).

Research on tourism work and workers has for example brought a degree of nuance to labour geographies around questions of mobility, emotion and subjectivity (Ioannides & Zampoukos, 2018). Service workers in tourism both provide the service and are a part of the consumed product (Sörensson, 2012). Jobs with emotion and care dimensions remain heavily gendered – women for example dominate hotel work-forces – with links to notions of domestic work (Figueroa-Domecq et al., 2015). Even when women and men perform similar tasks, their bodies are positioned and per-ceived differently. In Indonesia, class, gender, and colonial discourses intertwine; work-ers see themselves as 'providers of fun', and men are constructed as the norm against which women are compared (Sörensson, 2012). In other contexts, younger women with entrepreneurial dispositions that align with neoliberal governmentalities have gained power (Xu, 2018). In Kazakhstan, where yurt stays are on the rise (analogous to Airbnb in the West), women tend to assume responsibility for most of the tasks associ-ated with running the commercial home and hosting guests (Talinbayi, Xu, & Li, 2018). Old inequalities are reproduced while others shift.

Again, here, we return to questions of limits and excesses. Bodies, emotions and affects are governed within frames understood to have 'political limits' (Mostafenezhad, 2014). Orchestrations and choreographed tourism experiences set such limits, and obscure absences and violence (Alderman, 2018; Devine, 2017). But what happens in the micro-spaces of encounter also matters, when events don't go to script. Everyday encounters, and the moral, ethical, and political dimensions that inform them, unfurl within what Mostafanezhad calls 'everyday geopolitical encounters between 'hosts' and 'guests''. As de Jong (2017, p. 128) has argued, discourses and commodification processes are contingent and embodied; 'spatially and socially specific to the moment of encounter'. In North Korea, for example, where tourism is utterly choreographed to the point that it has become 'political theatre', glitches and slippages in the performance of tourism nevertheless elicit emotional and critical perspectives among tourists, that 'involved simultaneous empathy with and detachment from 'ordinary' Koreans, despite a sense of difference' (Connell, 2019, p. 38).

Such 'off-script' slippages remind critical tourism scholars that while a systemic worldview is essential, critiques of the persistence of hegemonic colonial and capitalist power relations ought not fix the frames of analysis such that subjects in tourist places (locals, visitors, nonhuman beings, etc.) are denied agency to resist and transform them (Everingham, 2016). Tourism mobilities are themselves a means to embody and perform activism (de Jong, 2016). *Paris Noir* for example is an effort among French-African activists to re-inscribe black experiences and make visible colonial legacies – framing a counternarrative 'to observe the effects of the racial denial mechanisms occurring in French society' (Boukhris, 2017, p. 684). Tourism makes possible unanticipated solidarities.

And coming full circle, among such potential solidarities are reconstituted emotional and affective relations with nonhuman others – 'modes of acknowledgement, humility, and discernment' (Hatley, 2012, p. 1) that are at the heart of an ethics for the Anthropocene. While tourism mobilities are a form of Anthropocenic excess, they nevertheless open possibilities to 'bear witness' to the Anthropocene's endangerments (cf. Dewsbury, 2003; Gibson, 2019; van Dooren, 2017), reconfiguring nonhuman others beyond fixed categories of 'nature', as 'earthly companion species … with rights and territories that require humans to behave in a new way and engage in new relationships' (Kristoffersen, Norum, & Kramvig, 2016, p. 108). For example, the arctic tourism cited above for its ecological excesses has fuelled increasing concern articulated for other species, such as whales, suggesting 'a possible ethical relationship with the cetaceans reflecting an attentive and hospitable ethos that can be aligned with current notions of the Anthropocene' (Kristoffersen et al., 2016, p. 94). Commercial imperatives and discourses govern touring bodies and emotions. Nevertheless, as Griffiths (2015, p. 628) has aptly put it, life 'constantly escapes'.

Volatile times: towards uncertain horizons

Tourism's Anthropocenic excesses and volatilities warrant urgent and compelling responses. Critical scholars are pursuing with pace the 'theoretical dethroning of humanity' (Sharp, 2017, p. 158). Awash with toxicity all the way from the cellular to the oceanic scales, our shared journey towards 'Destination Anthropocene' (Moore, 2018b, p. 1), means there is no going back to pre-capitalist times or (false) notions of 'wilderness' (Saarinen, 2018b).[1] Previously unthinkable questions now appear pertinent: how might tourists, governments and tour operators be discouraged from their 'obedience to the global tourism industry' (Zimmermann, 2018, p. 335)? How can ceaseless growth in aeromobility be justified, especially given that rapid increases in tourism demand are 'outstripping the decarbonisation of tourism-related technology' (Lenzen et al., 2018, p. 522)? Ought the notion of growth be abandoned altogether (Fletcher et al., 2018; Saarinen, 2018a)? My sense is that another conversation is needed about responsibility and responsibilising (to lift a somewhat unwieldy phrase from governmentality theory): who or what is responsible for tourism's excesses; who or what is responsible for responding and altering arrangements – and who or what is made responsible as power relations, decisions, and actions unravel (cf. Dean, 2010). One thing seems certain: as times become more volatile, market actors cannot ultimately be relied upon to deliver palatable, let alone sustainable, tourism. Tourism displacements, commodifications, exclusions, and exploitation warrant sustained scrutiny from critical tourism scholars. Identifying and understanding such forces is a necessary precursor to 'imagine and build less violent and more just and 'sustainable' forms of tourism' (Devine & Ojeda, 2017, p. 606).

With one eye squarely focussed on such forces, the other is drawn to the quieter, but perhaps no less significant, possibilities to learn, to address wrongs, and to extend empathy that emerge from tourism's everyday moments of encounter. Capturing the Anthropocene zeitgeist entails contemplation of multiple unknown futures well beyond the frames of capitalist modernity, beyond 'the human', and beyond our own human lifetimes, while enacting and relating differently in the everyday here and now (cf. Gibson & Warren, 2019; Norum & Mostafanezhad, 2016). Countering the 'disillusionment narrative' that characterises much critical scholarship (Woodyer & Geoghegan, 2013) are contributions by entangled subjects and marginal others to enacting different possibilities (Everingham, 2016). It is important not to lose sight of the messy details, the experiences and ethnographies – to pay 'close attention to how different forms of domination, exclusion and suffering are actually lived and understood, as well as contested and reworked' (Devine & Ojeda, 2017, p. 607). Power must be understood not as a simple capacity of which one has more or less, but rather, as strategies, practices, and techniques – fluid and contextual (Nepal, Saarinen, & McLean-Purdon, 2016, p. 5). More research is for example needed on the potential and limits of community anti-tourism campaigns. Consumer agency is another evolving theme. The #flygskam ('flying shame') social media movement championed by Swedish Olympic champion Bjorn Ferry involves people pledging not to use planes and to take fewer trips, or use trains instead. Quickly it has since spread to Germany (#flugscham), and the Netherlands (#vliegschaamte) (Norman, 2019). As well as more formal activisms and resistances are possibilities in the micro-spaces of tourism encounters for everyday Anthropocenic ethics to emerge, navigating, and negotiating positionalities into new forms of communality. Beyond Pollyannaish interpretations of 'hopeful tourism' (Higgins-Desbiolles &Whyte, 2013, p. 428), it remains true that

> ... the ambiguity of emotional connections and embodied encounters with local people opens up spaces for hopeful possibilities ... embodied encounters cannot be predetermined within a specific outcomes-based model, nor are all these encounters hopelessly subsumed within 'capitolcentric' discourses' (Everingham, 2016, p. 521).

'It cannot be', argues Mark Griffiths (2015, p. 629), 'that neoliberal sleights of hands are surreptitiously pulling all the strings'.

Notwithstanding a sense of urgency surrounding the Anthropocene, exactly how to live differently amidst volatility, to construct and reconfigure society and economy, theoretically and concretely, remain open agendas (Sharp, 2017). A sense of grief and loss accompanies the Anthropocene's extinguishment of the stabilities and equilibria underpinning Holocene modernity (Head, 2016). Profound disruptions to the normal ways of things appears almost certain. Whatever may come, tourism and its analogues – moving, visiting, witnessing – will be a part of it, and must therefore hold a prominent place in our research reckonings.

And on the subject of visiting and witnessing, I conclude by returning to the dying forests and melting glaciers of the Canadian Rockies. Hiking through mountainsides of dead spruce into ex-glacial valleys (through which ice a thousand feet thick once moved), the gravity of the Anthropocene is clear. What is required is nothing short of a 'new earth' (Saldanha & Stark, 2016, p. 427), permanently altered, populated with 'citizens of the Anthropocene' (Head, 2016, p. 142) who live differently with re-evaluated and reconfigured relations with nonhuman others. With full admission of my own family's privilege and our vacation's dependence on colonial-capitalist circuits and systems, I'm nevertheless glad my children got to witness melting glaciers and talk about our planet, as we walked among dying trees.

Notes

1. Discourses of impending environmental doom also perform geopolitical and geoeconomic purposes (Norum & Mostafanezhad, 2016, p. 159), among which are the spurring of 'last chance tourism' (Piggott-McKellar & McNamara, 2016). Some destinations are thus likely to *benefit* commercially from climate change (cf. Tervo-Kankare, Kaján, & Saarinen, 2018).

Disclosure statement

No potential conflict of interest was reported by the authors.

ORCID

Chris Gibson ⓘ http://orcid.org/0000-0002-7242-8255

References

Alderman, D. H. (2018). The racialized and violent biopolitics of mobility in the USA: An agenda for tourism geographies. *Tourism Geographies*, *20*(4), 717. doi:10.1080/14616688.2018.1477168

Bianchi, R. V. (2009). The 'critical turn' in tourism studies: A radical critique. *Tourism Geographies, 11*(4), 484–504. doi:10.1080/14616680903262653

Blakie, P. (2016). Foreword. In S. Nepal & J. Saarinen (Eds.), *Political ecology and tourism* (pp. xi–xiii). London: Routledge.

Boukhris, L. (2017). The Black Paris project: The production and reception of a counter-hegemonic tourism narrative in postcolonial Paris. *Journal of Sustainable Tourism, 24*, 684–702. doi: 10.1080/09669582.2017.1291651

Britton, S. (1991). Tourism, capital, and place: Towards a critical geography of tourism. *Environment and Planning D: Society and Space, 9*(4), 451–478. doi:10.1068/d090451

Büscher, B., & Fletcher, R. (2017). Destructive creation: Capital accumulation and the structural violence of tourism. *Journal of Sustainable Tourism, 25*(5), 651–667. doi:10.1080/09669582.2016.1159214

Castree, N. (2016, August 30). An official welcome to the Anthropocene epoch - but who gets to decide it's here? *The Conversation.* pp. 1–5.

Connell, J. (2019). Tourism as political theatre in North Korea. *Political Geography, 68*, 34–45. doi: 10.1016/j.polgeo.2018.11.003

Connell, J., & McManus, P. (2019). Flower viewing from horseback? New directions in Chinese tourism in Australia. *Australian Geographer, 50*, 333–347. doi:10.1080/00049182.2019.1591568

de Jong, A. (2016). Rethinking activism: Tourism, mobilities and emotion. *Social & Cultural Geography, 18*, 851–868. doi:10.1080/14649365.2016.1239754

de Jong, A. (2017). Unpacking pride's commodification through the encounter. *Annals of Tourism Research, 63*, 128–139. doi:10.1016/j.annals.2017.01.010

de Jong, A., & Varley, P. (2017). Food tourism policy: Deconstructing boundaries of taste and class. *Tourism Management, 60*, 212–222. doi:10.1016/j.tourman.2016.12.009

Dean, M. (2010). *Governmentality: Power and rule in modern society.* London: Sage.

Debbage, K. (2018). Economic geographies of tourism: One 'turn' leads to another. *Tourism Geographies, 20*(2), 347–353. doi:10.1080/14616688.2018.1434816

Devine, J. A. (2017). Colonizing space and commodifying place: Tourism's violent geographies. *Journal of Sustainable Tourism, 25*(5), 634–650. doi:10.1080/09669582.2016.1226849

Devine, J. A., & Ojeda, D. (2017). Violence and dispossession in tourism development: A critical geographical approach. *Journal of Sustainable Tourism, 25*(5), 605–617.

Dewsbury, J. D. (2003). Witnessing space: 'Knowledge without contemplation. *Environment and Planning A: Economy and Space, 35*(11), 1907–1932. doi:10.1068/a3582

Dickson, A. (2015). Distancing asylum seekers from the state: Australia's evolving political geography of immigration and border control. *Australian Geographer, 46*(4), 437–454. doi:10.1080/00049182.2015.1066240

Duval, D. T. (2013). Critical issues in air transport and tourism. *Tourism Geographies, 15*(3), 494–510. doi:10.1080/14616688.2012.675581

Everingham, P. (2016). Hopeful possibilities in spaces of 'the-not-yet-become': Relational encounters in volunteer tourism. *Tourism Geographies, 18*(5), 520–538. doi:10.1080/14616688.2016.1220974

Figueroa-Domecq, C., Pritchard, A., Segovia-Péreza, M., Morgan, N., & Villacé-Molinero, T. (2015). Tourism gender research: A critical accounting. *Annals of Tourism Research, 52*, 87–101.

Fletcher, R. (2018). Ecotourism after nature: Anthropocene tourism as a new capitalist "fix". *Journal of Sustainable Tourism, 27*, 522–535. doi:10.1080/09669582.2018.1471084

Fletcher, R., Blanco-Romero, A., Blázquez-Salom, M., & Murray, I. (2018, March 8). Tourism and degrowth: Impossibility theorem or path to post-capitalism? Entitle. Retrieved from https://entitleblog.org/2018/03/08/tourism-and-degrowth-impossibility-theorem-or-path-to-post-capitalism/

Fletcher, R., & Neves, K. (2012). Contradictions in tourism: The promise and pitfalls of ecotourism as a manifold capitalist fix. *Environment and Society, 9*, 60–77. doi:10.3167/ares.2012.030105

Fox, L. (2018). Tonga to start paying back controversial Chinese loans described by some as 'debt-trap diplomacy'. ABC News (Australia). Retrieved from http://www.abc.net.au/news/2018-07-19/tonga-to-start-repaying-controversial-chinese-loans/10013996

French, S., Leyshon, A., & Wainwright, T. (2011). Financializing space, spacing financialization. *Progress in Human Geography*, *35*(6), 798–819. doi:10.1177/0309132510396749

Freytag, T., & Bauder, M. (2018). Bottom-up touristification and urban transformations in Paris. *Tourism Geographies*, *20*(3), 443–460. doi:10.1080/14616688.2018.1454504

Gardner, B. (2016). *Selling the Serengeti: The cultural politics of safari tourism*. Athens, GA: University of Georgia Press.

Gibson, C. (2008). Locating geographies of tourism. *Progress in Human Geography*, *32*(3), 407–422. doi:10.1177/0309132507086877

Gibson, C. (2009). Geographies of tourism: Critical research on capitalism and local livelihoods. *Progress in Human Geography*, *33*(4), 527–534. doi:10.1177/0309132508099797

Gibson, C. (2010). Geographies of tourism: (Un)ethical encounters. *Progress in Human Geography*, *34*(4), 521–527. doi:10.1177/0309132509348688

Gibson, C. (2019). A sound track to ecological crisis: Tracing guitars all the way back to the tree. *Popular Music*, *38*(2), 183–203. doi:10.1017/S0261143019000047

Gibson, C., & Warren, A. (2019). Keeping time with trees: Climate change, forest resources, and experimental relations with the future. *Geoforum*, doi:10.1016/j.geoforum.2019.02.017

Gonzalez, V. V. (2013). *Securing paradise: Tourism and militarism in Hawai'i and the Philippines*. Durham: Duke University Press.

Goodsman, D. W., Grosklos, G., Aukema, B. H., Whitehouse, C., Bleiker, K. P., McDowell, N. G., … Xu, C. (2018). The effect of warmer winters on the demography of an outbreak insect is hidden by intraspecific competition. *Global Change Biology*, *24*(8), 3620–2628. doi:10.1111/gcb.14284

Gren, M. & Huijbens, E. H. (Eds.). (2016). *Tourism and the Anthropocene*. London: Routledge.

Griffiths, M. (2015). A compelling and flawed story of power-body relations. *Tourism Geographies*, *17*(4), 627–629. doi:10.1080/14616688.2015.1053975

Groundwater, B. (2017, May 27–28). After a fashion. Sydney Morning Herald, pp. 16–21.

Gurran, N. (2018). Global home-sharing, local communities and the Airbnb debate: A planning research agenda. *Planning Theory & Practice*, *19*, 298–304. doi:10.1080/14649357.2017.1383731

Gurran, N., Searle, G., & Phibbs, P. (2018). Urban planning in the age of Airbnb: Coase, property rights, and spatial regulation. *Urban Policy & Research*, *36*, 399–416. https://doi.org/10.1080/08111146.2018.1460268

Hall, C. M. (2015). Tourism and biological exchange and invasions: A missing dimension in sustainable tourism? *Tourism Recreation Research*, *40*(1), 81–94. doi:10.1080/02508281.2015.1005943

Hall, C. M. (2016). Loving nature to death: Tourism consumption, biodiversity loss and the Anthropocene. In M. Gren & E. H. Huijbens (Eds.), *Tourism and the Anthropocene* (pp. 52-73). London: Routledge.

Hall, C. M., James, M., & Wilson, S. (2010). Biodiversity, biosecurity, and cruising in the Arctic and sub-Arctic. *Journal of Heritage Tourism*, *5*(4), 351–364. doi:10.1080/1743873X.2010.517845

Hamilton, C. (2016). The Anthropocene as rupture. *The Anthropocene Review*, *3*(2), 93–106. doi:10.1177/2053019616634741

Hampton, M., & Bianchi, R. (2018). *Towards a political economy of coastal tourism development in South-East Asia*. Paper Presented at the Re-Centering Critical Tourism Studies Conference, Gadgah Mada University, Yogyakarta Indonesia, March 4.

Hampton, M., & Hamzah, A. (2016). Change, choice, and commercialization: Backpacker routes in Southeast Asia. *Growth and Change*, *47*(4), 556–571. doi:10.1111/grow.12143

Haraway, D. (2015). Anthropocene, capitalocene, plantationocene, chthulucene: Making kin. *Environmental Humanities*, *6*(1), 159–165. doi:10.1215/22011919-3615934

Harvey, D. (2011). *The Enigma of capital and the crises of capitalism*. Oxford: Oxford University Press.

Hatley, J. (2012). The virtue of temporal discernment: Rethinking the extent and coherence of the good in a time of mass species extinction. *Environmental Philosophy*, *9*(1), 1–12. doi:10.5840/envirophil2012912

Head, L. (2010). Cultural ecology: Adaptation - retrofitting a concept? *Progress in Human Geography*, *34*(2), 234–242. doi:10.1177/0309132509338978

Head, L. (2016). *Hope and Grief in the Anthropocene: Re-conceptualising human-nature relations*. London: Routledge.

Head, L. (2017). Why stop at response? *Dialogues in Human Geography*, *7*(2), 203–206. doi:10.1177/2043820617720094

Higgins-Desbiolles, F., & Whyte, K. P. (2013). No high hopes for hopeful tourism: A critical comment. *Annals of Tourism Research*, *40*, 428–433. doi:10.1016/j.annals.2012.07.005

Hoogendoorn, G., & Fitchett, J. M. (2018). Tourism and climate change: A review of threats and adaptation strategies for Africa. *Current Issues in Tourism*, *21*(7), 742–759. doi:10.1080/13683500.2016.1188893

Ioannides, D., & Zampoukos, K. (2018). Tourism's labour geographies: Bringing tourism into work and work into tourism. *Tourism Geographies*, *20*(1), 1–10. doi:10.1080/14616688.2017.1409261

Jedd, T. M., Hayes, M. J., Carrillo, C. M., Haigh, T., Chizinski, C. J., & Swigart, J. (2018). Measuring park visitation vulnerability to climate extremes in U.S. Rockies National Parks tourism. *Tourism Geographies*, *20*(2), 224–249. doi:10.1080/14616688.2017.1377283

Kaján, E., & Saarinen, J. (2013). Tourism, climate change and adaptation: A review. *Current Issues in Tourism*, *16*(2), 167–195. doi:10.1080/13683500.2013.774323

Kristoffersen, B., Norum, R., & Kramvig, B. (2016). Arctic whale watching and Anthropocene ethics. In M. Gren & E. H. Huijbens (Eds.), *Tourism and the Anthropocene* (pp. 94–110). London: Routledge.

Lee, D., Hampton, M., & Jeyacheya, J. (2015). The political economy of precarious work in the tourism industry in small island developing states. *Review of International Political Economy*, *22*(1), 194–223. doi:10.1080/09692290.2014.887590

Lejano, R. P. (2017). Assemblage and relationality in social-ecological systems. *Dialogues in Human Geography*, *7*(2), 192–196. doi:10.1177/2043820617720093

Lenzen, M., Sun, Y.-Y., Faturay, F., Ting, Y.-P., Geschke, A., & Malik, A. (2018). The carbon footprint of global tourism. *Nature Climate Change*, *8*(6), 522–528. doi:10.1038/s41558-018-0141-x

Ley, D. (2010). *Millionaire migrants: Trans-Pacific life lines*. Oxford: Blackwell-Wiley.

McPherson, G. (2017). Floating on a sea of funny money: An analysis of money laundering through Miami real estate and the Federal Government's attempt to stop it. *University of Miami Business Law Review*, *26*, 159–189.

Millar, L. (2017, August 12). Amsterdam bands new souvenir shops and live streams crowds to fight tourist invasion. *Australian Broadcasting Corporation*. Retrieved from https://www.abc.net.au/news/2017-11-08/amsterdam-fights-fight-tourist-invasion/9127518.

Minoia, P. (2016). Venice reshaped? Tourist gentrification and sense of place. In N. Bellini & C. Pasquinelli (Eds.), *Tourism in the city* (pp. 261–274). New York: Springer.

Montezuma, J., & McGarrigle, J. (2018). What motivates international homebuyers? Investor to lifestyle 'migrants' in a tourist city. *Tourism Geographies*, *21*(2), 214–234. doi:10.1080/14616688.2018.1470196

Moore, A. (2015). Islands of difference: Design, urbanism, and sustainable tourism in the Anthropocene Caribbean. *The Journal of Latin American and Caribbean Anthropology*, *20*(3), 513–532. doi:10.1111/jlca.12170

Moore, A. (2016). Anthropocene anthropology: Reconceptualizing contemporary global Change. *Journal of the Royal Anthropological Institute*, *22*(1), 27–46. doi:10.1111/1467-9655.12332

Moore, A. (2018a). Selling Anthropocene space: Situated adventures in sustainable tourism. *Journal of Sustainable Tourism*, *27*(4), 436–451. doi:10.1080/09669582.2018.1477783

Moore, A. (2018b). *Destination Anthropocene*. Berkeley: University of California Press.

Mostafenezhad, M. (2014). *Volunteer tourism: Popular humanitarianism in neoliberal times*. New York, NY: Routledge.

Mostafenezhad, M. (2018). The geopolitical turn in tourism geographies. *Tourism Geographies*, *20*, 343–346.

Mostafanezhad, M., & Norum, R. (2018). Tourism in the post-selfie era. *Annals of Tourism Research*, *70*, 131. doi:10.1016/j.annals.2017.11.008

Nadeau, B. (2009, September 1). Why are the Venetians fleeting Venice? *Newsweek*. Retrieved from https://www.newsweek.com/why-are-venetians-fleeing-venice-76751

Nadeau, B. (2012, October 31). Are cruise ships damaging Venice? *Daily Beast*. Retrieved from https://www.thedailybeast.com/are-cruise-ships-damaging-venice

Neef, A., & Grayman, J. H. (2018). *The tourism-disaster-conflict nexus*. Bingley: Emerald Publishing.

Nepal, S., Saarinen, J., & McLean-Purdon, E. (2016). Political ecology and tourism – concepts and constructs. In S. Nepal & J. Saarinen (Eds.), *Political ecology and tourism* (pp. 1–15). London: Routledge.

Nisbett, M. (2017). Empowering the empowered? Slum tourism and the depoliticization of poverty. *Geoforum, 85*, 37–45. doi:10.1016/j.geoforum.2017.07.007

Nofre, J., Giordano, E., Eldridge, A., Martins, J. C., & Sequera, J. (2018). Tourism, nightlife and planning: Challenges and opportunities for community liveability in La Barceloneta. *Tourism Geographies, 20*(3), 377–396. doi:10.1080/14616688.2017.1375972

Norman, J. (2019, March 28). 'Flying shame' has spread across Europe - Are Australians feeling it too? *SBS News Australia*. Retrieved from https://www.sbs.com.au/news/flying-shame-has-spread-across-europe-are-australians-feeling-it-too

Norum, R., & Mostafanezhad, M. (2016). A chronopolitics of tourism. *Geoforum, 77*, 157–160. doi:10.1016/j.geoforum.2016.10.015

Novy, J. (2018). Destination' Berlin revisited: From (new) tourism towards a pentagon of mobility and place consumption. *Tourism Geographies, 20*, 418–442. http://www.tandfonline.com/doi/abs/10.1080/14616688.2017.1357142 doi:10.1080/14616688.2017.1357142

Ooi, N., Duke, E., & O'Leary, J. (2018). Tourism in changing natural environments. *Tourism Geographies, 20*(2), 193–201. doi:10.1080/14616688.2018.1440418

Pang, S. F. H., McKercher, B., & Prideaux, B. (2013). Climate change and tourism: An overview. *Asia Pacific Journal of Tourism Research, 18*(1-2), 4–20. doi:10.1080/10941665.2012.688509

Piggott-McKellar, A., & McNamara, K. (2016). Last chance tourism and the Great Barrier Reef. *Journal of Sustainable Tourism, 25*, 1–19. doi:10.1080/09669582.2016.1213849

Rogers, D., & Koh, S. Y. (2017). The globalisation of real estate: The politics and practice of foreign real estate investment. *International Journal of Housing Policy, 17*(1), 1–14. doi:10.1080/19491247.2016.1270618

Rowen, I. (2014). Tourism as a territorial strategy: The case of China and Taiwan. *Annals of Tourism Research, 46*, 62–74. doi:10.1016/j.annals.2014.02.006

Rowen, I. (2018). Tourism as territorial strategy in the South China Sea. In J. Spangler (Ed.), *Enterprises, localities, people, and policy in the South China Sea* (pp. 61–74). New York: Springer.

Ruddick, S. M. (2017). Rethinking the subject, reimagining worlds. *Dialogues in Human Geography, 7*(2), 119–139. doi:10.1177/2043820617717847

Ryall, J. (2018, September 11). Japan tourism: The polite Japanese are finally getting sick of tourists. *Sydney Morning Herald*.

Saarinen, J. (2018a). Beyond growth thinking: The need to revisit sustainable development in tourism. *Tourism Geographies, 20*(2), 337–340. doi:10.1080/14616688.2018.1434817

Saarinen, J. (2018b). What are wilderness areas for? Tourism and political ecologies of wilderness uses and management in the Anthropocene. *Journal of Sustainable Tourism, 27*(4), 472–487. doi:10.1080/09669582.2018.1456543

Saldanha, A., & Stark, H. (2016). A new earth: Deleuze and Guattari in the Anthropocene. *Deleuze Studies, 10*(4), 427–439. doi:10.3366/dls.2016.0237

Salmon, G. (2018, January 22). Environmental message lost as Insta-traffic takes its toll on Tasmania's natural wonders. *ABC News* (Australia). Retrieved from http://www.abc.net.au/news/2018-01-22/instagram-trophy-hunters-beating-destructive-path-in-tassie/9344444

Schmude, J., Zavareh, S., Schwaiger, K. M., & Karl, M. (2018). Micro-level assessment of regional and local disaster impacts in tourist destinations. *Tourism Geographies, 20*(2), 290–308. doi:10.1080/14616688.2018.1438506

Sharp, H. (2017). Spinoza and the possibilities for radical climate ethics. *Dialogues in Human Geography, 7*(2), 156–160. doi:10.1177/2043820617720063

Smet, K. (2016). Housing prices in urban areas. *Progress in Human Geography, 40*(4), 495–510. doi:10.1177/0309132515581693

Sörensson, S. (2012). Providing fun in the 'world of tourism': Servicing backpackers in Indonesia. *Gender, Place & Culture, 19*, 670–685. doi:10.1080/0966369X.2011.625083

Steffen, W., Persson, A., Deutsch, L., Zalasiewicz, J., Williams, M., Richardson, K., … Svedin, U. (2011). The Anthropocene: From global change to planetary stewardship. *Ambio, 40*(7), 739–761. doi:10.1007/s13280-011-0185-x

Talinbayi, S., Xu, H., & Li, W. (2018). Impact of yurt tourism on labor division in nomadic Kazakh families. *Journal of Tourism and Cultural Change, 7*, 1–17. doi:10.1080/14766825.2018.1447949

Tervo-Kankare, K., Kaján, E., & Saarinen, J. (2018). Costs and benefits of environmental change: Tourism industry's responses in Arctic Finland. *Tourism Geographies, 20*(2), 202–223. doi:10.1080/14616688.2017.1375973

van Dooren, T. (2017). Making worlds with crows: Philosophy in the field. *RCC Perspectives: Transformations in Environment and Society, 1*, 59–66.

Weaver, A. (2005). Spaces of containment and revenue capture: 'Super-Sized' cruise ships as mobile tourism enclaves. *Tourism Geographies, 7*(2), 165–184. doi:10.1080/14616680500072398

Woodyer, T., & Geoghegan, H. (2013). (Re)enchanting geography? The nature of being critical and the character of critique in human geography. *Progress in Human Geography, 37*(2), 195–214. doi:10.1177/0309132512460905

Xu, H. (2018). Moving toward gender and tourism geographies studies. *Tourism Geographies, 20*(4), 721. doi:10.1080/14616688.2018.1486878

Young, M., & Markham, F. (2019). Tourism, capital, and the commodification of place. *Progress in Human Geography*, 030913251982667. doi:10.1177/0309132519826679

Yrigoy, I. (2014). The production of tourist spaces as a spatial fix. *Tourism Geographies, 16*(4), 636–652. doi:10.1080/14616688.2014.915876

Yrigoy, I. (2016). Financialization of hotel corporations in Spain. *Tourism Geographies, 18*(4), 399–421. doi:10.1080/14616688.2016.1198829

Yrigoy, I. (2018a). Transforming non-performing loans into re-performing loans: Hotel assets as a post-crisis rentier frontier in Spain. *Geoforum, 97*, 169–176. doi:10.1016/j.geoforum.2018.11.001

Yrigoy, I. (2018b). Rent gap reloaded: Airbnb and the shift from residential to touristic rental housing in the Palma Old Quarter in Mallorca. *Urban Studies*, 004209801880326.doi:10.1177/0042098018803261

Yrigoy, I., & Cañada, E. (2019). Fixing creditor-debtors' tensions through labor devaluation: Insights from the Spanish hotel market. *Geoforum, 98*, 180–188. doi:10.1016/j.geoforum.2018.11.012

Zimmermann, F. M. (2018). Does sustainability (still) matter in tourism (geography). *Tourism Geographies, 20*(2), 333–336. doi:10.1080/14616688.2018.1434814

What western tourism concepts obscure: intersections of migration and tourism in Indonesia

Kathleen M. Adams

ABSTRACT

Classic Anglo-European definitions of tourism as recreational travel have hindered more nuanced locally-grounded understandings of travel phenomena elsewhere in the world. Moreover, contemporary global labor and educational mobility have produced novel travel forms and behaviors that straddle the Western categories of "tourist" and "migrant." The purpose of this analysis is to examine Toraja (Indonesia) perspectives on travel which can be instructive for correcting the binary divides between tourism and migration that have long plagued dominant Western models of travel. Drawing from data culled from long-term qualitative fieldwork and online research, I convey three ethnographically-grounded stories of Toraja migrants on return visits to their homeland in order to destabilize Western-centrism in tourism studies. Research findings underscore contemporary travel understandings and practices that do not fit neatly with Western mutually exclusive categories of "tourism" and "migration." These Toraja practices encompass local historical patterns of travel for experiential/financial enrichment (*merantau*), migration *and* tourism. This study also advances tourism scholarship by highlighting the importance of local knowledge and demonstrating the value of ethnographic storytelling as a scholarly strategy for destabilizing orthodox Western-centric theoretical understands of tourism. The global significance of this place-based research is that tourism studies can be enriched by widening our lenses to also consider emigrants on return visits to their homelands.

摘要

欧美地区旅游的经典定义为休闲旅行, 这阻碍了人们对世界其他地区的旅行现象更细微的、在地性的理解。此外, 当前全球劳动力和教育流动产生了跨越西方"游客"和"移民"范畴的新型旅游形式和行为。本研究的目的是检验印度尼西亚托拉雅族人对旅游的看法, 这有助于纠正长期困扰西方主流旅行模式中的旅游和迁移之间的二分法。从长期的定性田野调查和在线研究中收集的数据中, 我传达了三个以民族志为基础的故事, 是关于托拉雅移民为了颠覆旅游研究中的西方中心主义而回访故乡的故事。研究结果强调了当前有关旅行的理解和实践并不完全符合西方相互排斥的"旅游"和"迁移"类别。这些托拉雅族人的实践包括出于充实旅游体验和财务自由、迁移和旅游的当地悠久的旅行模式。本研究还通过强调地方知识的重要性和展示民族志叙事作为打破正统的西方中心的旅游理论理解的学术策略的价值, 推进了旅游学术的发展。这

种以地方为基础的研究的全球意义在于, 旅游研究可以通过拓宽
我们的视角来丰富, 从而也可以为虑侨民回访他们的祖国。

Culturally-grounded travel stories, critical tourism studies and kudzu

The recent critical turn in tourism studies underscores the need for more systematic examinations of how both the practice of tourism and our analyses of it are embedded in asymmetrical power relations and hegemonic discourses (Ateljevic et al., 2007; 2012; Chang, 2019; Swain, 2009; Wijesinghe & Mura, 2018). Growing numbers of scholars are now calling for a decolonization of tourism studies (e.g. Hollinshead, 2016; Wijesinghe et al., 2019), or for approaches that embrace cultural plurality and difference (e.g. Coles et al., 2006, 2016; Hollinshead, 2010; Mura & Wijesinghe, 2019; Yamashita, 2019). With the aim of interweaving some of these threads of critical tourism studies, this article highlights some of the subtle ways in which our understanding of tourism in the Indonesia hinterlands continues to be filtered through inherited, largely Western-generated, scholarly constructs. Moreover, this article advocates for a particular methodological approach as a strategy for fostering more nuanced understandings of travel behaviors that can ultimately help dislodge Western/Anglo-centrism in tourism studies: this approach entails foregrounding local knowledge and anthropological story-telling.

Drawing on thirty years of ethnographic research in the Toraja homeland of Sulawesi, Indonesia, online discussions, and face-to-face interviews with Torajan migrants who have journeyed home, this article spotlights Torajans' lived experiences pertaining to travel. My aim is to showcase how culturally-grounded understandings of travel can offset and challenge classic Western models which tend to define tourists as outsiders engaged solely in leisured pursuits. I offer analysis of several case studies representing broader patterns observed over the course of decades of field research. Some of these ethnographically-grounded cases are presented in story-like fashion, which I argue is a potent but underrecognized avenue for conveying insights, fostering empathy and decentering the field. My overarching aim is to integrate locally-grounded cultural understandings of travel with broader political-economic factors coloring their experiences (e.g. Gillen & Mostafanezhad, 2019; Mostafanezhad, 2013; Sin, 2009; Su et al., 2018). The 'critical turn' has generally focused more on symbolic and cultural aspects of tourism (e.g. stressing hegemonic cultural imagery and discourses), often foregoing more penetrating analyses of "the asymmetries of power and divisions of labour that have grown under conditions of neo-liberal capitalism and globalization, and how these … manifest in specific tourism locations" (Bianchi, 2009, p. 487). This article seeks to interweave these two dimensions.

Before addressing the ethnographic material, a foray into the landscape of Asian critical tourism studies (A-CTS) is warranted. The idea of a "critical Asian tourism scholarship" is a recent endeavor (Chang, 2015, p. 84). Chang (2015) outlines two broad, sometimes-intersecting currents in A-CTS, one that interrogates the usefulness of orthodox Anglo-American constructs of Asian tourism for understanding Asian tourism dynamics (Winter, 2009; Winter et al., 2009), and a second, more radical current underscoring the need to Asianize the field (King & Porananond, 2014; Teo & Leong, 2006). As Chang observes, both currents signal not only desires to use scholarly knowledge to make a difference, but also recognition of the importance of foregrounding the diversity of "Asian" identities, perspectives, and experiences *vis á vis* tourism.

However, we might liken A-CTS to the kudzu vine—a creeping plant mythologized in American urban legends as an imported species that rampantly spread along rural highways and railroad embankments in America's Deep South. Like kudzu, from a certain standpoint, A-CTS approaches seem increasingly ubiquitous on the landscape. Yet, just as American scientists slowly came to realize that kudzu's all-enveloping spread was exaggerated, A-CTS's roots have not fully penetrated all realms of Asian tourism scholarship. As some have observed, A-CTS have flourished more along some byways than others (Mura & Wijesinghe, 2019). For instance, Mura and Pahlevan Sharif (2015) note that while critiques of Western-centric tourism scholarship are increasingly rampant in "Western/Anglo" realms, "sadly these voices have been less incisive within the "non-western"/"colonized" tourism academic world, including Asia" (Mura & Wijesinghe, 2019, p. 1). However unlike kudzu, which is generally disdained as a harmful invasive species, A-CTS offers great promise for yielding more insightful and (positively) impactful research.

Organizing framework: gates and yellow brick roads

This article draws on two framing images that are 'good to think' (*'bonnes à penser'*), to paraphrase Claude Lévi-Strauss's classic reference to forms that are productive for exploring broader social structures and formations (Lévi-Strauss, 1962). My first image is that of a gate, or gateway. As travelers, we are familiar with an assortment of gates, ranging from the gates of Jerusalem to welcoming gateway arches marking entry to tourist zones such as Chicago's Chinatown or Jakarta's Beautiful Indonesia Miniature Park. Yet these structures are not simply welcoming markers of inclusion: they can also contain human-made barriers. The ancient gates of once-walled cities like Jaipur and the contemporary gates of resorts and planned vacation communities remind us that gates not only mark "home" and "home communities"—that is, places of belonging—but also serve as edifices of control, regulating flows of people. In this vein, it pays to remember that gates are tethered to walls, the ultimate structures marking borders between insiders and outsiders. Recently, geographers have highlighted growing fixations on walls and wall-building as reactions to the "uncontrolled movement of individuals and non-state actors … [which] may be understood as a response to the decline of sovereign power in a globalized world … [where] 'enemy-others' … materialize in the figure of terrorists, … irregular (and errant) migrants" (Minca, 2017, p. 5). But as the Toraja case demonstrates, 'migrants' can also share terrain with the contrasting Western category of 'tourists' who are courted by nations and happily ushered through welcoming gateways.

Gates are more than sedentary markers. The very notion of a gate hinges on mobility: gates can open or close, controlling our moves between insider and outsider realms, and managing active congress between diverse peoples. Drawing on the framing image of gates enables contemplation of tourism's entwinement with privilege and inequality, migration and displacement. First, if we consider gates as material structures–ranging from bamboo to fortified steel–that are implanted in specific, legally dictated sites, we encounter politics, power and economics. Our scholarly perspectives on power set us apart from everyday tourists, enabling us to lend incisive insights into pressing issues. As Errington and Gewertz (1989) observed over 25 years ago, "If [we are] ... to have anything of importance to say ... we need to develop ... a voice as politically informed as that of Jamaica Kincaid" (1989, p. 52). In short, critical tourism studies can offer crucial insights into how entitlements and exclusions of nationality, class, ethnicity, gender, and religion inform the social practices that undergird human lives. The Toraja cases presented here illustrate how legal status, cultural and financial capital heavily color the experiences of Toraja migrants returning for homeland visits. While poorer Torajan migrant laborers are less likely to have the ability to return home for leisure visits, wealthier Torajan return visitors often enjoy the cocooned comfort of air-conditioned hotels, interspersing sometimes stressful family visits and ritual duties with extensive touristic activities (see Adams, 2019).

We can also reflect on gates and gateways as betwixt-and-between liminal zones, where one leaves behind one realm to enter another. Arnold Van Gennep (1909) coined the term 'liminality' to describe ritual transitions or passages from one culturally-defined state to another: this ritual stage involves "creation of a tabula rasa, through the removal of previously taken for granted forms and limits" (Szakolczai, 2009, p. 148). Liminality removes novitiates from earlier roles, ways of thinking, and habitual practices, ultimately transforming them. Thus, as markers of liminal zones, gates remind us of a key contribution of critical tourism studies, and this article: to challenge deeply engrained binaries between North and South, East and West, tourists and migrants, home and away. This is an enterprise undertaken by ever-growing numbers of tourism scholars over the past two decades (e.g. Ashtar et al., 2017; Cohen & Cohen, 2015; 2019; Franklin & Crang, 2001). Yet, in an era characterized by simplistic "we"/"they" worldviews, it bears underscoring that scholars can play a vital role in moving students, policy-makers, and publics away from reductionist thinking that can often obscure deeper understandings.

Finally, just as gateways and their gates are implanted in soil, a distinctive strength of ethnographically-based tourism scholarship is its grounded-ness in people's everyday lives, in the mundane and momentous moments that offer insights into broader issues. Tourism ethnographers' abilities to convey and amplify the stories, textures, tastes, and feelings of those in whose lives we partake offers the promise of engendering empathy in an increasingly intolerant world. Recently, the field of anthropology (where ethnographic methodologies originated) has witnessed a small flurry of books pushing for an expanded vision of "the anthropologist as storyteller" (Gottlieb, 2016, p. 93; also see Narayan, 2007, 2012; Wulff, 2016). Anthropology's 1980s post-modern "crisis of authority" ultimately prompted anthropologists to "sharpen their writing tools" (Wulff, 2016, p. 3), paving the way for a wider range of acceptable genres for imparting ethnographically-honed insights, ranging from graphic novels to creative-nonfiction. Such story-oriented interventions have yet to hit tourism studies. While tourism scholars have examined storytelling as a creative, improvisational guiding practice merging entertainment and education (Wynn, 2005), as an avenue for–or challenge to–destination-branding (Martin & Woodside, 2011) or as central to tourist practices (Bruner, 2005; Chronis, 2005), less attention has been directed to storytelling's value as a scholarly strategy for destabilizing orthodox theoretical understandings. Via the stories of traveling and returning Torajans presented here, I suggest that ethnographic storytelling is one important yet under-explored pathway towards fostering a de-centering of Western-Anglo-centrism in tourism studies.

A second, subsidiary image framing this article is that of the 'yellow brick road.' For film buffs, these twin images of 'gates' and 'yellow brick road' will evoke the 1939 Hollywood film *The Wizard of Oz*, a tale of a Kansas farm girl, Dorothy, swept away by a tornado to the distant Land of Oz. There, Dorothy and her newfound friends follow a meandering, perilous, yet promise-filled yellow brick road to reach the gates of the Emerald City, where they hope to realize their dreams. (For Dorothy, the dream is a return home to Kansas). Like Oz's proverbial 'yellow brick road,' tourism development was initially envisioned as a pathway to economic empowerment and nation-building in Indonesia (Soekarno n.d.; Picard, 1997; Adams, 1997, 2018). Yet just as Oz's yellow brick road was plagued by unanticipated twists, turns and detours, so too has been the story of tourism in these nations. Anchoring a *Tourism Geographies* article in images from an old musical about a displaced person may seem odd, but this film (centering on a voyage) can be instructive as a starting point for problematizing the persistent categories structuring our field.[1]

As tourism scholars, we've inherited sets of traditional Western-centric disciplinary and topical boundaries: we study tourists, be they "domestic" or "international," and we study touristic places. Also, depending on our disciplinary emphases, we study how tourists perceive, interact with, or transform host communities, be it economically, culturally or ecologically. Yet, Oz's Dorothy, like many travelers, can blur our categories: Though a displaced person facing possible permanent exile with no clear path home from Oz, Dorothy occasionally adopts touristic gazes and enacts touristic behaviors (as when admiringly the Emerald City from a distance and later enjoying the city's novel delights that contrast with her grey, depression-era rural Kansas home). How might expanding our traditionally narrow focus on 'tourists' to encompass a broader range of (im)mobilities–displacement, migration, exile–yield new insights? Some scholars have begun exploring this question in other parts of the world (e.g. Bloch, 2017; Coles & Timothy, 2004, Tie and Seaton, 2013), yet this merits further examination in Indonesian contexts. As I will suggest, many Torajans' travel practices and conceptions accommodate more overlap between tourism and migration than is typically captured via our Western categorizations.

I argue that a "Toraja-centric" perspective highlighting the blurriness between tourism/migration can be instructive for correcting the one-size-fits-all binary divides between tourism and migration that have plagued classic Western models of travel. The Western academic history of narrowly defining tourism primarily in terms of recreational travel and excluding returning migrants from our lenses has hindered more nuanced understandings of contemporary travel dynamics in Toraja (and elsewhere in Indonesia). Moreover, contemporary post-colonial global labor and educational mobility have produced novel travel forms and behaviors that are betwixt and between the categories of "travel" and "migrant" return visits that western scholarship has classically treated as mutually exclusive. As I will illustrate, for first and second-generation Torajan migrants', homebound trips entail interwoven activities (attending family funerals *and* visiting local tourism sites) that cannot neatly fit into the Western rubrics of "returning migrant" or "visiting friends and relatives" (VFR). Examining Toraja travel activities and attitudes from the ground-up underscores that travel patterns vary in different societies and are dynamic. In short, the Toraja case offers a corrective to classic, narrower Western-centric definitions of who is a tourist and what constitutes tourism.

The story of Oz is instructive in yet another way. While Dorothy and her friends focused on reaching their destination, the Emerald City, it turns out their journey *along* the yellow brick road was as important as their destination. With some exceptions, many of us in tourism studies habitually focus our lenses on tourist destinations, yet the Wizard of Oz reminds us that *the pathw*ays to and from those destinations–*not just the destinations themselves*–merit our scholarly attention. Thus, a secondary theme in this article entails exploring insights we might gain by focusing our lens on the trails and gateways to the ultimate destination, rather than squarely on the ultimate destination itself. Such a focus disrupts our habitual spotlighting of the dynamics at play in tourist zones, obliging us to better appreciate and more carefully attend to tourism's reverberations in out-lying zones.

Methodology

This study employed a mixed methods approach (Bernard, 2006): data collection involved a combination of ethnographic methods, in which participant observation (Spradley, 2016) figured prominently. Beginning in 1984–85 with biennial returns through 1998, and 2007, 2012, and 2017 visits, my participant observation in South Sulawesi spans three decades and informs my understanding of tourism and broader travel dynamics in and beyond the region. This long-term, grounded field research also lent insights into local ideas surround travel and locals' experiences of the overlapping terrain of tourism and migration in the Toraja highlands. My participant activities ranged widely. While residing with a rural highland Toraja family, I apprenticed with tourism carvers and souvenir vendors, attended government tourism planning meetings, guide workshops, cultural festivals, and over 70 Toraja rituals (which also drew return migrant participants who spent free hours touring their homeland with their offspring). I also joined formal tours alongside visiting return-migrants and participated in Toraja family trips to local tourism destinations (initiated when relatives returned for visits). These local destinations included Londa and Lemo's burial sites, Sa'dan weaving village, Mt. Sesean, Buntu Burake's giant Jesus statue, and Lolai, a "land above the clouds" sunrise vista site.

While some quantitative researchers bemoan ethnographic methods as too focused and specific to be replicated (Alder & Alder, 1994, cited in Hollinshead, 2004), many observe that ethnographic methods are particularly valuable for illuminating people's often emotionally-charged experiences with challenges, disruptions, and change (Ellen, 1984). Thus, these methods are especially suitable for this study. Particularly when researching disenfranchised minorities or those with precarious legal statuses (such as tourist visa overstayers), reliable data is notoriously difficult to obtain via question-naires and surveys. However, long-term participant observation fosters trust and more candid responses, thereby facilitating insights into topics not openly discussed in for-mal interview settings (Adams, 2012; Cole, 2004).

Due to the centrality of mobility as a research theme, data collection necessitated multi-sited ethnography (Marcus, 1995). While multi-sited (a.k.a. multi-local) research has its deepest roots in migration studies (e.g. Watson, 1977), and has also been adopted by geographers (e.g. Conran, 2011; Olwig, 2012), this technique has become prominent in tourism scholarship in recent decades (e.g. Adams, 2006; Bruner, 2005; Haldrup & Larsen, 2010). My research sites ranged from the Toraja highlands, to the road-stops along the nine-hour drive from Makassar (the lowland capital) to Toraja, to first- and second-generation Toraja migrant homes in Makassar and Eastern Indonesia (Figure 1). I also visited Jakarta and Chicago migrant Torajans who recounted their returns to the ancestral homeland. Some of my field sites were not in the physical world but in the internet cloud. As a key meeting space where far-flung Torajan migrants share memories and photos of travels home, the cyberworld of social media proved to be a fertile space for enriching my understanding of Torajan migrants' diverse experiences vis á vis their homeland. Although Indonesia is hailed as one of the world's most social-media active nations (Lim, 2013; Yogaswara, 2010), and had the third highest number of Facebook users in 2019,[2] scholarly examinations of Indonesian travelers' social media use to navigate experiences during homeland visits remain rare (but see Hamzah, 2013). This is surprising, since the cyber-world of social media offers a potentially fertile data source for these inquiries, particularly when used in triangulation with other data pools.

Formal and informal interviews (Gubrium & Holstein, 2001) were also employed. I selected participants via a purposive sampling strategy (Cresswell & Plano Clark, 2011). This technique entails identifying interviewees with extensive experience in the realm of one's study (Bernard, 2006). Interviewed with first and second generation Toraja migrants who had returned for visits to their ancestral homeland were in person and online (private and group interviews). A call for participants on Toraja migrant social media pages drew 59 interviewees. Most responded to an announcement on the larg-est Toraja migrant Facebook page–"Worldwide Toraja Migrants United" (*Persatuan Perantau Toraja Seluruh Dunia*)—which boasts over 40,000 members. These interviews included three private group discussions in 2015 and 2016 consisting of 25, 15, and 7 migrants who shared their motivations for and experiences while returning home for visits. An additional 12 were interviewed privately via Facebook messenger and phone, to ensure validity of group interview findings and elicit more detailed commentary.

After the administrator of the "Worldwide Toraja Migrants United" Facebook group announced my presence and research goals, I scanned page postings for a three-month period (in 2015–2016). These routine scans attuned me to overarching themes in migrant homeland returns and helped inform subsequent formal-interview ques-tions. However, I refrain from quoting directly from these Facebook postings due to ethical concerns stemming from the unlikelihood that all 40,000 members spotted periodic posts announcing my presence as a researcher. This caution was reinforced by Harng Luh Sin's (2015) observations concerning social media fieldwork ethics. As

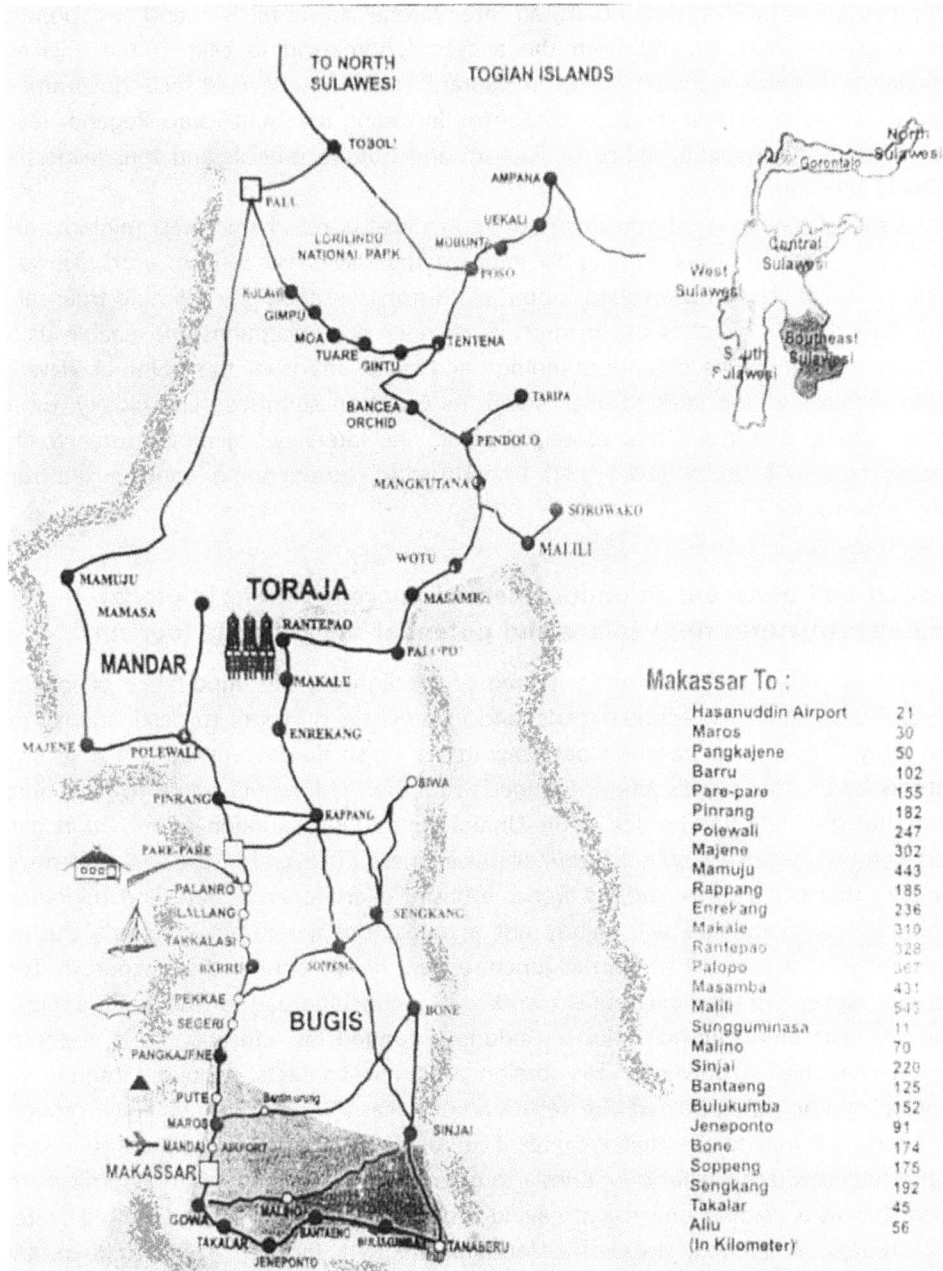

Figure 1. Map of South Sulawesi, indicating the Toraja highlands, Makasar and Stanis and Katrin's restaurant. Drawn by the late Stanis Sandarupa and used with permission from Dirk Sandarupa.

Sin observes, social media blurs the traditional divide between private and public spaces offering new data feasts yet also serving up an ethical "can of worms" (2015, p. 680).

In 2017, an additional 22 informal face-to-face interviews were conducted with first- and second-generation Toraja migrants visiting tourist sites in the Toraja highlands,

Rantepao cafes, local funeral rituals, an international music festival, and two popular rest stops en route to and from the ancestral homeland (a café in the adjacent regency and a Toraja-owned coastal restaurant). I also interviewed local government officials whose sectors encompassed tourism (including the Tana Toraja Regency leaders, North Toraja Regency Office of Tourism and Culture officials and four leaders in districts with tourist sites).

Additional data derived from focused life histories of returning Toraja migrants and tourists, as well as those hoping to improve their lives via tourism work. Howard Becker's (1970) classic observations about life history methodology still hold true: carefully collected life histories can disrupt researchers' preconceptions and enable us to reframe our categories of understandings and questions from the point of view of those whose lives we seek to understand. As Goodson surmises, "Life history, by its nature, asserts and insists that power[ful actors and interviewers] should listen to the people it claims to serve" (2001, 131). I now turn to recount some stories culled from this fieldwork.

Betwixt and between: an undocumented migrant caught in global political structures finds solace and potential salvation via tourism

In keeping with the Wizard of Oz theme of spotlighting the importance of looking beyond destinations to address roads (and experiences en route to destinations), my first story begins on a traveler's pathway, in the harsh fluorescent light of a security gate in Japan. There, a distraught Indonesian mother, Elsi, awaits return to her homeland after spending half her life in the United States. Elsi's situation–at an airport gate between two worlds—resulted from her involvement in several of the varied forms of mobility that characterize today's world: tourism, overseas education, and migration. While at this moment, Elsi is clearly not a tourist but a returning migrant, the life experiences that brought her to this juncture were tightly entwined with tourism. Two decades earlier, Elsi's father's tourism work was a launching pad for the family's migration. His part time tour-guide job in Indonesia funded his education at a respected regional Indonesian university. His foreign academic contacts, acquired through his tour guiding activities, helped him secure an overseas Ph.D. fellowship, which brought him and his family to the United States. Eventually, Elsi's father completed his degree and returned with his wife to Indonesia to operate a tour company and assume a professorship at a leading university, leaving Elsi behind to study in the United States. Instead, love intervened: Elsi wed a Honduran migrant, birthed several children, and began building a life in the undocumented Latino community of her American town, all the while pining for her parents and siblings in Indonesia. As she explained, she "longed to show her husband and kids the beauty of her [ancestral] homeland," but without legal papers, a visit home was too risky.

Some ten years after her parents' return to Indonesia, Elsi learned of her father's imminent death. Suddenly, the structural impediment that had kept her immobile – her undocumented status – seemed irrelevant. Using her old Indonesian passport and rush-order American passports for her three toddlers, Elsi and her children flew to Japan. At Narita's transit gate, however, a routine passport check revealed her expired

American visa. Tearful and fearful as Japanese airport officials interrogated her, she clutched her children tightly, waiting to learn if they would be permitted to pass through the gates and on to her homeland. Elsi sent me several anguished texts between interrogations, wondering if she had "done wrong" by trying to quench her "longing" to see her father one last time.

Eventually, they were allowed passage through Narita's transit gate and, upon landing in Indonesia, Elsi's toddlers' American passports were stamped not with the permanent entries afforded returning migrants but with two-month tourist visas. The security gate delay cost Elsi dearly: she reached home too late for a final embrace with her father. Instead, Elsi and her toddlers spent two bittersweet months with her mother, acquainting the children with their Indonesian kin as well as the tastes, smells and textures of their heritage, even visiting ancestral villages and tourist sites in her parents' homeland. While Elsi was a migrant who had returned home for her father's funeral, many of her activities in these months over-lapped with those of tourists (discussed below). It was only when Elsi attempted to book a flight back to the United States that she learned her undocumented status rendered return impossible. Her husband's probationary status in the United States offered neither legal means for him to bring them home, nor the ability to join them in Indonesia.

As the months of separation wore on, Elsi and her children moved in and out of despondency, feeling trapped at 'home' but not 'at home' in Indonesia. In our conver-sations during these limbo months, Elsi shared stories of long-weekend trips with her children that combined village kin visits with tours of Toraja touristic destinations (orchestrated by her highland relatives, as is common on such visits). While primarily based in Makassar (the lowland provincial capital where she was reared), Elsi, her chil-dren, and her mother spent weeks in her father's natal highland village preparing his funeral and in her mother's highland home town. She recounted how, on this first visit and subsequent highland visits, visiting Toraja tourist sites became a source of solace and reconnection with her heritage. In her words,

> Our first arrival in Toraja, it was all culture shocked (sic) for the kids. But they love more Toraja than Makassar. One of my kids loves animals and was not afraid of the pigs, chickens, wild dogs. The other love (sic) the adventures. They played with their cousins but just with laughter and smiles and hand signals with resulting adventures and wonderful memories for them. We were at OMG I forgot the name … [Simbuang] Batuallo [a Toraja gravesite and tourism destination] … We also went to Singki [a hill with a tourist pavilion overlooking Rantepao, North Toraja's capital], walking up the stairs. Passing the ancient grave[s] … [There] my mom and brother explained a bit about the Toraja culture. [My daughter] had so many questions …

On another Toraja highland visit, local relatives took Elsi and her children to visit Toraja's newest tourist destination inaugurated in 2015: a 40-meter-tall statue of Jesus hailed as the world's tallest Jesus statue. Via Facebook, Elsi shared photos of herself and her children posing at the statue's touristic gateway, explaining how her local cousins shared the cultural importance of saying *tabe* (excuse me) as they passed Toraja graves along the newly bulldozed tourist road to the site. She also sent photos of the sweeping vistas from the statue's base, describing her "pride" derived from "shar[ing] the beauty of [her parents' rural] homeland" with her American-born children. Elsi's experiences during her ancestral homeland visit on the occasion of her father's death are not atypical: Toraja first- and second-generation migrants' returns home are often sparked by desires to reunite with local kin or attend family rituals, but these visits also entail touristic activities and cultural heritage education for children reared abroad. This Toraja mélange of tourism and migrant return visits for family rites is perhaps best encapsulated by a 2012 Facebook post made by a visiting migrant Toraja during her return home for a relative's funeral. Sharing a photo of herself in traditional Toraja funeral attire standing in the ritual arena, she offers the caption "Being a tourist in my (own) home." (For additional details, see Adams, 2019) (Figure 2).

Elsi's subsequent experiences while in limbo in her homeland, unable to return home to her husband, illustrate how tourism can also serve as a "pathway" enabling migrants to regain equilibrium as they contemplate their next move. Using English and Spanish skills acquired as an undocumented migrant, Elsi toiled for almost a year in a tourist-oriented café in Sulawesi hoping that tourism-derived income and contacts might pave a path back to America, just as they had once done for her father during his tour-guiding years. In the meantime, Elsi's Facetime app becomes a nightly cyber gateway to her husband, making her fractured nuclear family momentarily whole again, albeit electronically.

The betwixt and between-ness of Elsi's situation is hardly unique in the contemporary world. While tourists from affluent nations pass easily through airport gates, economically marginalized migrants seeking labor or love beyond their homeland borders can find themselves locked out. Writing on the intersections of Indonesian labor migrants and Singaporean sex and drug tourists on Batam Island in the Asian Growth Triangle, Johan Lindquist has movingly chronicled these 'anxieties of mobility' (Lindquist, 2009). Elsi's story, however, indicates different ways in which migration and tourism are co-entwined: Either embodied in the activities of a *single* individual (Elsi), or in the ironic passport statuses of three half-Indonesian toddlers residing in Indonesia with the legal status of foreign 'American tourists.' The toddlers' official status as foreigners poses a further economic burden on Elsi, as they must exit the country every three months to reactivate their tourist visas in order to remain with their mother. Elsi explains that these costly trips are not optional, as without valid identity papers, the children would have to forego schooling. So, for now, Elsi's children retain their status as American "tourists," and Elsi waits tourists' tables hoping to stave off a plunge into Indonesia's underclass. Stories like Elsi's remind us of ethnography's value in problematizing entrenched legal, political, and behavioral classificatory binaries such as "tourist" versus "migrant" that obscure far more complex realities.

Din (2017) recently explored the blurring between migrant return visits and tourism in the Malay world (encompassing Indonesia and Malaysia) via an auto-ethnographic examination of the Malay *balik kampung* tradition. *Balik kampung* is a Malay expression that refers to migrants' returns to ancestral villages for holidays. Din ultimately suggests that this Malay migrant rite of holiday return visits to homeland villages constitutes a neglected form of tourism. His autoethnographic work offers an additional,

Figure 2. A Toraja migrant visitor photographing a funeral. Photo by the author.

related example of how Western orthodox categories of "tourist" versus "migrant" fail to capture on-the-ground realities in the Malay world (which includes Indonesia).

More broadly, it merits underscoring that the Indonesian/Malay term for migration (*merantau*) does not precisely map onto the terrain covered by the English term. *Merantau* is a widely used term generally defined as "leaving one's cultural territory voluntarily whether for a short or long time, with the aim of earning a living or seeking further knowledge or experience, normally with the intention of returning home" (Naim, 1976, p. 149–150). Today, *merantau* has become an Indonesian cultural institution (Lindquist, 2009): Indonesians use the term *merantau* to cover a wide array of mobilities, encompassing not only migration, but studying abroad and even long-term around-the-world type travels to gain experience and knowledge. Some scholars (Colombijn et al., 2012) gloss *merantau* as "wanderlust," which captures the historical salience of physical mobility in the Malay world. But I believe the notion of the 'quest' better conveys this cultural notion of travel as involving (a) undertaking movement in order to achieve some sort of transformation and (b) a vision of a final, permanent return home following achievement of the goal. (Although the Western notion of 'quest' does not assume that broad swaths of the population engage in this form of travel). Various scholars (e.g. Adams, 2016; Forshee, 2000) note that the overlay between *merantau* and tourism invites comparisons. Perhaps not surprisingly, the concept of *merantau* is "inscribed in Indonesian and Malay touristic culture, as [attested by] the existence of tourist-oriented lodgings [bearing the term] Merantau attest" (Adams, 2016, p. 19).

The practice of *merantau* was not widely adopted by Torajans until the 1970s and 1980s, when the suppression of regional Muslim armed rebellions, improved infrastructure, and new work opportunities in foreign-owned mines and timber industries made it viable for highlanders to seek fortune outside the homeland (Volkman, 1984, p. 158). Volkman characterized classic Toraja society as highly 'centripetal': in the pre-1980s era, most Torajans avoided travel and remained anchored to sacred ancestral sites and 'centers' in their homeland. She writes,

> The most significant center was the ancestral house known as 'the place where the umbilical cord is buried' (*inan larnunan lolo*). The 'planting' of a newborn child's umbilical cord in the earth on the eastern side of the house metaphorically rooted the person to a particular place. Rituals reinforced the centripetal tendencies of Toraja life by periodically reconcentrating dispersed family members at the center, whether ... [for] a funeral or a house-roofing celebration. The centripetal ideal was expressed in kinship as well: preferred marriages took place within the family and were known as 'returning to the house' [*sulle langan banua*]" (Volkman, 1984, p. 157–158).

Perhaps most significantly, Volkman observes that

> A centripetal, homeward-bound orientation is expressed even in the Toraja term for merantau: *ma'lemba kalando*; roughly, 'the long haul.' Ma'lemba means to carry something lashed to a bamboo pole across one's shoulder, and it evokes an image of a man returning from the fields with newly cut *padi* [rice stalks], bringing home the life-sustaining harvest" (Volkman, 1984, p. 158).

The paired Toraja conceptions of (1) family identity as rooted in physical places and (2) the life-sustaining role of returns home offer clues as to why contemporary migrant Torajans feel such strong affinities towards their homeland, and why they are drawn to travel home as return migrants cum tourists. Understanding these uniquely Torajan ideas regarding the magnetic pull of homeland hamlets also offers a corrective to the Western cookie-cutter divides between "tourism" and "migration." I turn now to another gateway that offers an entrée for exploring additional dynamics of migrant Torajan vacations in the homeland.

Gateway to Toraja ancestral tourism: a migrant return festival embodies economics, politics, and heritage-lite

Just as Dorothy and her rag-tag companions experienced euphoria as they neared the gates of the Emerald City (the destination that offered the promise of realizing their desires), many visiting Torajans express similar emotions upon arriving at the monumental cement and wood gateway arch marking formal entry into the Toraja highlands (Figure 3). Hundreds of thousands of travelers have passed through this gateway since it was erected (originally in simpler form in the early 1980s, as part of the government's efforts to both mark district boundaries and welcome tourists). Over 800,000 Torajans live in the mountainous homeland, working predominantly as farmers or in civil service and tourism. Over one million more Torajans reside outside the homeland, many in cities and towns elsewhere in Indonesia or Malaysia, and others in more distant countries. Toraja out-migration began in the 1970s and soon thereafter migrant remittances began fueling ritual inflation, as successful first-generation migrants raise family status by funding elaborate homeland funerals, costing up to

Figure 3. Gateway Arch marking entry to the Toraja highlands. Photo by the author.

several hundred thousand dollars (Volkman, 1984). Remittances also inflate land and livestock prices, driving still more to emigrate (de Jong, 2013).

Yet, just as the spiraling costs of Toraja rituals propel some Torajans to become migrants, these very rituals and the region's spectacular landscape also draw tourists.

(Table 1). Indonesia actively promoted international tourism to the Toraja highlands throughout the 1980s and 1990s, as tourism was envisioned both as a revenue generator and an avenue for development promising local job opportunities and raised income. At tourism's initial mid-1990s pinnacle, approximately 200,000 domestic tourists and 50,000 international tourists were visiting the Toraja highlands yearly. These internationally-derived categories of traveler types, long obscured awareness that a large percentage of those labeled "domestic tourists" were Toraja diaspora members returning home for visits (Adams, 1998). By the early and mid-2000s, however, the Indonesian economic and political crisis, ethno-religious tumult, Bali bombings, and other natural disasters slowed tourism to a trickle, puncturing the Toraja tourism economy. Thus, in 2006, local officials looked to a new, more reliable source of visitor-driven revenues. Ultimately, the successful strategy they embraced was one they were well-positioned to appreciate as Torajans: courting new pools of Toraja migrants and their families back for vacations. Here we glimpse how "local knowledge," culled from growing up in a cultural milieu, enables insights into tourism dynamics and potentials (that are often obscured by externally imposed categorizations), a theme addressed in more detail in the final case.

In December of 2006, the local government staged its first annual migrant oriented 'Longing to Return Home to Toraja' (*Toraja Mamali*) festival. Using government funds and donations from wealthy Toraja migrants, the festival featured fireworks displays, migrant welcoming parades, traditional foods, dances, and hotel discounts for returnees and their families. Heavily promoted on Facebook and conventional news outlets, the festival was a tremendous success, dramatically boosting hotel bookings and restaurant revenues during tourism's slow season. As one Torajan migrant observed, "If you don't stay in the village (with local kin), it's hard to find a hotel room. So many migrants return for the festival ... " Another second-generation Torajan migrant commented, "Though lots of Torajan [migrants] still visit home in the summer, the [December] *Toraja Mamali* Festival is really *ramai* (crowded/lively). Lots of us come home at Christmastime and attend the festival."

Further contextualization of the history of Torajan migrant return visits is necessary here. These visits fit under the broader Indonesian/Malay-based notion of *pulang kampung* (a.k.a. *balik kampung*) (re Din, ibid). Torajan homeland return visits typically occur in July and August during school vacations, and most highlanders schedule funerals and house rituals for these months to facilitate maximal familial participation (Torajans practice delayed burial, holding funerals months or years later). As Toraja kinship is structured around ancestral houses (all descendants of a particular ancestral house's founder are kin), maintaining membership in one's house-based families requires making potentially debt-inducing contributions to familial house-based rituals. For wealthier first-generation migrants, returns for ceremonies offer not only occasions to amass prestige in the homeland via material and livestock contributions but also opportunities to tutor their foreign-born children in Torajan practices. As witnessed with Elsi, such heritage tutorials frequently transpire via touristic visits to Torajan cultural sites (which are generally interdigitated with family visits and funeral or house ritual participation). As a financially-comfortable migrant residing in Jakarta offered, "Most Torajans who migrate (*merantau*), if they have a chance to return home to Toraja their goal is

Table 1. Tourist arrivals in Toraja.

Year	Foreign	Domestic	Total
1983	9,007	57,957	66,964
1984	12,547	84,338	96,885
1985	15,325	70,987	82,312
1986	19,726	113,590	133,316
1987	22,108	168,985	191,903
1988	25,308	154,865	180,173
1989	32,566	152,927	185,493
1990	39,700	171,689	211,389
1991	40,695	174,542	215,237
1992	46,799	171,172	217,971
1993	51,259	195,544	246,803
1994	56,565	204,987	261,552
1995	59,388	176,849	236,237
1996	42,123	32,930	75,053
1997	41,586	42,578	84,164
1998	22,624	30,597	53,221
1999	30,397	31,415	61,812
2000	37,805	32,207	70,012
2001	37,142	34,218	71,360
2002	30,058	32,638	62,696
2003	15,385	27,520	42,905
2004	5,762	21,802	27,564
2005	5,385	19,933	25,318
2006	5,321	20,817	26,138
2007	4,989	13,102	18,091
2008	n.a.	n.a.	25,583
2009	5,499	34,716	40,215
2010	5,627	12,631	18,258
2011	21,027	40,037	61,064
2012	25,652	35,263	64,880
2013	35,956	70,128	112,223
2014	41,058	71,522	112,580
2015	43,575	87,462	131,037
2016	51,793	112,628	164,421
2017	62,356	223,210	285,566
2018	44,425	235,712	235,712

Sources: Office of Tourism (1983–1989) & Office of Statistics (1990–2010), Tana Toraja Regency, Indonesia. 2011–2018 data derive from North Toraja Regency Office of Statistics (reflecting Tana Toraja's division into two regencies).

to get together with family, and also [attend] *rambu solo* [funeral] or *rambu tuka'* [house consecration] rituals. But, Torajans also return for tourism, especially those who are married to non-Torajans, so that our partner can get to know Toraja."

However, for first-generation Torajan migrants without means, or for second-generation Torajans reared far from the homeland, the prospect of becoming entangled in ritual debt during summertime homeland visits can be distressing. A 44-year-old first-generation Torajan migrant's (whose remittances from cruise ship employment funded construction of his mother's house) comments capture the ambivalence many migrants attach to summer vacation/ritual season visits,

> I'm married now and support my wife and kids on Java and help my mother in the village [in Toraja]. I long to see my mother and my village, and for her to see my kids. I want my kids to know Toraja. But visiting in summer is too expensive. Not the travel, but she'll [mother] want me to give money for [extended family] rituals happening then. And how can I say 'no' to my mother who birthed me? It is better for my [nuclear] family that I stay with them [on Java, not visit Toraja in the summer ritual season]."

However, by scheduling the 'Longing to return home to Toraja' festival in December, when few Torajans hold funeral or house consecration rituals, Toraja migrants gained a festive framework for returning to their ancestral homeland strictly for leisure and heritage-exploration, without the economic burdens of ritual participation. A first-generation migrant's comments concisely capture the economic and cultural issues underlying the festival's appeal to migrants, "I don't want to oblige my kids to get tangled up in Toraja ritual customs [euphemism for getting encumbered with funeral debt], but the … festival interests me. There's art, Toraja clothing, bamboo music, traditional dances." As her comments suggest, the festival-based heritage experiences of visiting Torajans are generally aesthetic and sensual, yet anesthetized (Adams, 2019, p. 204). That is, the festival enables feasting on a smorgasbord of heritage, without the long-term indigestion entailed in funeral season visits, which inevitably and inextricably entwine returnees in the burdensome realms of funeral debts and local family political rivalries. In short, these migrant Torajan festivalgoers consume 'Heritage Lite' (Adams, 2019, p. 163).

While migrants' festival experiences enable them to sidestep the politics and economics normally entailed in return visits, the festival itself has not escaped these realms. After the successful 2006 "Longing to Return Home to Toraja Festival," the event became an annual affair. Two years later, in 2008, it was rebranded to attract a broader array of Christian Indonesian and international travelers. Renamed the "Lovely December" Festival, new advertising promoted the Christian Toraja highlands as *the* destination for ecumenical Christmas and New Year festivities in Muslim Indonesia.[3] This economically-motivated rebranding was not without politics. Externally initiated, funded and imposed by the non-Torajan lowland Makassarese governor of South Sulawesi, Syahrul Yasin Limpo, the rebranding prompted pleas from Toraja leaders in both highland Toraja districts to return to the festival's original migrant-homecoming focus.[4] As one Toraja official lamented,

"Before, there were so many migrants here for the festival that Rantepao traffic stretched from the Hotel Misiliana (at the city's southernmost edge) to the bridge over the Sa'dan river (at the city's northern end). And the city glowed with decorative lights! It was so *ramai* (crowded, bustling). Now, it is less interesting … "

Other Torajan officials echoed these sentiments, observing that the festival was at its liveliest and most bustling in its early years when its exclusive focus was welcoming home visiting migrants. Moreover, various Toraja officials stressed that, unlike the English language re-branding, the festival's original name reflected Toraja regional identity and Toraja "local wisdom" (Rante, 2016). A North Toraja Regency senator's response to the English "Lovely Toraja" festival captured many Torajans' sentiments, "It sounds so much more delicious (*enak*) to use *Toraja Mamali* (Longing to Return Home to Toraja)—the ingredient of Toraja local wisdom is palpable" (Rante, 2016). The trope of 'local wisdom' offers a tacit challenge to Torajans' historic position as provincial underdogs, historically dismissed as tribal others and disenfranchised from political power (Adams, 2006; Biglke, 2005). As an ethnic and religious highland minority in a province long dominated by lowland Bugis and Makassarese, many Toraja leaders had believed their tourist drawing power would grant them authority over the branding of their own festivals. However, the reality of their limited options given political and bureaucratic asymmetries at the provincial level was increasingly evident .

The 2018 election of a new non-Torajan governor eager to imprint his mark on the political and touristic landscape brought another external rebranding of the festival as 'Kemilau Toraja,' or Toraja Glitter (sheen)" (Fajriani, 2018). Possibly inspired by Indonesia's Ministry of Tourism and Culture's embrace of "glitter" imagery for touristic festivals elsewhere (e.g. Erb, 2009), the new externally imposed rebranding was accepted by one of the two Toraja districts (Tana Toraja). However, Northern Toraja District rejected "Toraja Glitter," opting to remain with the previously imposed 'Lovely December' Festival branding. As they reasoned, at least, Lovely December had achieved name-recognition amongst returning migrants, domestic and international tourists. While the festival's name may seem trivial, the struggle embodies both economic agendas and political rivalries at the local, regional and provincial levels. Moreover, the festival re-brandings also reveal a symbolic battle between Torajan officials and provincial-level tourism consultants regarding the merits of local versus outsider wisdom surrounding the visitor drawing power of festival labels. Returning to the theme of Oz and gateways, the touristic festival branding battles also remind us of the power gateways shaping whose branding imagery prevails.

Arung pala restaurant: learning from the road[5]

Turning now from "gateways" to the second Oz frame of "yellow brick roads," we shift to the road running between Makassar (dubbed the island's lowland "gateway to Toraja") and the Toraja highlands. My aim here is to showcase the value of looking at the journey–and stopping points along the road traveled–rather than focusing solely on travelers' destinations.

Voyaging to the highlands entails a wearying nine-hour car or bus journey. The roadway hugs the coast for the initial three and a half hours, traversing Muslim Bugis and Makassarese lands, bustling county seats and slower-paced towns, as well as shrimp hatcheries and rice farms. At the port city of Pare Pare, the road twists inland, winding north through gentle hills until it ultimately narrows and switchbacks its way up into the mountains. As the ribbon of land at the road's cliffside edge thins, each bend reveals increasingly dramatic vistas. Finally, the road plateaus briefly and one arrives at the decorative gateway marking entry to the Toraja highlands.

For many Christian Torajan travelers I interviewed, the lowland portion of the journey to the homeland was qualitatively distinct from the mountainous portion. Especially in the 1980s, 1990s and early 2000s, Torajan comments about the lowland sections of this road often conveyed mild unease. As one Torajan explained in 1986, "it is not until the bus turns inland and the air cools [with the ascent into the mountains] that I can relax." And as another Torajan recounted in 1998, "When we finally reach Bambapuang (an area in the rugged regency bordering Tana Toraja Regency), then I feel good [enak]. It's not like the earlier stops. There, I look forward to getting off the bus for a coffee and smoke. The Duri people there, they are a lot like us ... - even their sweets [depa' tori] resemble ours ... " These and other Torajan comments reflect historical tensions and even mistrust between lowland Muslim groups and the predominantly Christian Torajan highlanders. For many travelling Torajans, the bus stops and rest stops along 'the yellow brick road' to the homeland marked far more than kilometers, they also marked a movement from lowland Muslim-dominated spaces of unease to higher altitude stops of growing comfort, both somatic (in terms of cooler temperatures) and psychological.

While space does not permit a full examination of the various stops along the route, here I focus on the transformations of one spot on this road that has shifted sensibilities concerning both the specific location and the journey to and from the Toraja highlands. A gateway frames this final ethnographic story: This gateway is not in a tourist destination but in the South Sulawesi District of Barru. It marks entry to a restaurant on a lonely stretch of highway between Makassar and the Toraja highlands. The worn mosaic writing paving the restaurant's entryway reads 'ta pada salama' in Bugis Lontara script. The words bear multiple meanings—welcome, goodbye, and a prayer for safe travels. For local Muslim Bugis, they convey a prayer for the peace and safety of those entering and leaving. In 2000, in a period of inter-faith and inter-ethnic frictions elsewhere in Indonesia, Stanis Sandarupa, a Toraja professor-cum-travel agent who had migrated to Makassar as a young man, purchased the run-down café from its bankrupt owner, hoping to develop it into a destination restaurant. The site was far from his mountainous childhood Toraja homeland and from the city that had become his second home. Although two Bugis farming villages were nearby, the site promised no viable local clients. Moreover, despite the café's seaside backdrop, many South Sulawesi travelers found this stretch of the road "eerie" (sunyi). For most who plied the road, this was a not a destination, nor a rest-stop, but a 'pass though' zone.

Sandarupa's urban friends and realtives cautioned him against the purchase, warning him "Muslim Bugis locals would never buy food from a pork-eating Catholic family" and that (then-on-going) ethno-religious tumult elsewhere in Indonesia did not bode well for the venture. They feared he would be putting his family at risk as the only Christian Toraja entrepreneurs in the district. Yet Sandarupa held steadfast to his plan. He had studied ethnic relations in graduate school and, as he explained, he "knew ethno-religious violence rarely happened in places where one group was clearly dominant." This academically-based insight reinforced his culturally-honed instincts— or local knowledge– and went against the grain of commonly-taught Western tourism business strategies, which tend to emphasize 'location,' steady flows of tourists, supplies (for restaurants) and affordable labor.

Here, elaboration on the concept of local knowledge is needed. Generally, local knowledge refers to nuanced understandings of in situ sociocultural practices, be they spiritual, ecological, or economic. Such understandings are often intuitive rather than analytical and overlap with indigenous knowledge (Butler & Menzies, 2007, p.18). As Butler and Menzies elaborate, both local knowledge and indigenous knowledge "have a sub-altern relationship with Western 'modern' scientific knowledge" (ibid, 7).

Sandarupa's inspiration for his traveler-oriented restaurant, as well as his innate understanding of how to achieve success, were grounded in local knowledge. His own experiences as a Toraja migrant who routinely moved between the lowland capital of Makassar and the highlands, like countless others, taught him that migrants would continue to travel this road, even when tourists were scarce. Sandarupa was also well-acquainted with the vague unease and mistrust many highland Torajans felt when traversing Muslim Bugis lowlands. He knew many Torajans preferred packing their own meals for the long nine-hour journey, rather than eating 'risky' food prepared by Bugis in what many imagined—rightly or wrongly–to be unsanitary roadside kitchens. As his mother-in-law once warned me when I was a graduate student new to the area, "Be careful where you eat at the Bugis bus stops, they wash with dirty water and never dry the plates—if you must eat there, do what we Torajans do, and eat only instant noodles [*Indomie*]." Stanis Sandarupa and his wife intuitively understood that a Toraja-owned restaurant on this lowland stretch of road roughly midway between the highlands and Makassar would attract a constant flow of hungry Toraja travelers. In addition, Sandarupa had observed his foreign tourist clients' penchant for sea vistas and calculated that they would also find the spot irresistible, since few roadside Bugis restaurants offered sea views. As he explained, "When I was a guide, I'd always stop on that stretch of the road, so the tourists could stretch by the sea–they always took a lot of photos."

After the Sandarupa family purchased the restaurant, they relocated there and sold snacks while slowly remodeling, sleeping together in the main room. As their son recounted, "it was a creepy place—naughty people [drunks, thugs and sex workers—the prior owners' clients] came there at night and my parents had to start closing early. Once they threw stones at the restaurant when we closed early." Given the restaurant's prior associations with illegal mischief (*kenakalan*), locals and police viewed the Sandarupa family with suspicion. Stanis tackled the situation by rejecting the western 'time is money' business mantra. He began devoting long hours to daily visits, partaking in relaxed meandering conversations with local male villagers and authorities. These visits ultimately built trust and mutual respect. Stanis' wife, Katrin, also started spending days visiting the village women, exchanging recipes and coming to understand the texture of their lives. Both Stanis and Katrin were reared in rural areas and intuitively understood the cultural importance of these informal visits and leisurely conversations. Rather than foregrounding their still floundering restaurant's economic 'bottom line,' they nourished local social relations, buying vegetables and seafood from neighboring farmers and fishermen, selling villagers' baked goods at the café, going door-to-door with gifts of food on New Year's Day, donating to the mosque, and hosting community parties. As Stanis's wife Katrin summed up, "Now they trust us. Before, they did not want their teens to work for us, now they send them to seek work here." Gradually, they became valued members of the community; locals stage wedding receptions in their restaurant and officials now routinely bring guests to dine there.

By 2012, their café had blossomed into a thriving restaurant featuring locally-sourced food and the family felt secure enough to open small hotel adjacent to the restaurant. The site now bustles with transit buses and local customers, provides livelihood to twenty-six local employees, and is recognized by Barru Regency for its role in "transforming outsider perceptions of the area." As Sandarupa's son declares, they are becoming "Barru's tourist destination." For Torajan travelers, the restaurant is now a mini-destination and has eroded older concerns about the ardors of travel through the lowlands. One Toraja friend recalled the first time she stopped there to break up the trip, "We had a picnic with the food we packed, sitting on a [covered open-air] platform by the sea. It felt cooler there, with the sea breezes. Katrin brought us drinks and some fresh fish. It was delicious." And as another Torajan woman explained, "Before, we'd have to pack our own food for the trip, and only drank coffee at the rest stops, but now we know we can eat good food [there]…we don't have to worry."

This is a small story in an out of the way place, but it is instructive, as it illustrates how attending to travelers' experiences and sensibilities along the road can lend insights otherwise overlooked when we focus solely on touristic destinations. Moreover, it demonstrates how "local knowledge" can lay the foundations for a mini-destination restaurant catering to locals, domestic and even international tourists. As a migrant with a foot in tourism, Sandarupa drew on his cultural and academic savvy to transform an ordinary road leading to a distant tourist destination into a promise-filled yellow brick road. Finally, this story of a roadside restaurant that evolved into a Barru tourist stop also underscores additional ways in which tourism and migration are entangled.

Conclusion

This article illustrates how ethnographic storytelling can serve as an important pathway towards fostering a de-centering of Western-Anglo-centrism in tourism studies. Via ethnographically grounded stories of return migrant visitors, voyagers with problematic legal statuses, and travel-oriented café founders, I have highlighted Indonesians' lived experiences and locally-derived understandings of travel, thereby revealing some of the limitations of western definitions of tourism. As illustrated here, ethnographic storytelling has tremendous potential for not only conveying grounded insights about the ways in which culture, politics, and economics are entwined in specific tourist settings, but also for engendering empathy, an important but sometimes under-emphasized aspect of critical tourism studies.

Moreover, this study's findings problematize the deep-seated binaries that historically structured tourism research. Via the Oz-themed organizational framework of gates and yellow brick roads, I have highlighted some of the insights gleaned by adjusting our lens to focus less exclusively on inherited conventional categories like 'tourists,' 'tourism' and 'destinations.' When such generic categories are set aside and scholars embrace ethnography-based approaches foregrounding local knowledge, we gain more nuanced understandings of the culturally varied contours of travel. (As we saw, classic categories such as 'domestic tourist' versus 'international tourist' or 'visiting friends and relatives (VFR)' mask the more complex cultural themes, economic issues, and personal longings undergirding the homeland visits of Toraja migrants). Although a growing chorus of Asian critical tourism scholars has begun chipping away at these binaries, and more broadly at Western-generated one-size-fits all tourism models, categories, and understandings, there is still work to be done.

Notes

1. My selection of this Hollywood film captures some of the ironies and challenges inherent in efforts to "Asianize the field." I initially contemplated using an Asian film as an article "frame": the 2009 Indonesian film, *Merantau* (trans.: migration/quest). But *Merantau* is largely

unknown outside Indonesia and would have necessitated a lengthy plot recitation, whereas the *Wizard of Ox's* familiarity to international audiences rendered it more serviceable for this journal's readership. Ironically, this calculus reflects how Anglo-Western publication contexts can vex desires to Asianize the field.

2. See https://www.statista.com/statistics/268136/top-15-countries-based-on-number-of-facebook-users/
3. https://www.indonesia.travel/gb/en/event-festivals/lovely-december-festival-2017-in-the-scenic-highlands-of-toraja
4. http://makassar.tribunnews.com/2016/01/31/toraya-mamali-atau-lovely-toraja-tokoh-toraja-ingin-kembali-mamali
5. An earlier article (Adams & Sandarupa, 2018) concerning tourism entrepreneurship and local wisdom includes some of this data.

Acknowledgments

I thank the organizers of the first Critical Tourism Studies Asia-Pacific conference for inviting me to deliver a keynote lecture that ultimately became the basis for this article. I am especially grateful to the Toraja migrants and residents who generously shared their stories with me. My appreciation also goes to Dirk Sandarupa for providing the map, and for help with gathering additional data on the Sandarupa family's restaurant. Peter Sanchez, Andrew Causey, Jill Forshee, Harng Luh Sin, Alan Lew and the journal's anonymous reviewers offered helpful advice and suggestions, for which I am appreciative.

Disclosure statement

No potential conflict of interest was reported by the author.

References

Adams, K. M. (1997). Touting touristic "primadonas": Tourism, ethnicity, and national integration in Sulawesi, Indonesia. In M. Picard and R. Wood (Eds.), *Tourism, ethnicity and the state in Asian and Pacific societies* (pp. 155–180). University of Hawaii Press.

Adams, K. M. (1998). Domestic tourism and nation-building in South Sulawesi. *Indonesia* and the *Malay World*, *26*(75), 77–96. https://doi.org/10.1080/13639819808729913

Adams, K. M. (2006). *Art as politics: Re-crafting identities, tourism and power in Tana Toraja, Indonesia.* University of Hawai'i Press.

Adams, K. M. (2012). Ethnographic methods. In L. Dwyer, A. Gill and N. Seertaram (Eds.), *Handbook of research methods in tourism: Qualitative and quantitative methods* (pp. 339–351). Edward Elgar/Ashgate.

Adams, K. M. (2016). Tourism and ethnicity in insular Southeast Asia: Eating, praying, loving and beyond. *Asian Journal of Tourism Research*, *1*(1), 1–28. https://doi.org/10.12982/AJTR.2016.0001

Adams, K. M. (2018). Revisiting 'wonderful Indonesia': Tourism, economy and society. In R. Hefner (ed.) *Routledge handbook of contemporary Indonesia* (pp. 197–207). Routledge.

Adams, K. M. (2019). 'Being a tourist in (my own) home': Negotiating identities and belonging in Indonesian heritage tourism. In Leite, N., Casteñeda, Q. and Adams, K. M. (Eds.), *The ethnography of tourism: Edward Bruner and beyond* (pp. 148–165). Lexington Books/Rowman and Littlefield.

Adams, K. M., & Sandarupa, D. (2018). A room with a view: Local knowledge and tourism entrepreneurship in an unlikely Indonesian locale. *Asian Journal of Tourism Research, 3*(1), 1–26. https://doi.org/10.12982/AJTR.2018.0001

Alder, P. A., & Alder, P. (1994). Observational technique. In N. K. Denzin, and Y. S. Lincoln (Eds.), *Handbook of qualitative research* (pp. 377–392). Sage.

Ashtar, L., Shani, A., & Uriely, N. (2017). Blending 'home' and 'away': Young Israeli migrants as VFR travelers. *Tourism Geographies, 19*(4), 658–672. https://doi.org/10.1080/14616688.2016.1274775

Ateljevic, I., Pritchard, A., & Morgan, N. (2007). *The critical turn in tourism studies: Innovative research methodologies.* Routledge.

Ateljevic, I., Morgan, N., & Pritchard, A. (2012). *The critical turn in tourism studies: Creating an academy of hope.* Routledge.

Becker, H. (1970). *Sociological work: Method and substance.* Aldine.

Bernard, H. R. (2006). *Handbook of methods in cultural anthropology: Qualitative and quantitative methods.* AltaMira Press.

Bianchi, R. (2009). The 'critical turn' in tourism studies: A radical critique. *Tourism Geographies, 11*(4), 484–504. https://doi.org/10.1080/14616680903262653

Biglke, T. (2005). *Tana Toraja: A social history of an Indonesian people.* Singapore University Press.

Bloch, N. (2017). Barbarians in India: Tourism as moral contamination. *Annals of Tourism Research, 62*, 64–77. https://doi.org/10.1016/j.annals.2016.12.001

Bruner, E. (2005). *Culture on tour: Ethnographies of travel.* University of Chicago Press.

Butler, C., & Menzies, C. (2007). Traditional ecological knowledge and indigenous tourism. In R. Butler and T. Hinch (Eds.), *Tourism and indigenous peoples: Issues and implications* (pp. 15–27). Elsevier.

Chang, T. C. (2015). The Asian wave and critical tourism scholarship. *International Journal of Asia Pacific Studies, 11*(1), 83–101.

Chang, T. C. (2019). 'Asianizing the field': Questioning critical tourism studies in Asia. *Tourism Geographies*, 1–18. https://doi.org/10.1080/14616688.2019.1674370

Chronis, A. (2005). Co-constructing heritage at the Gettysburg storyscape. *Annals of Tourism Research, 32*(2), 386–406. https://doi.org/10.1016/j.annals.2004.07.009

Cohen, E., & Cohen, S. (2015). Beyond Eurocentrism in tourism: A paradigm shift to mobilities. *Tourism Recreation Research, 40*(2), 157–168. https://doi.org/10.1080/02508281.2015.1039331

Cohen, S., & Cohen, E. (2019). New directions in the sociology of tourism. *Current Issues in Tourism, 22*(2), 153–172. https://doi.org/10.1080/13683500.2017.1347151

Colombijn, F., Jaffe, R., & Klaufus, C. (2012). Mobilities and mobilizations of the urban poor. *International Journal of Urban and Regional Research, 36*(4), 643–654. https://doi.org/10.1111/j.1468-2427.2012.01119.x

Cole, S. (2004). Shared benefits: Longitudinal research in Eastern Indonesia. In J. Phillimore and L. Goodson (Eds.), *Qualitative research in tourism: Ontologies, epistemologies and methodologies* (pp. 292–310). Routledge.

Coles, T., Hall, C., & Duval, D. (2006). Tourism and post-disciplinary enquiry. *Current Issues in Tourism, 9*(4–5), 293–319. https://doi.org/10.2167/cit327.0

Coles, T., Hall, C., & Duval, D. (2016). Tourism and post-disciplinarity: Back to the future? *Tourism Analysis, 21*(4), 373–388. https://doi.org/10.3727/108354216X14679788636113

Coles, T., & Timothy, D. (2004). *Tourism, diasporas and space.* Routledge.

Conran, M. (2011). They really love me! Intimacy in volunteer tourism. *Annals of Tourism Research, 38*(4), 1454–1473. https://doi.org/10.1016/j.annals.2011.03.014

Cresswell, J., & Plano Clark, V. (2011). *Designing and conducting mixed methods research* (2nd ed.). Sage.

Din, K. (2017). Returning home: A reflection on the Malaysian practice of balik kampung. *Asian Journal of Tourism Research, 2*(1), 36–49. https://doi.org/10.12982/AJTR.2017.0002

Ellen, R. (1984). *Ethnographic research*. Academic Press Inc.

Erb, M. (2009). Tourism as glitter: Re-examining domestic tourism in Indonesia. In T. Winter, P. Teo and T.C. Chang (Eds.), *Asia on tour: Exploring the rise of Asian tourism* (pp. 170–182). Routledge.

Errington, J., & Gewertz, D. (1989). Tourism and anthropology in a post-modern world. *Oceania, 60*(1), 37–54. https://doi.org/10.1002/j.1834-4461.1989.tb00350.x

Fajriani, N. (2018, December). Sejarah Lovely December Program Andalan SYL, sekarang diganti Kemilau Toraja era NA (History of Lovely December SYL Mainstay Program, now replaces the Toraja Sparkle era NA). *Tribun Timur.* http://makassar.tribunnews.com/2018/12/03/tribunwiki-sejarah-lovely-december-program-andalan-syl-sekarang-diganti-kemilau-toraja-era-na

Forshee, J. (2000). Shifting visions: Along the routes of Sumba cloth. *The Asia Pacific Journal of Anthropology, 1*(2), 1–25. https://doi.org/10.1080/14442210010001705900

Franklin, A., & Crang, M. (2001). The trouble with tourism and travel theory. *Tourist Studies, 1*(1), 5–22. https://doi.org/10.1177/146879760100100101

Gillen, J., & Mostafanezhad, M. (2019). Geopolitical encounters of tourism: A conceptual approach. *Annals of Tourism Research, 75*, 70–78. https://doi.org/10.1016/j.annals.2018.12.015

Goodson, I. (2001). The story of life history: Origins of the life history method in sociology. *Identity, 1*(2), 129–142. https://doi.org/10.1207/S1532706XID0102_02

Gottlieb, A. (2016). The anthropologist as storyteller. In H. Wulff (Ed.), *The Anthropologist as writer: Genres and contexts in the twenty-first century* (pp. 93–117). Berghahn.

Gubrium, J., & Holstein, J. (2001). *Handbook of interview research*. Sage.

Haldrup, M., & Larsen, J. (2010). *Tourism, performance and the everyday: Consuming the orient.* Routledge.

Hamzah, Y. (2013). Potensi media sosial sebagai sarana promosi interaktif bagi pariwisata Indonesia. *Jurnal Kepariwisataan Indonesia, 8*(3), 1–9.

Hollinshead, K. (2004). Ontological craft in tourism studies: the productive mapping of identity and image in tourism settings. In J. Phillimore and L. Goodson (Eds.), *Qualitative research in tourism: Ontologies, epistemologies and methodologies* (pp. 83–101). Routledge.

Hollinshead, K. (2010). Tourism studies and confined understanding: The call for a "new sense" postdisciplinary imaginary. *Tourism Analysis, 15*(4), 499–510. https://doi.org/10.3727/108354210X12864727693669

Hollinshead, K. (2016). Postdisciplinarity and the rise of intellectual openness: The necessity for "plural knowability" in tourism studies. *Tourism Analysis, 21*(4), 349–361. https://doi.org/10.3727/108354216X14600320851613

de Jong, E. (2013). *Making a living between crises and ceremonies in Tana Toraja: The practice of everyday life of a South Sulawesi Highland Community*. Brill.

King, V., & Porananond, P. (2014). Introduction: Rethinking Asian tourism. In P. Porananond and V. T. King (Eds.), *Rethinking Asian tourism: Cultures, encounters and local response* (pp. 1–21). Cambridge Scholars Publishing.

Lévi-Strauss, C. (1962). *Le totémisme aujourd'hui. [Totemism today]*. Press Universitaire de France.

Lim, M. (2013). The internet and everyday life in Indonesia. *Bijdragen Tot de Taal-, Land- en Volkenkunde / Journal of the Humanities and Social Sciences of Southeast Asia, 169*(1), 133–147. https://doi.org/10.1163/22134379-12340008

Lindquist, J. (2009). *The anxieties of mobility: Migration and tourism in the Indonesian borderlands*. University of Hawai'i Press.

Marcus, G. (1995). Ethnography in/of the world system: The emergence of multi-sited ethnography. *Annual Review of Anthropology, 24*(1), 95–117. https://doi.org/10.1146/annurev.an.24.100195.000523

Martin, D., & Woodside, A. (2011). Storytelling research on international visitors: Interpreting own experiences in Tokyo. *Qualitative Market Research: An International Journal*, *14*(1), 27–54. https://doi.org/10.1108/13522751111099319

Minca, C. (2017). Walls! Walls! Walls! *Society and Space*. http://societyandspace.org/2017/04/18/walls-walls-walls/

Mostafanezhad, M. (2013). Getting in touch with your inner Angelina': celebrity humanitarianism and the cultural politics of gendered generosity in volunteer tourism. *Third World Quarterly*, *34*(3), 485–499. https://doi.org/10.1080/01436597.2013.785343

Mura, P., & Pahlevan Sharif, S. (2015). The crisis of the "crisis of representation": Mapping qualitative tourism research in Southeast Asia. *Current Issues in Tourism*, *18*(9), 828–844.

Mura, P., & Wijesinghe, S. (2019). Behind the research beliefs and practices of Asian tourism scholars in Malaysia, Vietnam and Thailand. *Tourism Management Perspectives*, *31*, 1–13. https://doi.org/10.1016/j.tmp.2019.03.009

Naim, M. (1976). Voluntary migration in Indonesia. In A. Richmond and D. Kubat (Eds.), *Internal migration: The new world and the old world*. Sage Studies in International Sociology. 4.

Narayan, K. (2007). Tools to shape texts: What creative nonfiction can offer ethnography. *Anthropology & Humanism*, *32*(2), 130–144. https://doi.org/10.1525/ahu.2007.32.2.130

Narayan, K. (2012). *Alive in the writing: Crafting ethnography in the company of Chekhov*. University of Chicago Press.

Olwig, M. (2012). Multi-sited resilience: The mutual construction of "local" and "global" understandings and practices of adaptation and innovation. *Applied Geography*, *33*, 112–118. https://doi.org/10.1016/j.apgeog.2011.10.007

Picard, M. (1997). Cultural tourism, nation-building, and regional culture: The making of a Balinese identity. In M. Picard and R. Wood (Eds.), *Tourism, ethnicity and the state in Asian and Pacific societies* (pp. 181–214). University of Hawai'i Press.

Rante, Y. (2016). Toraya Mamali atau Lovely Toraja? Tokoh Toraja ingin kembali Mamali. (Toraya Mamali or Lovely Toraja? Toraja leaders want Mamali back). *Makassar Tribune*. January. http://makassar.tribunnews.com/2016/01/31/toraya-mamali-atau-lovely-toraja-tokoh-toraja-ingin-kembali-mamali

Sin, H. L. (2009). Volunteer tourism: "Involve me and I will learn? *Annals of Tourism Research*, *36*(3), 480–501. https://doi.org/10.1016/j.annals.2009.03.001

Sin, H. L. (2015). "You're not doing work, you're on Facebook!": Ethics of encountering the field through social media. *Professional Geographer*, *67*(4), 676–685.

Spradley, J. (2016) [1980]. *Participant observation*. Waveland Press.

Su, R., Bramwell, B., & Whalley, P. (2018). Cultural political economy and urban heritage tourism. *Annals of Tourism Research*, *68*, 30–40. https://doi.org/10.1016/j.annals.2017.11.004

Soekarno. (n.d.). President Soekarno on economic defensibility/Tourism in Indonesia. Ministry of Information, Republic of Indonesia. Special issue No. 15.

Swain, M. (2009). The cosmopolitan hope of tourism: Critical action and worldmaking vistas. *Tourism Geographies*, *11*(4), 505–525. https://doi.org/10.1080/14616680903262695

Szakolczai, A. (2009). Liminality and experience: Structuring transitory situations and transformative events. *International Political Anthropology*, *2*(1), 141–172.

Teo, P., & Leong, S. (2006). A postcolonial analysis of backpacking. *Annals of Tourism Research*, *33*(1), 109–131. https://doi.org/10.1016/j.annals.2005.05.001

Tie, C., & Seaton, T. (2013). Diasporic identity, heritage, and "homecoming": How Sarawakian-Chinese tourists feel on tour in Beijing. *Tourism Analysis*, *18*(3), 227–243. https://doi.org/10.3727/108354213X13673398610538

Van Gennep, A. (1909). *Les rites de passage. [Rites of passage]*. Émile Nourry.

Volkman, T. (1984). Great performances: Toraja cultural identity in the 1970s. *American Ethnologist*, *11*(1), 152–169. https://doi.org/10.1525/ae.1984.11.1.02a00090

Watson, J. (Ed.). (1977). *Between two cultures*. Blackwell.

Wijesinghe, S., & Mura, P. (2018). Situating Asian tourism ontologies, epistemologies and methodologies: From colonialism to neo-colonialism. In Mura, P. & Khoo-Lattimore, C. (Eds.), *Asian* qualitative research in tourism: Ontologies, epistemologies, methodologies and methods (pp. 95–115). Springer.

Wijesinghe, S., Mura, P., & Culala, H. (2019). Eurocentrism, capitalism and tourism knowledge. Tourism Management, 70, 178–187. https://doi.org/10.1016/j.tourman.2018.07.016

Winter, T. (2009). Asian tourism and the retreat of Anglo-western centrism in tourism theory. Current Issues in Tourism, 12(1), 21–31. https://doi.org/10.1080/13683500802220695

Winter, T., Teo, P., & Chang, T. C. (2009). Asia on tour: Exploring the rise of Asian tourism. Routledge.

Wulff, H. (2016). The anthropologist as writer: Genres and contexts in the twenty-first century. Berghahn.

Wynn, J. (2005). Guiding practices: Storytelling tricks for reproducing the urban landscape. Qualitative Sociology, 28(4), 399–417. https://doi.org/10.1007/s11133-005-8365-2

Yamashita, S. (2019). Southeast Asian tourism from a Japanese perspective. In M. Hitchcock, V. King and M. Parnwell (Eds.), Tourism in Southeast Asia: Challenges and new directions (pp. 189–205). NIAS Press.

Yogaswara, A. (2010). The power of Facebook: Gerakan 1,000,000 Facebookers. Mediakom.

Decentring scholarship through learning with/from each 'other'

Hazel Tucker and Stuart Hayes

ABSTRACT

The contemporary tourism world is characterised by mobilities from the East and West. Yet despite this, tourism scholarship remains highly Anglo-Western-centric. How, then, might we go about engendering a decentring of Anglo-Western-centrism in tourism scholarship? In this paper, we respond to this question by discussing some 'situated' educational experiences during a tourism fieldschool programme in Pai, Thailand. The international mix of (trainee and trainer) researchers on the fieldschool provided a rich source for exploring the ways in which tourism dynamics are gazed upon and theorised about by researchers from various 'Western' and 'non-Western' backgrounds. Our shared interactions and reflections allowed each of us to experience being 'other' in relation to tourism knowledge and concepts. Moreover, we were afforded a heightened awareness of each 'other's' experiences of being 'other' in different contexts. Our experiences thus serve, on the one hand, to highlight the pervasive influence of Anglo-Western centrism in tourism knowledge creation. On the other hand, our somewhat serendipitous experiences and discoveries during the fieldschool demonstrate the potentiality of the 'international classroom' to produce a tourism scholarship which is more open and sensitive to diversified and culturally relevant knowledge(s); in other words, a decentred tourism scholarship.

摘要

当代旅游世界的特点是东西方之间的移动性增强。然而, 尽管如此, 旅游学问仍然高度地以英美为中心。那么, 我们该如何着手在旅游学术领域产生一种英美西方中心主义的去中心化呢? 在这篇论文中, 我们通过讨论泰国拜县旅游田野学校项目中的一些'情境'教育经验来回应这个问题。来自不同'西方'和'非西方'背景的研究人员如何看待和理论化旅游业的动力, 田野学校的国际混合(培训生和培训师)研究人员提供了丰富的探讨方式来源。我们共同的互动和反思让我们每个人都能体验到在有关旅游知识和概念中成为'他人'。此外, 我们对彼此在不同情境下成为'他人'的经历有了更高的认识。因此, 我们的经验一方面突出了英美西方中心主义在旅游知识生成中的广泛影响; 另一方面, 我们在田野学校的一些偶然的经验和发现, 显示了'国际课堂'产生旅游学问方面的潜力, 这种学问对多样化和与文化有关的知识更为开放和敏感; 换句话说, 是一份去中心化的旅游学问。

Introduction

Alatas (2006) remarks that: 'The social sciences are dominated by theories, concepts and categories that were developed in Europe and North America … This domination has been at the expense of non-European ideas and concepts' (p. 178). Here, Alatas is referring to the historical dominance of Anglo-Western categories and concepts to explain the 'whole' social world, an inevitable consequence of social sciences' Western origins. Similarly, Winter (2009) provides a useful definition of Anglo-Western centrism in the context of tourism scholarship:

> Anglo-Western centrism: the accepted norm of uncritically applying certain analytical and theoretical approaches conceived in particular historical circumstances to all forms of tourism everywhere. (p. 24)

Winter's definition is based on the assertion that nearly all the major concepts in tourism have been developed in the West (i.e. by Western scholars) and as a response to Western tourism phenomena. According to Alneng (2002, p. 137), the resulting 'exclusiveness of Tourist-as-Modern-as-Westerner' has led to a 'relative indifference towards, and sometimes complete denial of, non-Western tourism'. Tourism theory thus continues to remain largely rooted in the 'Western tourist gaze', the relevance of which to non-Western tourists and tourisms, and indeed non-Western tourism students, is being increasingly questioned (Chang, 2015; Cohen & Cohen, 2015; Tucker & Zhang, 2016; Winter, 2009; Zhang, 2018). Furthermore, as this relative dominance of Western-based theory is likely to carry through into current tourism education curricula, future tourism scholars are perhaps also inclined to continue to perpetuate the Anglo-Western-centrism that currently dominates the field.

Concurrently, the global tourism landscape continues to change, and one prominent change is that Asia has become an increasingly key tourist-generating region. Over the past three decades, the number of outbound tourists from the Asia and Pacific region increased from 59 million in 1990 to 376 million in 2016 (United Nations World Tourism Organisation (UNWTO), 2008; World Bank, 2018). According to a recent report from the UNWTO, in 2017 one out of four tourist trips originated in Asia and the Pacific (UNWTO, 2018). In the same year, China ranked number one in the world in terms of international tourism expenditure ($US257.7 billion), almost doubling that of the USA ($US135 billion). In response to these increased Asian mobilities, authors such as Winter (2009) have asked whether the Anglo-Western-centrism in tourism knowledge means that the field is, in fact, ill-equipped to respond to the changes in the global tourism landscape. Others are more hopeful, however, such as Chang (2015), who suggests that the 'concurrent critique of orthodox Western thought and the emergence of non-Western tourists presents tantalising possibilities to rethink age old understandings of tourism' (p. 84). Chang (2015) argues that this rethinking constitutes a growth in critical Asian tourism scholarship, which ranges from relativist/revisionist to radical approaches to 'Asianising' the tourism field.

Certainly, a gradual increase in expanding and diversifying the field is occurring. As Edensor and Kothari (2018, p. 706) remark, 'recent approaches have attempted to expand understandings about tourist motivations and practices, and diverse studies of non-Western tourists are increasing in number'. However, Edensor and Kothari (2018,

p. 706) also concede that the continued relative 'dearth of available non-Western tourist accounts is matched in academic study where contemporary theories of tourism are based on a cluster of paradigmatic assumptions'. Alatas (2006), too, suggests, although in relation to the broader social sciences, that a fuller de-centring really 'requires a *plurality* of philosophical and cultural expression' (p. 194). As well as the need to redress the imbalance in research and theory development itself, therefore, this calls for a more diversified and culturally inclusive approach in tourism education, because this is the space where the orthodoxy of tourism knowledge and paradigmatic assumptions are (re)produced. On this, and particularly in relation to the increasingly 'internationalised' context of postgraduate tourism education, Botterill and Gale (2005, p. 478) point out that there is a need to find ways to reformulate tourism knowledge so as 'to encompass a more diverse 'situatedness' and [so] that previously silent voices contribute to create 'new' tourism knowledge'.

The crucial question then becomes: How to do this in practice? In this paper, we discuss some 'situated' educational experiences during a tourism fieldschool programme in order to 'anchor' some of these broader ontological/epistemological/pedagogical questions regarding how we go about engendering a decentring of Anglo-Western centrism in tourism theory. This ethnographic fieldschool programme, which the authors lead, is popular with international students, many from China. Along with the fieldschool's setting in Pai, a small town in northern Thailand, in our experience the programme creates opportunities for students and teachers to learn from one another (Kitano, 1997) in the sharing and co-creation of situated diverse tourism knowledges. The point that we were all able to experience being 'other' in relation to the variety of tourism situations occurring gave rise to many serendipitous discoveries and cathartic interactions taking place between the students and teacher/researchers. In turn, these interactions and discoveries resulted in our students', and our own, ability to go some considerable way in unpacking and repacking normative viewpoints (Chang, 2015, p. 97) regarding Western-centricism in tourism theory, as well as Asian 'difference'. Before we go onto discuss the processes of co-creation of situated tourism knowledges, we will firstly review key literature concerned not only with the need for a de-centring of tourism scholarship, but also of the role that tourism education might play in this.

Anglo-Western-centrism in tourism knowledge

Anglo-Western-centrism in tourism studies is manifested both in the fact that the 'tourist' that research has mainly focused on is still very often conceived as coming from the western, industrialised countries (Keen & Tucker, 2012; Huang, van der Veen, & Zhang, 2014; Winter, 2009), and in the fact that the producers of tourism knowledge are largely Western scholars. Consequently, in the changing global tourism landscape, characterised in part by increased mobilities amongst Asian tourists, what we are now seeing is the uncritical application of Anglo-Western theory to explain such phenomena. As Edensor and Kothari (2018, p. 706) remark: 'In declaring that all contemporary tourists are impelled by particular motivations and carry out specific practices, such influential theories disregard(ed) tourism's multiplicity'. This approach, which according to Winter (2009) hangs its 'theoretical import on the hooks of replication and predict-ability', too often provides 'the basis for misguided claims of universality' (p. 23). Indeed, due to non-European tourists and tourism having remained largely invisible, normative assumptions about tourist motivation and behaviour have been formed, such as tourists' search for 'authenticity', which have consistently failed to acknow-ledge the particular cultural context and time from which they emerged (Edensor & Kothari, 2018; Towner, 1995). Clearly, as Keen and Tucker (2012) suggest, 'it is crucial to consider how the ways of thinking and being (epistemologies and ontologies) which are the true legacies of colonialism and Western imperialism are contested through tourism' (p. 102).

In his pointing out that 'there has been a widespread failure to look more closely and incorporate non-Western forms of leisure travel into mainstream discussions and theories about tourism', Winter (2009, p. 24) suggests incorporation of 'non-Western' knowledge and theory as the answer. Such calls to confront the tourism academy's 'Eurocentrism' are present more broadly within critical postcolonial tourism studies. For example, Pritchard and Morgan (2007) have argued that 'we must act to decentre the tourism academy and respond to the challenges and critiques being articulated by indigenous scholars so that we may begin to create knowledge centered on indigenous epistemolo-gies and ontologies' (p. 22). More recently, also, Chambers and Buzinde (2015) called for a de-linking from Western epistemologies in order to strive for 'an agenda for tourism's decolonization' (p. 9): 'Tourism scholars in and from the South … need to undertake an epistemic de-linking, which requires a rejection of Western epistemologies about tourism representing the "God-eye" view' (Chambers & Buzinde, 2015, p. 13). Following Swadener and Mutua (2008), they thus propose a 'de-linking project' which would involve the reclaiming and foregrounding of indigenous voices and epistemologies. Many of the growing number of studies focusing on non-Western tourists, and conducted by non-Western researchers (for example, Mkono, 2011; Ong & du Cros, 2012; Teo & Leong, Mkuno, 2006) might be considered examples of this 'de-linking' project.

Fully achieving such a de-linking from Eurocentric knowledge, however, is unlikely to be straightforward. As Zhang (2018) asks, for example, how straightforward is it for Chinese tourism researchers to 'be non-Western and do non-Western tourism research?' (p. 132). Or, for Western scholars, how easy is it in reality to 'embody a cosmopolitan paradigm' (Swain, 2016)? Similarly, Tucker and Zhang (2016) problem-atise the idea of encouraging alternative discourses, suggesting that such practices run the risk of further entrenching dualisms which, ultimately, may hinder the ability of the tourism field 'to proceed to more open and pluralistic dialogues' (p. 252). They argue that many of the responses to the call for a re-centring of tourism scholarship have been problematic in that attempts to 'decentre' and 'delink' can often function to reinforce conventional colonial binaries and ways of thinking. Indeed, the ongoing discussion about recentring and decentring often tends to revert to such binaries as centre and periphery, colonised and coloniser, and West and non-West. This is despite the 'new era' of tourism itself subverting these binaries, and despite the recognition that a new era of scholarship is needed to do the same.

Cohen and Cohen (2015) suggest using a mobilities paradigm to study the emerging markets in order to be more 'attentive to differences within and between countries in the emerging regions, as well as to similarities between some of these and Western ones' (p. 2). Chang (2015), too, provides a useful overview of different approaches that scholars have employed in order to challenge the Western ethnocentrism in tourism studies and to develop a more adequately nuanced understanding of Asian tourism. One approach is a 'geography-matters perspective', which emphasises the significance of place or locality in the development of various forms of tourism. Relatedly is the critical post-colonial approach to Asian tourism, which 'challenges the core assumptions in the way knowledge is conceived and codified' (Chang, 2015, p. 90). Relevant to the present article, Chang refers to Teo and Leong's (2006) work on backpacking in Thailand as an example of critical post-colonialism, in that the need 'to unpack Western assumptions of backpacking/backpackers, and to repack concepts to suit contemporary Asian conditions (where Asians are increasingly the backpackers themselves)' is asserted (Chang, 2015, p. 91). Chang (2015, p. 91) continues: 'The traditional view of the western drifter who shuns mass tourism and technology, and who pursues alternative services and local exotica (Cohen, 1973) may be accurate in a particular time-space, but cannot be accepted as a universal, immanent truth'. Thus, a critical post-colonial tourism studies aims to expose Western-centric assumptions as well as essentialist frameworks and, in the context of Asian tourisms, would tend towards developing new concepts, or at least reconfiguring 'old' ones, by paying close attention to different socio-cultural contexts and time-spaces.

A relevant example of this can be seen in a recent article by Hsu and Huang (2016) in which the authors attempt to provide an update on the current value system in China and outline the relationship between these values and Chinese travel behaviour. They identify a range of different Chinese cultural values and link these to the travel behaviours of Chinese tourists. For example, 'thrift' is identified as an instrumental value (i.e. a desired character trait) that can be linked to specific travel behaviours like shopping around for a good deal or choosing a destination based on its value for money. This is all well and good, but in failing to locate such values within a broader global values framework there is an implicit assumption that Chinese values (and thereby travel behaviours) are somehow (all?) different to Western/other cultural values. The reality of the situation is likely to be more nuanced than this of course, and many values and behaviours may be shared across different cultures. For example, are not 'thrifty' and 'budget-conscious' travellers (a common term used to describe 'Western' tourists) one and the same? The questions posed by Zhang (2018) therefore become relevant again here, and rather than just thinking in terms of 'difference' perhaps we must also ask 'are we really so different from one another? Are Chinese tourists really so different from other tourists?' (p. 132).

It would seem, therefore, that de/re-centring may perhaps be easier to talk about in theory than it is to do in practice. Chang (2015) offers some good advice:

> The emergent wave in Asian tourism has implications for how and what we research on. While producing more knowledge on Asian tourism is an appropriate start, what is more important than a mere quantitative increase in research is a sensitive scholarship that acknowledges the contextual uniqueness of the phenomenon and appropriate concepts to showcase it. (p. 87)

This notion of developing a 'sensitive scholarship' speaks to the matter of (tourism) education and its ability/potential to better develop critical/postcolonial scholars who are sensitive to contextual uniqueness and appropriate concepts. As noted above, for instance, a suggestion put forward by Botterill and Gale (2005) is that the 'international classroom', in particular, has the possibility to create 'new' tourism knowledge by allowing previously silenced voices to contribute and to embrace a more diverse, contextualised 'situatedness' in the reformulated tourism knowledge.

However, any discussion concerned with de/re-centring scholarship in higher education must be wary of how the internationalisation of higher (tourism) education itself may serve to perpetuate (and even magnify) colonial, Western-centric knowledge production. As Majee and Ress (2018, p. 4) note: 'Suspicion remains … the proliferation of internationalisation agendas represent yet another vehicle to promote Euro-American logics in the guise of the "global"'. Indeed, the problem remains that the global hierarchy of knowledge instilled during European colonialism persists through to today, continuing the situation whereby 'Asian scholars, for instance, consider themselves as secondary or mere imitators of Western knowledge' (Wijesinghe & Mura, 2018, p.104). As Wijesinghe and Mura (2018, p.104) continue: 'Universities, which are primary knowledge production centres, remain dominated by Western traditions and English language'. In this sense, the pursuit of Westernised formalised education (for instance, in New Zealand for Asian or Chinese students) may itself be a project of learning to adopt the pervasive Anglo-Western centrism (or Euro-American logic, as Majee and Rees put it) in understanding society and social behaviour (such as tourism).

This matter becomes a particular focal point at the postgraduate level (the context for this study) given the strong demand for Western-based postgraduate study amongst international students, in particular those from Asia. As Botterill and Gale (2005) argue:

> The orthodoxy of tourism … studies lies at the centre of the postgraduate curriculum. A community of mainly white, middle class, male Anglophone, western academics, have been the architects of tourism knowledge… In the international classroom, it is laid out before an audience of students in which such characteristics are marginally represented… The dilemma is that the student body largely expects to receive the orthodox, not to challenge it, exactly because it is the currency that has drawn them thousands of miles from their homes and supposedly will form the future of their professional careers and provide payback for their own [and] their families … investment. (pp. 478–479)

Clearly, then, and as pointed out by Rizvi, Lingard, and Lavia (2006, p. 257),

> education has a systematically ambivalent relationship to postcolonialism. On the one hand, it is an object of postcolonial critique regarding its complicity with Eurocentric discourses and practices. On the other hand, it is only through education that it is possible to reveal and resist colonialism's continuing hold on our imaginations.

Consequently, without deliberate pedagogical interventions to decentre scholarship, the internationalisation of higher education risks potentially becoming yet another means – a powerful one at that - by which colonial legacies are reproduced. *With* deliberate pedagogical intervention, on the other hand, such as that discussed in this

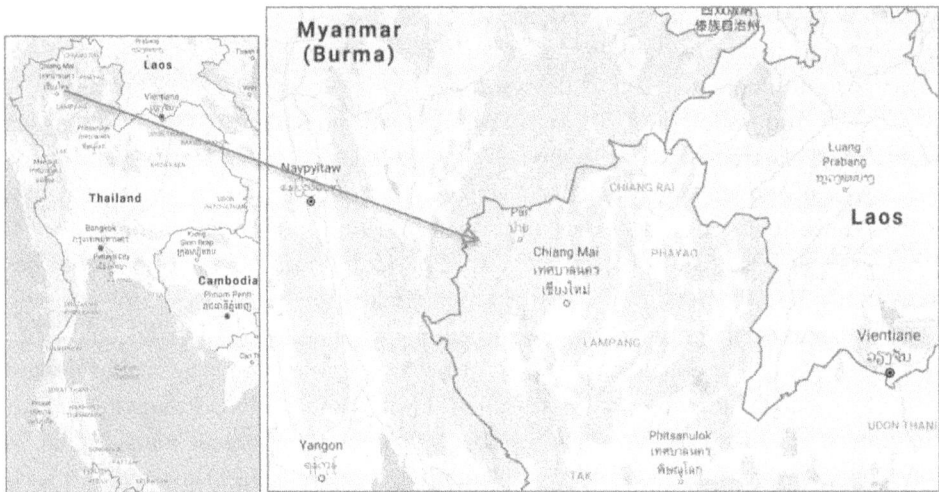

Figure 1. Field school location(s). Source: Google Maps

article, the international classroom does hold the possibility to create both 'new', diversified and situated tourism knowledge and a more sensitive tourism scholarship.

The remainder of this article focuses in on the teaching and learning interactions and discoveries experienced during the ethnographic fieldschool delivered by the authors as a part of a tourism Master's degree programme. Endeavouring to consider the ways in which the tourism dynamics in Pai are gazed upon and theorised about by researchers from various backgrounds, the fieldschool provided a valuable window through which to gaze at, and seek richer and more nuanced understandings of, the differences, multiplicities, similarities, cross-influences and mutations within tourist culture(s). A brief overview of the ethnographic fieldschool will first be presented.

Tourism and development ethnographic fieldschool

The Tourism and Development Ethnographic Field School is a residential, month long programme that takes place in Northern Thailand (Chiang Mai and the town of Pai, Figure 1).

The fieldschool is aimed at Master's students and has two main intended learning outcomes. These are:

- For students to have the ability to understand, design, perform and critically evaluate ethnographic research
- And for students to be able to critically analyse their own positionality as tourists and researchers.

The fieldschool has been running since 2012 and this particular research is based on the experiences of the 2017 (11 students and 2 staff) and 2018 fieldschools (10 students and 2 staff).[1] The 2017 and 2018 cohorts were made up of students and staff from different backgrounds, but primarily from China, India, and New Zealand. Most had relatively extensive travel backgrounds and were permanent residents of the aforementioned countries, with the exception of the two staff who are both UK-to-New Zealand migrants.

> All students wishing to enrol in the fieldschool must have completed the module 'Tourism Culture'. This module is: aimed at understanding tourists from a social and cultural perspective, including the scope and nature of tourism and tourist places", and providing students with "a critical understanding of tourists from a variety of perspectives and key theories relevant to tourist culture. (University of Otago, 2019, n.p)

The 'Tourist Culture' module therefore provides excellent academic grounding in the understanding of how and why tourists travel, and, given the different back-grounds of students (a consistent characteristic of Master of Tourism cohorts at the university), we are able to explore this from different cultural perspectives. Accordingly, and given that it takes place immediately following the module Tourist Culture, the experiential nature of the fieldschool provides students with the oppor-tunity to see 'tourist culture(s)' in action. Further preparation for the fieldschool involves students reading selected articles about ethnographic research methods and tourism in Northern Thailand.

The fieldschool itself takes place over a 4-week period in June during the University's mid-semester break. The first two weeks are spent in the regional capital Chiang Mai, where students receive orientation and instruction on how to conduct ethnographic research. During this phase, we concentrate more on the first intended learning outcome of the fieldschool which is to teach students how to 'Understand, design, perform and critically evaluate ethnographic research'. In the context of this particular intended learning outcome, students are instructed to undertake a variety of exercises. These exercises involve, for example, students practicing participant observa-tion and fieldnote-taking techniques. In this instance students are tasked with visiting a local attraction such as a market or temple to observe and take notes about touristic behaviour. Our instructions to the students are to 'notice what you notice', not only in terms of the actions of others but also in relation to the students' own internal responses to what they are seeing and experiencing (e.g. assumptions).

In practising ethnographic research techniques in this way, even during this phase of the fieldschool, students and staff are continually 'noticing' how tourist culture(s) manifest in Chiang Mai and this provides a rich source for group reflection and discus-sion. We constantly share with one another our 'noticings' and interpretations about tourist culture and reflect on what this might mean for us as individuals and/or as members of the fieldschool group/a cultural group/the tourism scholarly community. Indeed, the process of analysing our positionality as tourists and researchers through group discussions serves another important purpose in that it creates a relaxed envir-onment that continues throughout the fieldschool. This, we believe, is vital to the suc-cess of the fieldschool in allowing spaces for meaningful and open interactions among and between students and staff.

In Chiang Mai, students are also taught about and tasked with interviewing. Building on from participant observation, the students are by now beginning to iden-tify aspects of tourist culture that are of particular interest to them. Working in small groups (in which individual group members share a similar interest) we support students to develop research questions, followed by related interview questions. The groups then 'practice' interviewing a small number of tourists. Interviews are transcribed and we then work with students to identify themes and patterns in the data. Findings from each group are then presented in a symposium-style format.

The second part of the fieldschool then takes us to Pai, approximately 80 km northwest of Chiang Mai. The town of Pai itself provides a thought-provoking context for the students (and us, the 'teachers'). Pai, which was once a Western backpacker enclave (Cohen, 2006), has evolved over recent years to accommodate growing numbers of Asian tourists, in particular those from China. Although it remains popular with 'Western' backpackers, the town of Pai now plays host to a fascinating and contemporary mix of tourists (and resident, as well as resident-tourist for that matter), all of whom go about their touristic (and non-touristic) lives within a fairly small area. Tourists from various backgrounds go to Pai for a range of man-made and natural attractions and/or to experience a growing number of 'wellness' activities (e.g. meditation, yoga etc.). The town offers a variety of accommodation options, from budget 'huts' and dorms to upscale resort-style hotels. Many of the accommodations, as well as the excursions and activities that are sold, appear to specifically target, either intentionally or otherwise, certain segments of the visitor market. For example, a popular one-day tour aimed specifically at Chinese tourists includes being bussed around a number of different love-inspired attractions made famous by a popular Chinese film called *Lost in Thailand* (see Budde et al., 2013). In contrast, many of the wellness-oriented activities appear more deliberately intended for Western (or at least non-Asian) visitors.[2]

In this context, Pai offers the potential to read the tourism dynamics through the key binaries that pervade tourism studies – Western/non-Western, Asian/non-Asian, backpacker/mass tourist, etc. Moreover, the dynamics of the fieldschool group are also multicultural and could also be described using binary or essentialist categories, such as 'Asian' or 'Western'. Subsequently, Pai provides a fascinating backdrop to explore the different ways that multi-cultural researchers engage with contemporary, third millennium tourism dynamics. It is for the reasons mentioned above that Pai was specifically selected for the location of the second phase of the field school. For it is in this context that students undertake small group projects based on a subject which they find of particular interest based on observations during the first few days in Pai. Other locations could have been chosen for the fieldschool, and there are likely to be other places which have similarly developed into a destination for both Western 'backpackers' and Asian/Chinese independent tourists. However, Pai's 'tourism zone' offers a particularly vivid, and often contrasting, tourism dynamic within a fairly confined space, and has thus proven to be both a convenient and fruitful setting for the fieldschool.

Experiences of the fieldschool

We now turn to a discussion of our experiences in Pai vis-à-vis our 'unpackings' and 'repackings' of certain tourism-related cultural concepts and meanings. By way of introduction to this, it should be noted that, given that the majority of the fieldschool group were from New Zealand and China, the discussion here will focus mainly on the experiences and 'noticings' of 'Western' and 'Chinese' researchers. Due to this particular nationality mix, during the fieldwork experiences in Pai the group members' awareness became heightened in relation to particular ('Western' versus 'Chinese') aspects of cross-cultural tourist behaviour, including a heightened awareness of their own and each other's behaviour, as well as that of the predominant (Western and Chinese) cross-cultural mix of other tourists in Pai. As well as encouraging this awareness/noticing, following the recent 'trend' in tourism studies of focusing on 'emotions, the body, the senses (beyond ocular-centrism) and materialities' (Cohen & Cohen, 2017, p. 166), we also deliberately direct the student-researchers' attention, and our own researcher attention, on these particular 'beyond ocular-centrism' facets, or dimensions, of tourism phenomena. There are now many Chinese travellers visiting Pai, alongside the 'Western' backpackers who, as Cohen (2006) argued, the place had initially been 'made for' and 'made with'. In relation to how these different tourist groups construct meaning in relation to their experience in Pai, therefore, students' research questions are thus opened to encompass matters such as 'affective atmospheres' (Anderson, 2009), feelings and bodies/materiality (including clothing and smart-phones). This enables the students, as Kathleen Adams (2016) suggests, to 'see some of the promise of paying closer attention to the emotional realm of travel, and to how tourism-induced feelings relate to sensibilities about self and other, ethnicity, nation and humanity' (p. 17).

In discussing our experiences and presenting 'data', we draw on observations, learning and other aspects of 'noticing' from a variety of sources. These sources include our own and students' fieldnotes, students' written assessments, and an in-depth conversational interview between the authors and a mixed nationality group of students. In particular, we use 'vignettes' as a way of organising and presenting some key 'illustrative case material' (Yin, 2011, p. 260).

Essentialising dualisms in Pai

Vignette 1: Extract from a report written by a group of Chinese students exploring 'How Relaxation is Performed in Pai':

Western tourists in Pai prefer to go to the natural sites, whereas Chinese tourists are mainly interested in artificially-constructed sites, where they can take a picture as a 'keepsake'. A Western interviewee who had participated in a one-day tour said that she felt bored during the tour because many of the sites can be built anywhere and were nothing to do with Thai culture. She prefers to spend time trekking or soaking in the hot spring rather than 'idiotic' sightseeing' .... 'We interviewed a number of Chinese tourists for the research. Their imagination encouraged them to go to the places where the tourist photos are taken. One interviewee said that the purpose of visiting Pai is taking photos of the same scene she'd seen before on social media.'

The extract presented in **Vignette 1** helps to illustrate some of the inherent dualisms or binaries that pervade tourist culture and, thereby, tourism knowledge creation. It feels almost inevitable, from this extract, that conventional binaries of West/non-West are reverted to (Tucker & Zhang, 2016) in such a way that Western and Chinese tourists are separated and portrayed as somehow different beings. Perhaps without an adequate sense of ontological or epistemological security in 'reading' or interpreting their observations of different tourists, the Chinese students revert to the 'fall-back' position of comparing and contrasting different nationalities of tourists and grouping them into West vs non-West. It is often these 'differences' that are theorised about and highlighted in contemporary tourism scholarship. For example, in Yang et al.'s (2018) introduction to the edited work *Asian Cultures and Contemporary Tourism*, the notion of difference is clearly explicated whilst little attention is paid to how Western and Asian tourist behaviour may be similar: 'While Asian and Westerners are imagined and evolving labels, we argue that these labels are still meaningful social constructions. The labels identify the differences between *us* and *others* reified in different beliefs, norms and traditions, and such differences are known as "culture"' (Yang et al., 2018, p. 13). Just like the portrayal of Chinese and Western tourists in Vignette 1, Yang et al.'s comments illustrate how tourist culture(s) may be fixed, almost exclusively, to notions of difference. Thus, it seems, dualisms and essentialisms are normalised and there is little attempt to be attentive to the 'similarities' that likely exist, nor to the nuances between and within tourist groups.

This binary way of thinking was also shown in certain things that were said during an interview which we (the teachers on the program) conducted during the field-school with a multi-cultural group of the students in order to probe their experiences of being tourists/researchers in Pai. One thing that became clear from speaking with the students was the extent to which certain ways of thinking and seeing are normalised. For example, when asked about whether they were 'aware' of their nationality within the group, a student from New Zealand replied:

> Even today, this guy asked me what I was doing and I told him I was waiting for my friends over there, but then I was like, oh, he probably doesn't think I have any friends because there were no Westerner's there!

This extract could be seen as exposing a deeply ingrained binary (e.g. being Western/non-Western), and it highlights the idea that people automatically and unconsciously identify with and position themselves with one or other 'camp' (Tucker & Zhang, 2016). There is also the idea here that cross-cultural convergence is, or would at least be perceived to be, somehow alien (How could/why would a Western woman possibly be waiting for (or even have) a group of non-Western/Asian friends?). Again, it seems, given the extent to which normalised ways of binary thinking pervade everyday (tourist) life, the task of re-centring tourism scholarship is not an easy one to navigate.

Universalising 'good vibes' in pai

Vignette 2: Extract from a New Zealander's fieldnotes depicting the relaxed 'vibe' in Pai:
Sitting in cafe – VERY hippy, chilled, Westerners here along with hippy Thai guys that hang out, long hair, cotton clothing, reading books, a Thai hippy guy playing chess with older western guy. Young woman with feathers in her hair, sitting reading. When different people come in they greet each other with hugs. You have to take your shoes off at the entrance, so bare feet as you walk around inside the café, Indian music, incense burning, Thai hippies. At the next table, one tourist says to another 'you feel like Alice in Wonderland when you arrive in Pai'. On the menu it reads: 'Start the day with slow food'. We order herbal tea and coconut cheese cake. Posters on the walls display quotes from the Dalai Lama; adverts for yoga classes and macramé workshops; and one poster promotes the café's weekly 'Spoken Words night',

described in the poster as: 'the Pai focal point for local and travelling musicians, creative artists and free thinking people. A friendly, chilled out melting pot of pure creativity and sharing … .Bring your poem, story of your travels, a joke, an experience, instrument, voice, dance, talent, or simply chill out in the good vibes … .'

The extract in Vignette 2 shows that the New Zealander is quite comfortably able to describe the 'hippy' atmosphere of this particular Pai café, perhaps because she can identify with the experience on a cultural, as well as perhaps on a personal, level. Moreover, she would likely have a wealth of research into (Western) backpacker culture with which to draw upon when she comes to analysing her fieldnotes and theorising about the ways of performing relaxation in Pai (for example, as well as earlier 'backpacker' theory, this café scene fits with more recent trends as identified in Kathleen Adam's (2016) paper on 21st Century tourism in Southeast Asia). In this sense, the cafe represents not only a physical and culturally familiar 'comfort zone' for the New Zealand researcher (ontological security), but in 'knowing' how to interpret this place, and having the means to do so, this may also offer her a significant level of epistemological security. On the other hand, given that, according to our observations, neither non-Western tourists, nor our non-Western students, tend to visit that particular cafe, there is the risk that an analysis of Pai based on this experience would likely only capture a very narrow aspect of touristic 'relaxation' in Pai. Moreover, and following on from Chang (2015), there is an added risk that such analysis might even ignore the complex socio-cultural tourism context of this time and space and instead falsely reinforce narrow and traditional 'Western' views of 'backpackers', 'backpacker culture' and tourist behaviour and experience in Pai. In other words, in this context the New Zealand researcher is so ontologically and epistemologically secure in her observing and interpreting this kind of scene, there is risk of 'universalising' this café's, actually very particular, rendering of the 'good vibes' in Pai.

The difficult task of shifting tourism scholarship away from this very Western or Euro-centric view is further highlighted in the comments below from one of the Chinese students:

A lot of times when I was writing assignments and I was trying to interpret my feelings about Chinese tourists, it was so hard to find literature … to support my idea. It's like there is … always some part of ourselves which is not being explained by the literature. So, for me, writing academically, you've got to have something to support your idea, but I would always get puzzled about whether I should put something into my writing or not because there is no 'evidence', just my personal opinion.

This comment highlights the frustration that a lack of (non-Western) tourism theory that adequately considers Chinese values and beliefs (Mura & Sharif, 2015) hinders the capability of this student to fully express his understandings about tourists and tourism in written work. Not only is the student expected to write in English but he is also expected to express himself based on Western epistemologies (and indeed Western tourist behaviours), which are unlikely to provide a suitable theoretical anchor.

A different sort of frustration was highlighted in comments made by a New Zealand student. In the extract below the student is talking about her experiences of working with an Asian colleague on a group project about tourist's photography behaviour:

She wouldn't pick up on what we just saw. I'd be like 'Did you see that ... ?' And she'd be like 'No, I wrote something else' ... She would definitely miss something that I think would be important and write something else that I don't think is important, but could be important. Whereas I'll completely miss that.

Here, the student is noticing that even when observing the same tourist behaviour there might be differences between what different researchers from different cultural backgrounds 'see'.

Interestingly, though, it is this sort of multi-cultural encounter, such as students working together in a group, that may in fact be the key to *de/re*-centring tourism scholarship through the opening up of pluralistic dialogues (Chang, 2015; Huang et al., 2014; Tucker & Zhang, 2016). Simply by acknowledging that her colleague's observations 'could be important', this student is beginning to open herself up to be able to look at things in different ways. This student went on to say that she had now learned that she 'shouldn't be telling people what to notice and should value their different 'noticings''. There is, in this, affirmation of the 'others' way of seeing the world. It is reflexive realisations such as this that might, eventually, help in what Chambers and Buzinde (2015) call the 'dismantling of views that Western and non-Western knowledges are not compatible [and seeing that] ... although these two knowledge systems function independently, they can certainly benefit from creative interconnectivity' (pp. 10–11). For us, as well as our students, working on the fieldschool has enabled this 'dismantling' and realisation of creative interconnectivity to occur, albeit often serendipitously. In the following section we explore this idea by considering how we, as Western tourists/researchers, came to know the contemporary Chinese concept of 'Small Fresh' or 小清新 (xiǎo qīng xīn), through cross-cultural dialogue with our Chinese students.

Negotiating meaning – 'small fresh'

In **Vignette's 3** and **4** we focus in on a particular aspect of Chinese tourist behaviour that we now know to be 'Small Fresh', or 小清新 (xiǎo qīng xīn). Prior to the introduction of the fieldschool module as part of a Master of Tourism programme, neither of us were aware of the concept 'Small Fresh'. The process of our coming to understand this unfamiliar concept helps to further illustrate both the complexities surrounding Western-centric tourism scholarship, and the 'productive re-centrings' which the fieldschool interactions provided.

Vignette 3: Student report from 2017 fieldschool:
During the 2017 fieldschool one group report handed in for assessment, written by two students, one from China and the other from India, focussed quite broadly on the behaviours of Chinese tourists in Pai. Throughout the report there were continual references to 'Small Fresh', as in the following excerpt:
In Pai, tourist attractions like Tree House and Coffee in Love highly fit into the category of 'Small Fresh' as it's surrounded by nature with images of country side, artistic buildings, simple lifestyle. The 'get-away' image of these places provides tourists a platform to perform their quest for 'Small Fresh' lifestyle, that is freedom, nature, arts and romance. It's a lifestyle that is dreamed by the young generation who are struggling with a stressful life in the fast-growing economy in China; they long for running away from anxiety and stress and just living a simple life in a small town.
We spoke with the students while they had been writing the report, and at the time they were really frustrated. They, and in particular the Chinese student, knew that the tourist behaviour they were observing

was linked to the notion of xiǎo qīng xīn/'small fresh' but there was little, if any, relevant literature with which to ground their descriptions. It was suggested that if they could not find any specific academic literature about 'Small Fresh' that they look to mainstream tourism literature for something that might similarly describe what they were observing. They may as well have been asked to hammer a square peg in a round hole.

In the report the students tried hard to convey what 'Small Fresh' meant, and they did a really good job. They had even found one or two references, but these were mainly from non-academic sources. There were no references to 'mainstream' tourist behaviour literature, and because there were no 'familiar' anchors (i.e. the use of Western theory/concepts), the report was difficult to assess in that it appeared to lack academic rigour.

In this vignette, the students and teachers found themselves in a predicament that is likely to be familiar given the prevalence of Anglo-Western centrism in tourism knowledge creation. The students presented a final report in which they tried to make a culturally significant aspect of Chinese tourist culture understandable, yet with no culturally relevant theory available for them to draw upon. By working (or battling to work) within the narrow confines of Anglo-Western centric tourism knowledge creation, a scenario such as that presented in Vignette 3 carries with it the risk that non-Western researchers may come to view their own culturally relevant interpretations of the world as somehow less worthy, or valid, than those of their Western counterparts. This may be particularly problematic in cases where power differentials are at play (e.g. teacher/learner).

Yet despite the unfulfilling nature of the experience outlined in Vignette 3, the teachers and other class members had at least been exposed to this idea of 'small fresh'. The concept remained somewhat abstract until a serendipitous encounter occurred the following year during the 2018 fieldschool (see **Vignette 4**).

Vignette 4: Encounter at the Treehouse, Pai (2018 fieldschool)

As well as being an accommodation complex, the Pai 'Treehouse' is becoming extremely popular as a daytime visitor attraction, in particular with Chinese tourists. As part of the fieldschool, we visit the attraction every year to observe the behaviours of tourists in this place (including our own and each others' amongst the group). On this particular occasion, some Chinese students and the teachers were standing on a veranda overlooking the main 'treehouse' area. We were observing a scene that has become common in such places, but is one that, for the non-Chinese members of the fieldschool group at least, is difficult to fathom. The scene was a group of young and middle-aged Chinese women dressed in smart but summery outfits posing for portrait-style photographs. The women spent a seemingly inordinate amount of time getting themselves, and any 'props', in exactly the right position. Their flowery outfits were colourful and complemented the carefully applied make-up. There appeared to be a deliberateness to it all – the outfits, the make-up, the posing, the choice of the place; the women had come here to be photographed. The fieldschool teachers and non-Chinese students looked on, perplexed; it all looked a bit odd, a bizarre sort of collective tourist behaviour that we do not understand. From previous fieldschools, the teachers had learned (from Chinese students) that this sort of behaviour might be related to the Chinese term 'small fresh'/小清新 (xiǎo qīng xīn).

On this particular occasion of observing tourist behaviour, one of the teachers turned to a Chinese student and asked him: 'If I said to you 'small fresh', would that mean anything to you?' The student smiled, and immediately replied, 'Ah, yes, xiao qing xin, it is what is happening down there'. We then proceeded to talk about what the term might mean and, eventually, agreed that it equated to something like 'adult innocence'. This, of course, made more sense than 'small fresh' or 'xiao qing xin' and it felt as though we had been given a different lens through which to view, and better understand, this obviously significant touristic behaviour.

During this encounter the students and teachers had at least some point of shared reference. By talking together about what we were seeing, we were presented with an opportunity to reflect on our own and each other's culturally based interpretations. This enabled us to find a cross-cultural translation for the concept of small fresh, one that was comprehendible to all present whilst not detracting from the essence of its meaning.

It is relevant to note that the key rationale for developing the fieldschool was to provide students with the opportunity to ground complex theory in the everyday realities/practices of tourism. And whilst it is not the purpose of this article to discuss the merits, or otherwise, of experiential learning (for example, an overseas fieldschool), the encounter described in **Vignette 4** does highlight some of the benefits that may accrue when Western and non-Western scholars spend time together 'in'/'doing' tourism. The encounter documented in **Vignette 4** might be an example of the sort of 'horizontal dialogue' (Grosfoguel, 2006, p.179) which, according to Chambers and Buzinde (2015), is vital to recentring tourism scholarship.

This of course requires that our minds switch from our own to consider the 'other' and, in doing so, achieve a 'renewed mind' (Çakmak & Isaac, 2017, p. 76). Ultimately, 'other' knowledge, like that discussed in **Vignettes 3** and **4**, is likely to provide valuable, localised and culturally relevant insights into tourism. However if we are not prepared to make the mind switch that Çakmac and Isaac call for, then much of this knowledge and the insights that it provides may remain obscured from view. As the Chinese student from **Vignette 4** pointed out to us later, 'If I hadn't been asked about 'small fresh' I would never think about that, in that, it would feel so normal to me … I would understand these things from my own opinion'. This sort of situation is problematic since, by not consciously probing localised knowledge of certain phenomena, or by their not being deemed 'worthy', particular aspects of knowledge about tourism may fail to evolve into mainstream theory. Subsequently, certain tourist behaviour (for example, 'small fresh'-related behaviour of Chinese tourists) may be misconstrued, misinterpreted or missed altogether due to a lack of relevant understanding and theory, leaving researchers frustrated and the tourism knowledge system (Tribe & Liburd, 2016) lacking as a result. Moreover, relying on chance, or serendipitous, encounters as a means to 'renew our minds' may not be sufficient, and a more deliberate effort on the part of tourism researchers from different backgrounds to step out of their culturally familiar comfort zones may be needed. In the next section we discuss what occurred when we, the authors, made a conscious effort, during the 2018 fieldschool, to encourage more culturally relevant knowledge within the context of our research training exercises.

Beyond 'small fresh' – encouraging culturally relevant knowledge

Having witnessed first-hand some of the impacts of the Western-centrism prevalent in tourism knowledge, we decided during the 2018 fieldschool to actively encourage students to draw on localised, culturally relevant knowledge in their written work. An example of this is provided in **Vignette 5**. In this example the students, all from

China, base their interpretations of observed Chinese tourist behaviour on contemporary Chinese concepts.

Vignette 5: Student report from 2018 fieldschool:
Daka Travelers (打卡)
During our research, we have found a group of travellers who have come to Pai because it was … on their checklist for their Thai vacation … They all just wanted to skim Pai's attractions, and they referred to their behaviours as Daka (打卡). Daka literally means clock-in in Chinese … tourists [who] skimmed places and take pictures to prove that they have visited certain places
Internet celebrity (网红 wǎng hóng)
The prevalence of social media platforms, such as Weibo, Wechat moments, Tik Tok (抖音), has led to the emergence of internet influencers … known in China as internet celebrities. Named 'wang hong' in Chinese, these internet celebrities were commonly seen around the attractions in Pai … [T]he most common feeling that this group of people … described was 'small fresh' (小清新). 小清新 (xiǎo qīng xīn); its Chinese meaning is merely small fresh, but in the recent years, it has been a subculture in Chinese society. Typically, it refers to women who are artsy and graceful. Things like green fields, blue sky, and romantic sceneries are considered as essential elements to feel 'small fresh'. Pai has many attractions that have been surrounded by these things … Many people from this … group didn't want to interact with us, because they were [too busy] enjoying themselves taking photos and making arty poses. Sometimes they would spend a couple of minutes to edit and upload their pictures on their phones. Some of them explained to us as they were enjoying showing off their pictures on social media, because when they compete with others' photos, they feel satisfied and proud if their pictures are more beautiful than others'.

Due to the report quoted in **Vignette 5** we, the New Zealander teachers, were again exposed to different examples of Chinese concepts that will undoubtedly be helpful for us as Western scholars when attempting to explain particular touristic behaviours. In other words, examples such as 'Daka traveller' provide a basic foundation for observing Chinese tourist behaviour. These examples could also provide a theoretical 'anchor' when talking with Chinese students about Chinese tourist behaviour. In this sense, concepts such as these will allow us to move beyond the normalised 'monocultural focus' (Samuel & Burney, 2003, p. 95) of Western curricula and better contextualise tourist behaviour using concepts that are more readily understandable to non-Western students. This is particularly important given the significant numbers of students coming from Asia (and in particular China) to study tourism programmes in the 'West' (Hayes & Tucker, 2018).

Yet even within all of this, there is still the sense that we remain fixated on 'difference'. For example, whilst being exposed to concepts such as 'Daka travellers' or 'Internet celebrities' is undoubtedly useful, could these simply be particular cultural and linguistic interpretations of more universal tourist behaviours? Is it only Chinese tourists that 'tick off' places to visit, and is it only Chinese tourists who celebrate their experiences on social media? The answer of course is no, and such behaviours are indeed far more universal in contemporary tourism. Thus, what has become most apparent to us from our experiences of the fieldschool, and indeed throughout the process of writing this article, is that in order to move forward in terms of recentring the tourism field we must first somehow 'unpack' and loosen our ties to fixed/rooted notions of culture (our own and others'). More importantly, if we are to grasp the nuanced differences, multiplicities and similarities that exist within tourism culture(s) then, perhaps, we are better to adopt a *contemporary* lens rather than relying on notions of a 'Western' or 'Asian' lens, either of which will inevitably return us to deploying dualisms, essentialisms and fixed notions of 'difference'.

Chang's (2015) ideas about paying close attention to different socio-cultural contexts and time-spaces are highly relevant here. Let us look again at the example of 'small fresh' to explain this point. An important aspect of why the 'small fresh' tourists pose for these photographs, it seems, is to show off on social media (**see Vignette 5**). This sort of behaviour does not appear to be underpinned by traditional collectivist-oriented Chinese values but instead seems driven by other, more contemporary values such as ostentation (see Hsu & Huang, 2016), a value that has always been associated with individualistic cultures (i.e. Western cultures). As Fu, Cai, and Lehto (2017) reflect, 'the changing socio-economic dynamics in contemporary Chinese society should be acknowledged as an important contextual factor. China's engagement with the outside world contributed to salience of consumerist values, changing many aspects of tourist life' (p.166). What is important to remember here, though, is that the socio-economic-cultural dynamics in many/most societies, not just some societies, are changing in certain ways *together*. Therefore, as the very notion of culture becomes ever more dynamic and hybridised in contemporary 'globalised' society, so too does tourist behaviour. For example, the advent of mobile phone technology and social media has strongly influenced tourist behaviour in certain ways across the board, and subsequently, notions of difference, and similarity, become much more nuanced. To put it another way, there is perhaps no longer any clear 'other' in contemporary tourism.

Conclusions

Our experiences of the fieldschool serve, on the one hand, to highlight the on-going, pervasive, and at times damaging, influence of Anglo-Western centrism in tourism knowledge creation. Our student-scholars who are from Western countries, and particularly if English speaking Western countries, appear to be at an advantage from the start of their scholarship endeavours. For our non-Western student-scholars, the lack of culturally relevant tourism theory frustrates their attempts at interpreting what they are seeing and feeling in relation to the tourism world. There is also an implicit sense that what they may see as culturally relevant is somehow of less worth. On the other hand, however, our somewhat serendipitous experiences and discoveries during the fieldschool also demonstrate the potentiality of the 'international classroom' to produce a more diversified and situated tourism knowledge and a tourism scholarship which is more open and sensitive to culturally relevant knowledges(s).

A prominent example was how our ongoing exposure to the concept of 'small fresh' (as evidenced in Vignette's 3, 4 and 5) has enabled a deeper understanding of the importance of this concept and what it represents. The authors', and our non-Chinese students', prior ignorance of the concept (Vignette 3) was overcome by engaging in cross-cultural dialogue with, in this case Chinese, student-scholars (Vignette 4). This dialogue, in turn, led to further elaborations on the meaning of the concept (Vignette 5), so that 'small fresh' has evolved from a concept that previously meant nothing to the non-Chinese students and teachers to one that now provides a rich and culturally relevant window with which to view much contemporary tourist culture/behaviour. Of course, simultaneously, we must be careful about the situation, as noted previously by Tucker and Zhang (2016) and Zhang (2018), that in the current environment in which Chinese tourists are becoming an important growing market everywhere, there might be a tendency to expect Chinese researchers to take a certain role by somehow becoming essential mediators between the 'West' and 'East'. However, we believe that our discussion in this article of the interactions and discoveries taking place in the fieldschool offers one possible route, that is through active learning *with* and *from* 'each *other*', to de/re-centring and diversifying scholarship and theory for the ever-changing tourism landscape.

As Chang (2015) suggests then, perhaps the goal 'is not to discard extant theories and to reinvent the wheel so much as to reassess how knowledge is conceived in the first place and critically apply and/or create new conceptual lenses to represent this knowledge" (p. 87). In other words, it is not about replacing Anglo-Western tourism theory with non-Western theory. Indeed this would be to re-invent the wheel and could simply perpetuate the dualisms and binaries that currently exist. Instead, our representations of tourist culture must aim to better reflect the contemporary tourism world characterised by third millennium realities and increased mobilities (Adams, 2016), and in which Western and non-Western tourists likely have much in common. In this context, it is new/young scholars (regardless of background) who may become the best equipped and most able to de/re-centre the tourism field. Indeed, if actively encouraged to do so, it seems that the young students who take part in the fieldschool are capable of sensitively tuning in to the subtle differences and similarities in tourist culture. Furthermore, their, and our, continuous and shared interactions and reflections allowed all of us to each experience, at times, being 'other' in relation to tourism knowledge and concepts, as well as affording all of us a heightened awareness of each 'other's' experiences of being 'other' in different contexts. Ultimately, then, we, as teacher-scholars, have as much to learn from our student-scholars as they do from us; we can all benefit from being open to learning *with* and *from* each 'other'.

Of course, though, the interactions and discoveries discussed in this paper occurred somewhat serendipitously; the very nature of the fieldschool allowed for this. We should therefore be mindful that, whilst being highly conducive to the sort of cathartic interactions and discoveries outlined, fieldschools, and in particular international fieldschools, are likely to be the exception rather than norm when it comes to tourism curricula. This opens up the important question of how might we go about engendering similar interactions and discoveries within a more traditional classroom context? This, perhaps, represents a more challenging task and one that potentially requires an even more deliberate and planned pedagogical intervention. As we look to the task at hand, Leask (2008, p. 23) offers some pertinent questions which we might draw upon: How is what I will be teaching culturally constructed and shaped?; How is thinking in the [field] unique and culturally constructed?; What does this mean for the way I teach it?; What possibilities are there for students to explore the ways in which their own and each other's cultures organise knowledge? In considering these questions, we may become more aware of how pedagogy can open up new possibilities for de/re-centring tourism knowledge as well as the possibility of a more sensitive tourism scholarship.

Disclosure statement

No potential conflict of interest was reported by the authors.

Notes

1. To ensure rich, quality learning experiences/interactions, the fieldschool is limited at all times to 12 students.
2. It is interesting to note however that in recent years Asia has demonstrated the highest growth rate in terms of the number of wellness tourism trips taken (see Yang, Lee, & Khoo-Lattimore, 2018).

References

Adams, K. M. (2016). Tourism and ethnicity in insular Southeast Asia: Eating, praying, loving and beyond. *Asian Journal of Tourism Research, 1*(1) 1–28.

Alatas, S. F. (2006). *Alternative discourses in Asian social sciences: Responses to eurocentrism.* London: Sage

Alneng, V. (2002). The modern does not cater for natives: Travel ethnography and the conventions of form. *Tourist Studies, 2*(2), 119–142. doi:10.1177/146879702761936626

Anderson, B. (2009). Affective atmospheres. *Emotion, Space and Society, 2*(2), 77–81. doi:10.1016/j.emospa.2009.08.005

Botterill, D., & Gale, T. (2005). Postgraduate and Ph.D. education. In D. Airey & J. Tribe (Eds.), *An international handbook of tourism education* (pp. 469–482). Oxford: Elsevier

Budde, F., Tranter, P., Fechtel, A., Wise, A., Lui, V., & Milunsky, T. (2013). Winning the next billion Asian travelers—Starting with China. Retrieved from https://www.bcgperspectives.com/content/articles/transportation_travel_tourism_globalization_winning_billion_asian_travelers_starting_china/?chapter=3

Çakmak, E., & Isaac, R. K. (2017). A future perspective about tourism and power: A polyphonic dialogue in the agora. *Tourism, Culture & Communication, 17,* 75–77. doi:10.3727/109830417X14837314056979

Chambers, D., & Buzinde, C. (2015). Tourism and decolonisation: Locating research and self. *Annals of Tourism Research, 51,* 1–16. doi:10.1016/j.annals.2014.12.002

Chang, T. C. (2015). The Asian wave and critical tourism scholarship. *International Journal of Asia-Pacific Studies, 11,* 83–101.

Cohen, E. (1973). Nomads from affluence: Notes on the phenomenon of drifter-tourism. *International Journal of Comparative Sociology, 14*(1–2), 89. doi:10.1177/002071527301400107

Cohen, E. (2006). Pai—A backpacker enclave in transition. *Tourism Recreation Research, 31*(3), 11–27. doi:10.1080/02508281.2006.11081502

Cohen, E., & Cohen, S. A. (2015). A mobilities approach to tourism from emerging world regions. *Current Issues in Tourism, 18*(1), 11–43. doi:10.1080/13683500.2014.898617

Cohen, S. A., & Cohen, E. (2017). New directions in the sociology of tourism. *Current Issues in Tourism, 22*(2), 1–20.

Edensor, T., & Kothari, U. (2018). Consuming colonial imaginaries and forging postcolonial networks: On the road with Indian travellers in the 1950s. *Mobilities, 13*(5), 702–716. doi:10.1080/17450101.2018.1476020

Fu, X., Cai, L., & Lehto, X. (2017). Framing Chinese tourist motivations through the lenses of Confucianism. *Journal of Travel & Tourism Marketing, 34*(2), 149–170. doi:10.1080/10548408.2016.1141156

Grosfoguel, R. (2006). World-systems analysis in the context of transmodernity, border thinking, and global coloniality. *Review (Fernand Braudel Center), 29*(2), 167–187.

Hayes, S., & Tucker, H. (2018). Qualitative research skill training: Learning ethnography in the field. In P. Mura & C. Khoo-Lattimore (Eds.), *Asian qualitative research in tourism: Ontologies, epistemologies, methodologies, and methods* (pp. 139–151). Singapore: Springer

Hsu, C. H., & Huang, S. S. (2016). Reconfiguring Chinese cultural values and their tourism implications. *Tourism Management, 54*, 230–242. doi:10.1016/j.tourman.2015.11.011

Huang, S., van der Veen, R., & Zhang, G. (2014). New era of China tourism research. *Journal of China Tourism Research, 10*(4), 379–387. doi:10.1080/19388160.2014.952909

Keen, D., & Tucker, H. (2012). Future spaces of postcolonialism in tourism. In J. Wilson (Ed.), *The Routledge handbook of tourism geographies* (pp. 97–102). London: Routledge

Kitano, M. K. (1997). What a course will look like after multicultural change. In A. I. Morey & M. K. Kitano (Eds.). *Multicultural course transformation in higher education: A broader truth* (pp. 18–34). Toronto, ON: Allyn and Bacon.

Leask, B. (2008). Internationalisation, globalisation and curriculum innovation. In M. Helsten & A. Reid (Eds.), *Researching international pedagogies* (pp. 9–26). Dordrecht, Netherlands: Springer Science and Business Media.

Majee, U. S., & Ress, S. B. (2018). Colonial legacies in internationalisation of higher education: Racial justice and geopolitical redress in South Africa and Brazil. *Compare: A Journal of Comparative and International Education*, 1–19. doi:10.1080/03057925.2018.1521264

Mkono, M. (2011). African as tourist. *Tourism Analysis, 16*(6), 709–713. doi:10.3727/108354211X13228713394840

Mura, P., & Sharif, S. P. (2015). The crisis of the 'crisis of representation' – Mapping qualitative tourism research in Southeast Asia. *Current Issues in Tourism, 18*(9), 828–844. doi:10.1080/13683500.2015.1045459

Ong, C. E., & Du Cros, H. (2012). The post-Mao gazes: Chinese backpackers in Macau. *Annals of Tourism Research, 39*(2), 735–754. doi:10.1016/j.annals.2011.08.004

Pritchard, A., & Morgan, N. (2007). De-centring tourism's intellectual universe, or traversing the dialogue between change and tradition. In I. Ateljevic, A. Pritchard, & N. Morgan (Eds.), *The critical turn in tourism studies* (pp. 11–28). Oxford: Elsevier

Rizvi, F., Lingard, B., & Lavia, J. (2006). Postcolonialism and education: Negotiating a contested terrain. *Pedagogy, Culture & Society, 14*(3), 249–262. doi:10.1080/14681360600891852

Samuel, E., & Burney, S. (2003). Racism, eh? Interactions of South Asian students with mainstream faculty in a predominantly White Canadian university. *Canadian Journal of Higher Education, 33*(2), 81–114.

Swadener, B. B., & Mutua, K. (2008). Decolonizing performances: Deconstructing the global postcolonial. In N.K. Denzin, Y.S. Lincoln, & L. Tuhiwai-Smith (Eds.), *Handbook of critical and indigenous methodologies* (pp. 31–43). Thousand Oaks, CA: Sage.

Swain, M. (2016). Embodying cosmopolitan paradigms in tourism research. In A-M. Munar & T. Jamal (Eds.), *Tourism research paradigms: Critical and emergent knowledges* (pp. 87–112). Bingley, UK: Emerald Group Publishing Limited.

Teo, P., & Leong, S. (2006). A postcolonial analysis of backpacking. *Annals of Tourism Research, 33*(1), 109–131. doi:10.1016/j.annals.2005.05.001

Towner, J. (1995). What is tourism's history? *Tourism Management, 16*(5), 339–343. doi:10.1016/0261-5177(95)00032-J

Tribe, J., & Liburd, J. J. (2016). The tourism knowledge system. *Annals of Tourism Research, 57*, 44–61. doi:10.1016/j.annals.2015.11.011

Tucker, H., & Zhang, J. (2016). On Western-centrism and "Chineseness" in tourism studies. *Annals of Tourism Research, 61*, 250–252. doi:10.1016/j.annals.2016.09.007

United Nations World Tourism Organisation. (2008). UNWTO Tourism Highlights 2008 Edition. Retrieved fromhttps://www.e-unwto.org/doi/pdf/10.18111/9789284419876

United Nations World Tourism Organisation. (2018). UNWTO Tourism Highlights 2018 Edition. Retrieved fromhttps://www.e-unwto.org/doi/pdf/10.18111/9789284413560

Wijesinghe, S. N. R., & Mura, P. (2018). Situating Asian tourism ontologies, epistemologies and methodologies: From colonialism to neo-colonialism. In P. Mura & C. Khoo-Lattimore (Eds.), *Asian qualitative research in tourism ontologies, epistemologies, methodologies, and methods* (pp. 97–115). Singapore: Springer Nature.

Winter, T. (2009). Asian tourism and the retreat of Anglo-western centrism in tourism theory. *Current Issues in Tourism, 12*(1), 21–31. doi:10.1080/13683500802220695

World Bank. (2018). World development indicators: Travel and tourism. Retrieved fromhttp://wdi.worldbank.org/table/6.14

Yang, E. C. L., Lee, J. S. H., & Khoo-Lattimore, C. (2018). Asian cultures and contemporary tourism: Locating Asia, cultural differences and trends. In E. C. L. Yang & C. Khoo-Lattimore (Eds.), *Asian cultures and contemporary tourism* (pp. 1–20). Singapore: Springer

Yin, R. K. (2011). *Qualitative research from start to Finish*. New York, NY: The Guildford Press

Zhang, J. (2018). How could we be Non-Western? Some ontological and epistemological ponderings on Chinese tourism research. In P. Mura & C. Khoo-Lattimore (Eds.), *Asian qualitative research in tourism: Ontologies, epistemologies, methodologies, and methods* (pp. 117–136). Singapore: Springer

'Asianizing the field': questioning Critical Tourism Studies in Asia

T. C. Chang

ABSTRACT

The relationship between Critical Tourism Studies (CTS) and Asian tourism research is worthy of research. In particular, the challenges that CTS offers to Asian tourism scholarship and how the latter expands the scope of critical thought are areas for consideration. CTS deconstructs mainstream concepts and theorizations and gives voice to the marginalized in the real world of tourism industry and also in the research domain of knowledge creation. How CTS and Asian tourism scholarship are mutually implicated is considered through six exploratory questions relating to principle, language, authorship, concepts, emancipation and pedagogy. Drawing on select literature in Asia, the criticality of extant studies is examined and meaningful directions for further critical scholarship charted. The ultimate goal is to work towards a Critical Asian Tourism Studies (CATS) in which the mutually reinforcing relationship between Asian tourism scholarship and CTS is celebrated, and where research on Asian tourism can be made more theoretical, inclusive and emancipative.

摘要

批判性旅游研究与亚洲旅游研究的关系值得研究。特别是批判性旅游研究对亚洲旅游研究的挑战，以及亚洲旅游研究如何拓展批判性思维的视野，都是值得思考的领域。批判性旅游研究解构主流概念和理论，在旅游产业的现实世界和知识创造的研究领域，为边缘群体发声。本文从原则、语言、作者身份、概念、解放、教学法等六个探索性问题来探讨批判性旅游研究与亚洲旅游学术的相互关系。借鉴亚洲的部分文献，对现存研究的批判性进行了研究，并为进一步的批判性研究指明了有意义的方向。最终的目标是致力于批判性的亚洲旅游研究，在该研究中，亚洲旅游学术与批判性旅游研究之间的相互促进关系得到赞赏，从而亚洲旅游研究可以变得更加理论化、包容性和解放性。

1. Introduction

When King and Porananond (2014, p. 6) first used the phrase 'Asianise the field' in their book *Rethinking Asian Tourism*, they were urging for more indigenous scholarship in Asian tourism studies. It is their belief that Asian scholars bring useful insights that might differ from non-Asian perspectives in tourism conceptualizations and analyses.

Calls for an Asian-centric perspective have increased since the 2000s following the increasing numbers of Asians on tour (internationally and domestically) and greater sensitivity towards reflexive knowledge and tourism epistemologies. Alongside this surge in Asian tourists and indigenous scholarship is also an increasing attention paid to Critical Tourism Studies (CTS). The basic premise of CTS includes the application of critical social–cultural theory to research, the challenge to mainstream knowledge through focusing on the 'Other' (the marginalized, oppressed and indigenous groups among others) and social emancipation as an end goal (Ateljevic, Pritchard, & Morgan, 2007; Tribe, 2008). The mutual implications of CTS and Asian tourism knowledge are worthy of further contemplation.

Six questions are raised about the relationship between CTS and Asian tourism research. The questions centre around issues of principle, language, authorship, concepts, emancipation and pedagogy. By asking questions and suggesting possible answers, the properties of an emerging Critical Asian Tourism Studies (CATS) will become clearer. Indeed, the very act of asking questions is in line with CTS. As Wilson, Small, and Harris (2012, pp. 48–49) assert, the responsibility of critical scholars is to 'critique, question, unmask and understand … . [to] keep asking questions, challenging and accepting divergent views'. While the questions posed might appear rhetorical, the discussion problematizes each issue with an eye to clarifying what makes (and breaks) CATS.

The discussion is organized across three sections. First, the evolution of CTS as a knowledge field is presented and the implications for Asian tourism scholarship considered. As we shall see, 'critical' means various things over the years but a central concern has always been a critique on power relations. The second and key section poses six questions as we envisage the possibilities and challenges of critical approaches in and for Asia. Looking ahead, the conclusion contemplates the fulsome potential of CTS if the breadth and depth of Asian tourism knowledge are brought to bear upon its lofty universalist agenda.

2. Critical tourism studies and Asian scholarship

Before the term CTS came into popular usage in the 2000s, the most fundamental understanding of critical research was to scrutinize something carefully and to uncover faults through enquiry. The earliest critical tourism research emerged from non-government organizations (NGOs) in the 1980s voicing the experiences of local communities and tourism's impacts on them (Higgins-Desbiolles & Whyte, 2014). NGOs like Hawai'i Ecumenical Coalition, the Ecumenical Coalition on Tourism (based in Bangkok) and EQUATIONS (India) produced educational materials rather than academic research, an example being the Hawai'i Ecumenical Coalition's *Pacific Tourism as Islanders See It* (1980) which focused on the 'experiences and viewpoint from the inside looking out rather than the other way around' (cited in Higgins-Desbiolles & Whyte, 2014, p. 89). The roots of critical perspectives are therefore activist and community-based, emerging from 'people who have been harmed by various forms of tourism and the NGOs that have sought to represent them' (Higgins-Desbiolles & Whyte, 2014, p. 90).

A second understanding of critical research emerged amidst the reckoning of tourism's non-economic dimensions. The 'cultural turn' in the social sciences in the early 1990s saw increased attention paid to the representations of people and places in tourism. Critical research emerged as a direct challenge to mainstream approaches which were preoccupied with management and policy-dominated agendas, positivist methods and structuralist discourses. Influenced by social and cultural theories *ala* the Frankfurt School,[1] critical tourism sought to reject 'grand theories or "meta-narratives" that claim to be universally applicable... [but which] make no provision for spatio-temporal or cultural differences' (Gale, 2012, p. 38). By injecting social and cultural theories into analyses, tourism research sought to discuss issues of 'power, discourse, representation' while problematizing essentialist views founded on stark differences such as powerful Western guests and disempowered local hosts (Gale, 2012, p. 45). The launch of critical journals like *Tourist Studies* (2000) and *Tourism and Cultural Change* (2003) may be interpreted as part of this cultural turn.

Following closely behind and articulated more fully from the mid-2000s is the CTS that we know of today. A series of biennial conferences in Eastern Europe and publications saw tourism studies aspiring to values-led, humanist research approaches. Recognizing the 'power of sacred and indigenous knowledge' (Pritchard, Morgan, & Ateljevic, 2011, p. 949), critical research strove for 'just relations between researchers and tourism stakeholders' and participatory methodologies as a way for the researcher to be more reflexive and culturally sensitive (Higgins-Desbiolles & Whyte, 2014, p. 91). At the first Critical Tourism conference in Dubrovnik in 2005 and in its subsequent publication in 2007, co-organizers/editors Ateljevic, Pritchard, and Morgan (2007, p. 4) expressed CTS' goal to 'challenge the field's dominant discourse' which has traditionally been founded on 'masculine tradition of western thought'. More than just a 'way of knowing' (ontology of knowledge), CTS aspires to a way of asking where knowledge comes from (epistemology) and a 'way of being' that is committed to justice and equality (Ateljevic et al., 2007, p. 3).

A number of characteristics define the Euro-originated CTS of the late 2000s. They include: (i) a focus on researcher reflexivity and positionality; (ii) critical research methodologies (approaches that are post-modernist, feminist, post-colonial and indigenous; and methods that include auto-ethnography, grounded theory and participatory action research); (iii) critical pedagogy and curriculum; and (iv) inclusion of long-ignored voices (Higgins-Desbiolles & Whyte, 2014). On the last point, it argues that tourism research has too long been 'focused narrowly on the motivations and needs of tourists and the tourism business sectors' (Higgins-Desbiolles & Whyte, 2014, p. 92.), ignoring the voices of subaltern peoples of disabilities, age and alternative sexualities as well as members of the host community.

Whether we agree (or not) on the existence of a CTS school, a consistent thread across the scholarship has been an enquiry on power, particularly who has power, how power is used and how to reclaim it in the 'real world' and 'research domain'. In the 'real' world of industry, the critique focuses on who benefits/loses in tourism development, and how profits and costs are shared across governments, multinational corporations, local enterprises and communities. In the research domain, the enquiry focuses on who is heard and who is silenced, and who holds the power in knowledge production and dissemination. Critical tourism perspectives seek to 'expose whose interest are served and the exercise of power and the influence of ideology in the research situation and the research itself' (Tribe, 2007, p. 30). Far removed from the management concepts and tourism morphological models of the late twentieth century, CTS applies critical approaches and brings critical theory to bear on tourism studies and practices.

The focus on power and critical theory has implications for Asian tourism research and four concerns are apparent. The first is that as we give voice to local communities and indigenous researchers who represent them, we must remember the intent is not to displace existing knowledge and approaches but to supplement them. CTS does not seek to replace one perspective (positivist) with another (critical post-structural) but to have them alongside each other as 'plural epistemological bases and multi-dimensional approaches' (Pritchard & Morgan, 2007, p. 21). Speaking directly to Asian tourism, Winter (2009, p. 321) argues against 'analytical nativism' that seeks a total rejection of Western knowledge in favour of an Asian one. In as much as a 'West without the Rest' perspective is parochial, an 'Asia without the West' approach is equally unhelpful in apprehending tourism realities.

Second, we should question the usefulness of categories like 'Asian' and 'Western' and be mindful that labels are themselves constructs that can co-produce the inequalities that we lament. What then is the analytical merit of using the term 'Asian' beyond being a convenient proxy for 'Western' or 'Eurocentrism'? At the outset, the essentialist trap of interpreting Asia as a monolith and 'somewhere or something that is fundamentally "different"' from the West must be avoided' (Winter, Teo, & Chang, 2009a, p. 8). The adjectival 'Asian' is therefore emphasized here as a heuristic choice that distinguishes the literature from that which claims to be universal but is often not. We also note that the focus on Asia does not mean the discussions are representative of the entire continent nor exclusive to it. As the largest continent, there are bound to be differences within it and also variations within national borders. The debates in and on Asia are therefore meant to speak to 'Other' regions of the world while also speaking back to Western perspectives. Debates on silences, marginalization and 'othering' will hopefully resonate with the emerging tourism regions of Africa, Central America, South America, Middle East and the South Pacific (Cohen & Cohen, 2015).

A third concern regards the value of geographical categorizations. We must keep in mind that terms like 'Asian', 'regional', 'Anglo-American', etc. constitute one axis along which knowledge is constructed. Ethnic social categorizations (e.g. 'Chinese research' or 'Indonesian scholarship', e.g. Bao, Chen, & Ma, 2014; Oktadiana & Pearce, 2017) as well as class and gender are other axes through which knowledge may be categorized and interpreted. The fact that geographical categorization is emphasized does not mean this is the best or only way to critique the literature. Given that this publication is in *Tourism Geographies*, a spatial perspective is prioritized and particular attention is drawn to the role of geography and geographers in CTS debates. By asking questions on the geographical provenance of authors or how emancipatory goals play out differently in different settings, for example, the influence of geography is foregrounded.

Finally, attention is called to the dangers of self-inscribed othering. The choice to focus on Asia might be criticized by some as a process of self-othering – an act that divides rather than brings the tourism studies field together. This seems to go against the spirit of CTS which celebrates collaboration and knowledge co-production (Aitchison, 2007). For example, when the literature is reviewed to be overly Western-oriented, or when critics differentiate between Western and non-Western tourists, one wonders why a West-versus-non-West demarcation needs to take place at all. Why cannot Asian tourism be studied on its own terms and for its own sake without a Western (othering) referent? These questions go to the very heart of the CATS enterprise and are worthy of further discussion. In the next section, six questions are asked as we consider the contours of an emerging CATS and the value of an Asianizing approach to achieving this.

3. Critical Asian Tourism Studies (CATS): six questions

Six questions are raised with the goal of clarifying the characteristics of an emerging CATS. The questions include:

- Principle: is it possible to be critical without acknowledging CTS and its proponents?
- Language: must one use English to be critical?
- Authorship: can a non-Asian pursue critical Asian research?
- Concepts: must critical scholarship always critique concepts?
- Emancipation: is emancipation possible in and through Asian research?
- Pedagogy: must CTS be accompanied by critical teaching and learning?

While the questions are phrased for a 'Yes/No' response, the deliberations are far more contested. The first four 'PLAC' questions address the *process* of critical research, while the final two 'EP' issues interrogate the *product* or intended outcomes of research. Together, the 'PLAC-EP' ensemble is threaded through with an interrogation on power as we strive to raise awareness of the *process* as well as *outcome* of critical tourism scholarship.

In the spirit of critical research, the author's positionality is declared at the outset. As an Asian geographer with an interest in epistemological issues, the six questions here are but a personal and subjective interpretation of critical scholarship. They are not necessarily universal issues attending CTS but a specific enquiry aimed at fashioning the emergent field of CATS. On the one hand, the author identifies closely with King and Porananond's (2014, p. 6) mission to 'Asianise the field', and on the other undertakes a geography-oriented analysis of the literature. It is this ideological standpoint and disciplinary alignment that shape the questions and discussions to follow.

(a) Principle: being critical without CTS?

The first question asks whether it is possible to engage in critical research without acknowledging CTS and its proponents. The answer is 'yes, so long as critical practices are respected'. These practices include a scepticism of assumptions, being attentive to power relations and giving voice to the marginalized among others. CTS does not hold a monopoly on critical thought, and critical tourism research does not necessarily have to speak the CTS language or reference any of its key texts.

Consider the attempts by some writers to retheorize tourism's cultural impacts in order to better contextualize indigenous situations. Thirumaran (2009) has argued, for example, that the cultural commoditization thesis on Balinese dances is essentially framed from a Western-tourist perspective. By shifting the empirical focus to Indian Hindu tourists, he counter proposed that 'cultural affinity' and 'shared heritage' are more meaningful concepts. Field interviews with Balinese dancers and Indian Hindu tourists (from India and Singapore) reveal that guest/host relations are familial and intimate, something that does not happen between Western visitors and local hosts. He argues that cultural commoditization represents a partial, Western orientation that is relevant to Western travellers, but does not address the relationship between Asian travellers and local hosts. Along the same line, Jenkins and Romanos' (2014) research on visual artists in Bali refutes the concept of 'cultural authenticity'. They contend that authenticity is a tourist-centric idea reflecting tourists' preoccupation with exoticism and 'untouched' cultures. From the artists' end, what is more important is how their work provides them with a quality of life and the wherewithal to raise a family. Seen from this perspective, 'artistic well-being' rather than 'authenticity' is the key issue. Depending on where the academic gaze is directed, therefore, the ruling concept shifts. Both studies on Bali demonstrate that an academic gaze on the local community refocuses our analysis and the corresponding conceptual language to explain the analysis.

In Asia where domestic travel and intraregional tourism are booming, 'Asian tourists in Asia' also constitutes an important focus of research. Yamashita's work is illustrative of critical scholarship because regardless of the site – from Bali to Bangkok, and other Asian places in between – his focus is 'from the Japanese perspective' (Yamashita, 2009). Not presuming that as a Japanese academician and anthropologist, he can ever speak outside his Japanese persona, Yamashita argues that the Japanese gaze is highly particular, informed by the nation's historical connections to and political agendas within Asia. Popular concepts like 'exoticism', 'authenticity' and 'uniqueness' pervading the literature he argues have little currency in Japanese academic discourse. Instead, the Japanese often interpret Southeast Asia as a site of nostalgia, healing and cosy familiarity, a contrast to contemporary urbanized Japan (Yamashita, 2009).

From a CTS vantage point, the works mentioned above do not fall neatly into its fold. They do not deploy the linguistic terminologies or concepts of CTS, nor do they reference key CTS proponents. When they do use the term 'critical', they do so without invoking the traditions of social–cultural theory. However, it is difficult to deny their critical credentials. On the one hand, they speak to and advocate on the aspirations of local agents (e.g. dancers, artists), and on the other, the authors articulate a reflexive positionality in the field (either as a Hindu tourist or Japanese researcher). They are also sceptical of concepts that espouse a Western-touristic viewpoint. In many ways therefore, these works are eloquently critical without the CTS baggage.

It is possible to be critical without CTS because many critical principles have disseminated far beyond their starting points. While researchers owe a debt to critical traditions for debates on power and representations of the 'Other', not all are necessarily aware of the provenance of these ideas. Challenging dominant discourses and debunking stereotypes are what any sceptical researcher does. Rather than fly the CTS flag, what we can encourage instead is 'standpoint research'. By this term, Humberstone (2004, cited in Tribe, 2008, p. 364) means the declaration of one's ideological position in the production of knowledge. He observed that the development of theory and knowledge has been 'largely based upon Eurocentric research and the idea of mainly white middle class' and there is a neglect of research from the standpoint of non-whites or the disabled (Humberstone, 2004, cited in Tribe, 2008, p. 364). Along with a healthy scepticism of popular concepts, Humberstone encouraged a more robust awareness of one's ideological standpoint and positionality in research (to be further discussed under 'Authorship'). Not only will this make for more nuanced scholarship, it also raises consciousness of one's responsibility and positionality as a knowledge creator.

(b) Language: tyranny of a universal language?

A second question relates to language as we enquire whether English is a requisite in critical thought. This might appear as a rhetorical question because language has never been a barrier to tourism research. However, as a universal language, one must be conscious of the power and tyranny exerted through the use of English. While any language can carry critical thought, some ideas seem to travel faster when expressed in a universal tongue.

The universalism of English surfaces a number of concerns, the first being the 'trap of the Other'. By writing in a foreign language, the non-native English user inadvertently pays homage to the originator of the language, instead of writing in a native tongue which might allow for better self-projection (see Ngugi on the issue of African literature, cited in Tribe, 2006, p. 371). Chambers and Buzinde (2015) observed how some African academics have thus reverted to local languages to express the African experience even as they are aware of English as a global medium for ideas. Avoiding the use of English is thus a strategic refusal to be 'Other'.

A related concern is the tyranny exerted by English when it is used to translate indigenous thought. In the Thai context for example, one way of interpreting place is to understand the people occupying it, specifically its community life and politics (Jiraprasertkun, 2009). Unlike the West, the Thais have only recently conceived of the notion of 'community' (*chumchon*) as the Western world knows it. The idea of 'searching for identity' or 'loss of community identity', so popular in Western critical thought, is alien to the Thais. 'Since, typically, Thais were not specifically interested in defining their own qualities, the subject of how Thai space and place have been characterised was more of a concern to foreign scholars' (Jiraprasertkun, 2009, p. 276). Critiques of urbanization and tourism leading to 'identity loss' are thus Western-inclined criticisms. If such issues do not concern the Thais and find no expression in their language, do they really exist? Jiraprasertkun (2009, p. 276) argues that there is a need for 'a specific language and vocabulary' instead of 'using inapplicable Western concepts to explore Thai space and its constitutive places'. Only the local tongue can capture the Thai situation, 'the way Thai people, see, behave in and perceive their world' (Jiraprasertkun, 2009, p. 276).

A similar linguistic trap also surfaced in Alneng's (2002) study of Dalat, Vietnam. In Western representations of Dalat as a tourist destination, descriptions abound of it as dreary, not exotic, an 'uninspired imitation of European alpine atmosphere' and 'Vietnamese kitsch' (Alneng, 2002, p. 132). Alneng asks: 'How can Dalat be "the final word in Vietnamese kitsch" when Vietnamese does not even have a word equivalent to kitsch' (2002, p. 132). Contrary to Western perceptions, Vietnamese visitors regard the hill resort as exotically French and a perfect place for honeymoon and family vacations. For what the Westerners call kitsch, the Vietnamese considers romantic, desirable and 'seem to have no anxious preoccupations with authenticity' (Alneng, 2002, p. 133). In critical tourism thought, we must be aware of who is imputing ideas, concepts and idioms – the English-writing analysts or the host community/local author. The use of English as a universal academic language of/for critique thus carries its own worldview and linguistic predilections. In Asian tourism studies, linguistic cognizance is essential if we are to be sensitive to differing viewpoints across the cultural divide.

(c) Authorship: can a non-Asian speak?

Closely allied to language is the issue of authorship, and the question is whether a non-Asian can engage in critical Asian discourse. This may appear to be a moot query because Western researchers are the ones leading the (self-)critique on ethnocentrism and knowledge creation. In Asian tourism, King (2015, p. 518) observed that the charge of ethnocentrism 'has come primarily from Western social scientists or social scientists in Western institutions' and he went on to name Winter, Cohen and Alneng among the key proponents. Academics are often quick to seize on a literature gap and in this case, the ethnocentric charge was identified by self-reflexive Western writers (Chambers & Buzinde, 2015). To the question 'Can a non-Asian speak about Asian tourism?', the answer is 'Yes' and this is already being done. However, greater authorial reflexivity should be further encouraged.

In the preceding discussion on principle, the value of standpoint research was affirmed. This idea is revisited here as we discuss the importance of declaring one's cultural positionality in critical scholarship. A good example is the CTS concept of 'entanglement'. Harris, Wilson, and Ateljevic (2007, p. 44) define entanglements as 'forces that influence, constrain and shape the act of producing and reproducing knowledge within academic structures', and these include ideologies that shape research, systems of accountability that decide what is acceptable and positionality as embodied by the researcher's own 'lives, experiences, worldviews'. They maintain that critical researchers should acknowledge his/her 'position as an outsider to the indigenous colonised experience', writing as a privileged Westerner (Denzin, 2005 cited in Harris, Wilson, & Ateljevic, 2007, p. 46). The outsider researcher recognizes him/herself as an 'allied other', a 'fellow traveller of sorts ... an insider who wishes to deconstruct the Western academy and its positivist epistemologies from within' (Denzin 2005, cited in Harris, Wilson, & Ateljevic, 2007, p. 46).

We should note, however, that reflexive self-awareness is not exclusive to CTS, and post-colonial tourism writers have also espoused the value of positionality and standpoint research. Hollinshead (1999), for example, spoke of the need for researchers to be 'self-aware' yet 'other-regarded' (p. 7) in the hope that this will give rise to plural, even competing interpretations of tourism places and people. Other writers of a post-colonial stripe have similarly demonstrated the value of positionality in helping readers contextualize tourism development issues. For example, Teo and Leong (2006) wrote about Asian backpacking while declaring that their interpretations are filtered through their Asian, female, backpacker perspectives, while Amoamo (2011) had discussed her insider/outside/marginal position as a Māori researcher speaking on Māori tourism affairs. While these post-colonial writers may not explicitly self-identify with CTS, their awareness of ethnicity resonates closely with the CTS' tenor on authorial reflexivity.

It is with this awareness of one's ethnicity that King and Porananond (2014) spoke of the need to 'Asianise the field' believing that Asian scholars will bring enlightened insights that differ from non-Asian perspectives. In a quantitative count of Asian authors contributing to Asian tourism books, Chang (2015) recorded differing degrees of his 'Asianizing' trend. In Porananond and King's (2014) edited collection of 17 chapters by 21 authors, there were a total of 18 authors of Asian ancestry, a count of 85.7 per cent. This compares favourably to other collections such as Hitchcock, King and Parnwell's (2009) co-edited collection (16 chapters contributed by 16.7 per cent of Asian authors) and Winter, Teo, and Chang's (2009b) work (23 chapters comprising 52.5 per cent of Asian contributors) (Chang, 2015, p. 89). A recent volume on Asian qualitative tourism research by Mura and Khoo-Lattimore (2018) reflected a positive trend; of the 21 contributors, a total of 11 authors are of Asian descent. The proportions are much lower in tourism companions compiled by Blackwell and Routledge. The Blackwell companion by Lew, Hall, and Williams (2004) had three Asian authors out of 58 contributors; a decade later in the 2014 Wiley-Blackwell edition, there were four Asians out of 69 authors (Lew, Hall, & Williams, 2014). In the Routledge Companion by Wilson (2012), two out of 38 authors are Asians based in Asia. In the field of Indonesian tourism, Oktadiana and Pearce (2017, p. 1103) revealed that 79 per cent of publications were by 'bule' (white foreigner) researchers, with 51 per cent of from three western countries (Australia, the U.S. and U.K.).

These statistics are but a very blunt gauge of how 'Asianized' the Asian tourism field is. Nevertheless, the snapshot reveals a dearth of indigenous voices in Asian tourism particularly on matters of critical research. A perusal of individual chapters from the above edited books reveal that most of the works are empirical or focused on concepts relating to impacts, carrying capacity and tourist/local perceptions. Considerations of power and epistemological matters were absent except in the collections by Winter et al. (2009b) and Mura and Khoo-Lattimore (2018) (see also Mura and Sharif, 2015).

Moving forward, instead of obsessing over empirical data and findings ('what knowledge'), critical tourism scholarship should be concerned about 'who' and 'how' knowledge – asking questions about epistemologies and knowledge creation (Teo, 2009). 'Asianizing the field' does not mean that non-Asians should be held back from working on Asian research. Rather it is a call to encourage more ethnically Asian researchers to speak up critically on regional and local issues from 'within'.[2] It is this Asian consciousness that is manifested in Yamashita's (2009) work 'from a Japanese perspective'; Porananond's 'sense of Thai-ness' (Porananond & King, 2014) and Thirumaran (2009) research 'from an Indian-Hindu standpoint'. In the field of Indonesian tourism research, Oktadiana and Pearce (2017) proposed the 5Ps of patronage, partnerships, professionalism, pathways and patriotic pride as ways to support local Indonesian involvement in research vis-à-vis *bule* scholars (white foreigner). Sensitivity to the epistemology and ontology of knowledge is a first step to realizing the quantitative inequalities that pervade the knowledge field and the need to address them by supporting indigenous research.

(d) Concepts: reinventing the wheel?

When Pritchard and Morgan spoke about stasis in tourism studies, they were concerned about its inability to 'break new epistemological, *conceptual* or ethical ground' because of the lack of alternative, dissenting voices (2007, p. 12, emphasis added). The question before us is whether Asian tourism research needs to break new conceptual ground in order to be considered critical, or is it possible to be critical without refining, refuting and (re)inventing concepts?

The CTS is not foremost an exercise in conceptualization. While it is true an author may be described as 'critical' if s/he critiques concepts and (re)invents new ideas, this is not necessarily CTS' *raison d'etre*. Conceptualizations may be regarded as truly critical when they are a product of marginal voices, accompanied by the empowerment of peripheral groups that leads ultimately to social emancipation. Critical conceptualizations are therefore accompanied by activist commitments and social responsibility. As Tribe (2008, p. 251) maintains, CTS 'question taken-for-granted recipes and responses and lead to a greater engagement with aims and ends. It can illuminate tourism's blind spots'. Based on abstract database and keyword searches from 1974 till the late 2000s, Tribe (2008) concludes that very few tourism studies engage in critical theory (1.87 per cent of 35,194 texts). Instead, the dominant research paradigm remains positivism and where concepts are critiqued, the focus has mainly been on ideology, discourses and social exclusions (Tribe, 2008).

The above holds true in Asian tourism too. Critical Asian research must also be accompanied by social goals, otherwise it is merely conceptual discourse. Indeed, it has become fashionable in some Asian tourism writings to use verbs like 'rethink', 'revisit', 'reframe', 'unpack' and 'challenge' (in book and article titles) to suggest critical thought. With greater reflexivity and growing academic confidence since the 2000s, Asian tourism researchers have been at the forefront of 'adopting and *refuting*, endorsing and *challenging*' Western-centric concepts (Winter et al., 2009a, p. 6, emphasis added). Critical works must continue this tradition of scrutinizing assumptions, questioning concepts and debunking clichés with empirically sensitive accounts. However, to be truly critical, research must go beyond conceptual reflections to also include social action and change.

(e) Emancipation: possibility of an end goal?

The preceding questions have mainly considered the *process* of critical research. For the final two questions, attention turns to the desired *product* or outcome of CTS: social emancipation and critical pedagogy. 'Emancipation' is closely associated with power, and the role of critical research is to question power structures and dismantle unjust systems in order to 'set people free'. According to Tribe (2008, p. 246), critical tourism is about transforming the way people perceive and act in the world, 'unleashing human agency and autonomy' that will lead to 'better production and consumption of tourism'. The question for CATS is how might this work in an environment where tourism is sometimes the cause of social injustice and inequality.

A large part of social emancipation depends on governance. Tribe (2008, p. 250) views governance as a 'complex set of agencies … involved in the planning, regulation, and control of tourism itself, and the political environment in which it operates'. Critical research helps to check on whether tourism is planned and regulated ethically, if its benefits are shared among stakeholders, and whether practices are pro-local and community empowering. That research is taking place in Asia is not the issue; whether the right research questions are being asked, problems surfaced and socially just solutions proposed and carried out – these are the key concerns in critical scholarship.

The contrast between critical scholarship in the West and pro-development research in the developing world is often stark. Bao et al. (2014) have compared the blue-skies, theoretical research of the developed West with industry-sponsored research in China (and other parts of Asia where research is often in service of industry). To better appreciate this contrast, the different historical trajectories of countries and their research needs and funding opportunities should be contextualized. The advanced West/North has had a longer history of mass tourism and research has since progressed from serving early development needs to incorporating academic, theoretical and critical issues today (Butler, 2004). The liberal academic environment in Western institutions also differs from that in the developing East/South where research and universities are trying to 'catch up'. Practical, applied, industry-geared research is often needed in the latter countries to aid development. While it is possible to hasten critical research in select Asian countries with strong and at times liberal educational institutions (e.g. Hong Kong, Taiwan), the ground is uneven across Asia in terms of research expectations, government funding and university incentive systems.

In their assessment of Chinese tourism research, Bao et al. (2014, p. 179) critiqued existing scholarship as 'naive empiricism' focused on 'summaries and simple explanations of certain tourism phenomena, with no philosophical positions or hypothesis'. This critique is also relevant to other parts of Asia. Surveying the tourism literature on Southeast Asia, Mura and Sharif (2015) note that the majority of works focus on quantitative data and very few deal with ideological debates, reflexivity and researcher introspection (e.g. motives for and social agendas of research). Earlier, we had spoken about the need to advance Asian research from its preoccupation with 'what knowledge' (focusing on data) to 'who' and 'how' research (questioning the epistemologies and ontologies of knowledge). It is imperative to also include 'why' research as we question the role of social emancipation as an end goal in CATS.

A challenge surrounding the emancipation question is who exactly is research sup-posed to emancipate? While the obvious answer is the local community, often the focus veers to the 'unfree' researcher too. In a wave of critiques on Colonialism's stranglehold on non-Western researchers in Asia and elsewhere, much has been said about the need to decolonialize research methods and knowledge productions sys-tems (e.g. Chambers & Buzinde, 2015; Cohen & Cohen, 2015; Mura & Sharif, 2015; Wijesinghe, Mura & Bouchon, 2017; Wijesinghe, Mura, & Culala, 2019). In what is termed 'academic activism', the decolonializing agenda is to identify 'factors and struc-tures controlling knowledge production and dissemination and raise the need for decolonization of knowledge' (Wijesinghe, Mura, & Bouchon, 2017, p. 13). While this is a laudable epistemological goal, the social emancipatory intent of CTS appears to be side-lined.

Critiquing the 'Academy of Hope', Higgins-Desbiolles and Whyte (2013, p. 430) argue that CTS appears to be more interested in improving researcher and publication conditions than 'transforming the actual oppressive circumstances faced by commun-ities for which tourism bears at least some significant responsibility'. Debates on how the 'Other' may be better represented on editorial boards and knowledge-production circles, and discourses on inclusivity and Third-World authorial voices (Pritchard & Morgan, 2007) are all part of a researcher-oriented emancipatory project. Similarly, our earlier attention on language and 'Asianizing the field' (Porananond & King, 2014) may also be construed as projects to emancipate researchers rather than local commun-ities. Whether 'emancipation' should serve the community or the colonized researcher and whether it can do both simultaneously are questions worthy of further contemplation.

(f) Pedagogy: teaching critical thought in tourism?

The final issue to consider is critical pedagogy, or the practice of teaching and learn-ing critically. There are two questions: what should be taught to students in CATS, and is it possible to teach critical tourism without a social agenda? Both are sensitive questions because institutions, state agencies and business organizations in Asia (some of which fund education outfits and research projects) might construe said agendas to be potentially destabilizing. The two questions – the first, general and descriptive (focusing on *taught content*) and the second, specific and prescriptive (focusing on *thought contest*) – are foundations of any critical curriculum aimed not just at inculcating knowledge but also skills and sensitivities.

It was earlier argued that CTS is 'more than just a conceptual exercise'; the same is true of its pedagogical expectations. Teaching content is the first step to ensure stu-dents are aware of the possibilities and problems in tourism; at a higher level, they should also learn about contested thoughts in tourism. Such a gradation of taught/ learnt material demands a curriculum that moves learners from the basics to more sophisticated issues, and from mere 'knowing' to 'doing' and 'being' (Ateljevic, Pritchard, & Morgan, 2007). Whether higher levels of teaching and learning embrace such critical thought is often a matter of institutional vision and individual resources (e.g. availability of faculty, inclination of teaching staff, goals of institution, funding demands). Academics and educationists know only too well that what they *should* be teaching is not often what they end up doing for reasons often beyond their control.

Sofield's (2000) assessment of Asian tourism research is applicable to pedagogy. He identifies four platforms of research: the advocacy platform advocates for tourism's economic prospects and social–cultural value; the cautionary platform warns of tourism's negative effects; the adaptancy platform stresses alternative forms of tourism (e.g. eco-tourism); and the knowledge-based platform studies the tourism knowledge field in a post-disciplinary manner (identifying structures, gaps and agendas). These research platforms can also serve to scaffold critical pedagogy in Asian tourism. Lower-level students can start with learning about tourism's positive/negative impacts (advocacy) and development concerns (cautionary). At higher levels, they should learn about social responsibility and empowerment strategies (adaptancy), while at the highest level, they should be taught academic abstractions and the epistemologies/ontologies of knowledge (knowledge platform). Sofield concludes that 'There is a need for more research into tourism in Asia, taken from the knowledge-base platform with greater objectivity and incorporating local perspectives on leisure and travel' (2000, p. 55). How many institutions in Asia actually teach the knowledge platform from a local (rather than an exclusively global) perspective is not known.

The challenges of critical pedagogy are evident in other ways. In Fullagar's experience of teaching tourism and hospitality in Hong Kong, she spoke of the challenge of teaching leadership in a cross-cultural context (Fullagar & Wilson, 2012). Instead of 'preaching' Western concepts of leadership to Asian students, a starting point is for students to deconstruct their assumptions of leadership and to reconstruct models based on personal experiences. What is helpful is to get students to share reflexively about their practicum experiences particularly their observations of 'everyday dynamics of power' as well as leadership in the tourism workplace (Fullagar & Wilson, 2012, p. 55). Pedagogy employing a 'critically reflective approach' based on personal experience rather than one that exports concepts from elsewhere is rewarding for both educators and learners (Fullagar & Wilson, 2012, p. 55).

Pedagogical issues also concern the graduate student community. With more Asians pursuing graduate studies, tourism pedagogy can manifest in student–supervisor relationships too. In the Chinese tourism context, Huang et al. (2014) note an increasing pace in Asian-Western collaborative publications, much of which between educators and learners. Two pedagogically inclined relationships are highlighted, the first being the 'Chinese student-Western supervisor dyadic' in which co-learning and knowledge co-production lead to 'cross-cultural dialogues and interpretations' (Huang, van der Veen, & Zhang, 2014, p. 381). The second concerns Chinese students being trained in Western universities and who return to their homeland with post-positivist, ethnographic approaches and social agendas. The latter is particularly healthy if the methods and experiences acquired in the West are shared with others in an Asian environment that contextualizes local educational needs and societal realities.

In summary, for Asian tourism studies to be critical, what is needed is a pedagogy born from mindful teaching about the possibilities/problems of tourism (taught content), and the politics of knowledge creation and dissemination (thought contest).

Critical pedagogy should not be confined to the classroom as learning opportunities come from doing/being outside the classroom, and knowledge is best acquired through reflexive personal experiences. Neither should critical pedagogy be limited to undergraduates as co-learning and knowledge co-production occur at the graduate level, benefitting both students and supervisors.

4. Conclusion

In as much as Asian tourism scholarship can benefit from the opportunities availed by CTS, Asian insights can also enrich CTS in productive ways. In closing, the fulsome potential of CTS is contemplated as the breadth and depth of Asian knowledge are brought to bear on its lofty agenda. First, CTS's celebration of non-Western knowledge and perspectives is well matched by the expanding body of works that profess a distinct Asian consciousness. This is a healthy trend allowing Asian writers to declare a standpoint in their scholarship while broadening the CTS corpus by testing the boundaries of reflexivity and values-based research. As Pritchard and Morgan (2007, p. 25) aver, to 'decent[er] tourism's intellectual universe' and create new knowledge, 'we must be willing to learn from every knowledge tradition, from Africa, Asia and from indigenous peoples around the world'. Incorporating insights from Asia can make for a more robust understanding of tourism phenomena and impacts.

The value of standpoint research brings us to a second productive tension and that is an understanding of how tourism 'makes the world'. In Hollinshead's (2007) conception of 'worldmaking', he asserts that tourism creates but also destroys the world through places (tourism landscapes), people (social representations and livelihoods) and pasts (memories and heritage). The tourism world is perpetually made/unmade/remade by planners, practitioners and marketers, but we must also add in the context of this paper writers and researchers who 'make/unmake' through the written text. Since the turn of millennium, writers have become particularly practised at contesting 'Eurocentric worldmaking gazes' and devising new 'interpretations from various/significant multiple standpoints' (Hollinshead, 2007, p. 187). This task of unmaking dominant viewpoints and remaking them with fresh perspectives is what CATS aspires to. The task at hand, however, is not necessarily to unmake namely discarding past works but to remake through refining inherited concepts and perspectives. Only by embracing 'Other' ideas of people, places and pasts, is it possible to envision a post-colonial, post-industrial, post-Occidental tourism world? (Hollinshead, 2007, p. 190).

The mutual implication of Asian tourism and CTS produces what Bao et al. (2014, p. 178) have called 'tourism research based on Asian values'. Without entering the fray of what constitutes Asian values, we must appreciate the attempt here to strive for something uniquely Asian (or in their case, Chinese). In their quest for culturally sensitive research, they argue that one should start with the issue of *language* before proceeding to *research context* and *agenda* setting. On the matter of language, their paper is written in English and serves as a review of all tourism research on China in the Chinese language, written by Chinese working/living in China over a 30-year period (Bao, Chen, & Ma, 2014). Precisely because this is a literature inaccessible to most, the English review is itself a noteworthy contribution. By research context, they explain that different countries have different conditions under which research is conducted. These include the needs of industry, funding agencies (government, corporations, research institutes) and university incentive systems. To give meaning and direction to research, agenda setting is essential. To make an intellectual mark, Chinese researchers are encouraged to pursue a dual-pronged agenda. Scholarship that addresses distinct national/local problems (e.g. tourism real estate) as well as topics of international interest (e.g. tourism's effects on minorities) makes for values-driven scholarship with a Chinese flair (Bao, Chen, & Ma, 2014).

Ultimately CTS is about deconstructing power and engendering social change. With much of Asia designated as 'developing', Asian tourism research can contribute critically to spotlighting areas of abuse and identifying grounds for care and responsibility. With rich empirics on ethical and responsible tourism along with a groundswell of pro-local rights and social change (e.g. UNWTO, 2011), Asian research can help to document, critique and realize the outcomes to which critical scholarship is intended. The framework of six questions presented here – on principle, language, authorship, concepts, pedagogy and emancipation – is but an outline of the emerging contours of CATS. Hopefully, it also serves as a road map for a more sensitive and reflexive understanding of tourism phenomena, the knowledge academy and our place in this changing world.

Notes

1. We note that the origins of a critical paradigm lie in the work of Frankfurt's Institute of Social Research established in 1923. The term 'critical' came from 'dialectical critique of political economy', an approach by which to critique modern society (Chambers, 2007) which also extended to a critique of traditional positivist theories as being incapable of addressing societal issues. It is argued that the goal of research is to give people, especially marginal social groups, a voice thereby emancipating them from structural exploitation (Gale, 2012).
2. In the field of Indonesian tourism research, Oktadiana and Pearce proposed the 5Ps of patronage, partnerships, professionalism, pathways and patriotic pride as ways to increase local-Indonesian involvement in research vis-à-vis *bule* scholars (white foreigner) (see 2017, p. 1106 particularly).

Disclosure statement

No potential conflict of interest was reported by the authors.

References

Aitchison, C. C. (2007). Marking difference or making a difference: Constructing places, policies and knowledge of inclusion, exclusion and social justice in leisure, sport and tourism. In I.

Ateljevic, A. Pritchard, & N. Morgan (Eds.), *The critical turn in tourism studies. Innovative research methodologies* (pp. 77–90). Oxford: Elsevier.

Alneng, V. (2002). The modern does not cater for natives: Travel ethnography and the conventions of form. *Tourist Studies, 2*(2), 119–142. doi:10.1177/146879702761936626

Amoamo, M. (2011). Tourism and hybridity: Re-visiting Bhabha's third space. *Annals of Tourism Research, 38*(4), 1254–1273. doi:10.1016/j.annals.2011.04.002

Ateljevic, I., Morgan, N., & Pritchard, A. (2007). Editor's introduction: Promoting an academy of hope in tourism enquiry. In I. Ateljevic, A. Pritchard, & N. Morgan (Eds.), *The critical turn in tourism studies. Innovative research methodologies* (pp. 1–8). Oxford: Elsevier.

Ateljevic, I., Morgan, N., & Pritchard, A. (2011). *The critical turn in tourism studies: Creating an academy of hope.* Abingdon: Routledge.

Bao, J., Chen, G., & Ma, L. (2014). Tourism research in China: Insights from insiders. *Annals of Tourism Research, 45*(1), 167–181.

Butler, R. (2004). Geographic research on tourism, recreation and leisure: Origins, eras, and directions. *Tourism Geographies, 6*(2), 143–162. doi:10.1080/1461668042000208453

Chambers, D. (2007). Interrogating the 'critical' in critical approaches to tourism research. In I. Ateljevic, A. Pritchard, & N. Morgan (Eds.), *The critical turn in tourism studies. Innovative research methodologies* (pp. 105–119). Oxford: Elsevier.

Chambers, D., & Buzinde, C. (2015). Tourism and decolonialisation: Locating research and self. *Annals of Tourism Research, 51*(1), 1–16.

Chang, T. C. (2015). The Asian wave and critical tourism scholarship. *International Journal of Asia Pacific Studies (IJAPS), 11*(1), 83–101.

Cohen, E., & Cohen, S. A. (2015). Beyond Eurocentrism in tourism: A paradigm shift to mobilities. *Tourism Recreation Research, 40*(2), 157–168.

Fullagar, S., & Wilson, E. (2012). Critical pedagogies: A reflexive approach to knowledge creation in tourism and hospitality studies. *Journal of Hospitality and Tourism Management, 19*(1), 1–57.

Gale, T. (2012). Tourism geographies and post-structuralism. In J. Wilson (Ed.), *The Routledge handbook of tourism geographies* (pp. 37–45). London and New York: Routledge.

Harris, C., Wilson, E., & Ateljevic, I. (2007). De-centring tourism's intellectual universe, or traversing the dialogue between change and tradition. In I. Ateljevic, A. Pritchard, & N. Morgan (Eds.), *The critical turn in tourism studies. Innovative research methodologies* (pp. 42–56). Oxford: Elsevier.

Higgins-Desbiolles, F., & Whyte, K. P. (2013). No high hopes for hopeful tourism: A critical comment. *Annals of Tourism Research, 40*, 428–433.

Higgins-Desbiolles, F., & Whyte, K. P. (2014). Critical perspectives on tourism. In A. Lew, C. M. Hall, & A. Williams (Eds.), *A companion to tourism* (pp. 88–97). Malden, MA: Blackwell Publishing.

Hitchcock, M., King, V., & Parnwell, M. (2009). *Tourism in Southeast Asia: Challenges and new directions.* Copenhagen: NIAS Press.

Hollinshead, K. (1992). White' gaze, 'red' people – shadow visions: The disidentification of 'Indians' in cultural tourism. *Leisure Studies, 11*(1), 43–64. doi:10.1080/02614369100390301

Hollinshead, K. (1999). Surveillance of the worlds: Foucault and the eye-of-power. *Tourism Management, 20*(1), 7–23. doi:10.1016/S0261-5177(98)00090-9

Hollinshead, K. (2007). Worldmaking' and transformation of place and culture: The enlargement of Meethan's analysis of tourism and global change. In I. Ateljevic, A. Pritchard, & N. Morgan (Eds.), *The critical turn in tourism studies. Innovative research methodologies* (pp. 165–193), Oxford: Elsevier.

Huang, S., van der Veen, R., & Zhang, G. (2014). New era of China tourism research. *Journal of China Tourism Research, 10*(4), 379–387.

Jenkins, L. D., & Romanos, M. (2014). The art of tourism-driven development: Economic and artistic well-being of artists in three Balinese communities. *Journal of Tourism and Cultural Change, 12*(4), 293–306. doi:10.1080/14766825.2014.934377

Jiraprasertkun, C. (2009). Reading Thai community: The processes of reformation and fragmenta-tion. In T. Edensor & M. Jayne (Eds.), *Urban theory beyond the west. A world of cities* (pp. 273–294). Abingdon: Routledge.

King, V. (2015). Encounters and mobilities: Conceptual issues in tourism studies in Southeast Asia. *Journal of Social Issues in Southeast Asia, 30*(2), 497–527. doi:10.1355/sj30-2f

King, V., & Porananond, P. (2014). Introduction: Rethinking tourism. In P. Porananond & V. King (Eds.), *Rethinking Asian tourism: Culture, encounters and local responses* (pp. 1–21). Newcastle-upon-Tyne: Cambridge Scholars Publishing.

Lew, A., Hall, C. M. & Williams, A. (2004). *A companion to tourism.* Malden: Blackwell Publishing.

Lew, A., Hall, C. M. & Williams, A. (2014). *The Wiley Blackwell companion to tourism.* Oxford: John Wiley and Sons.

Mura, P., & Khoo-Lattimore, C. (2018). *Asian qualitative research in tourism. Ontologies, epistemolo-gies, methodologies, and methods.* Singapore: Springer.

Mura, P., & Sharif, S. P. (2015). The crisis of the 'crisis of representation' – Mapping qualitative tourism research in Southeast Asia. *Current Issues in Tourism, 18*(9), 828–844. doi:10.1080/13683500.2015.1045459

Oktadiana, H., & Pearce, P. L. (2017). The *"bule"* paradox in Indonesian tourism research: Issues and prospects. *Asia Pacific Journal of Tourism Research, 22*(11), 1099–1109. doi:10.1080/10941665.2017.1374987

Porananond, P., & King, V. (2014). *Rethinking Asian tourism: Culture, encounters and local responses.* Newcastle-upon-Tyne: Cambridge Scholars Publishing.

Pritchard, A., & Morgan, N. (2007). De-centring tourism's intellectual universe, or traversing the dialogue between change and tradition. In I. Ateljevic, A. Pritchard, & N. Morgan (Eds.), *The critical turn in tourism studies. Innovative research methodologies* (pp. 11–28), Oxford: Elsevier.

Pritchard, A., Morgan, N., & Ateljevic, I. (2011). Hopeful tourism. A new transformative perspec-tive. *Annals of Tourism Research, 38*(3), 941–963. doi:10.1016/j.annals.2011.01.004

Sofield, T.H.B. (2000). Rethinking and reconceptualizing social and cultural clashes in Southeast and South Asian tourism development. In C.M. Hall and S.J. Page (Eds.), *Tourism in south and southeast Asia: Issues and cases* (pp. 45–57), Oxford: Butterworth Heinemann.

Teo, P. (2009). Knowledge order in Asia. In T. Winter, P. Teo, & T.C. Chang (Eds.), *Asia on tour. Exploring the rise of Asian tourism* (pp. 34–51), London and New York: Routledge.

Teo, P., & Leong, S. (2006). A postcolonial analysis of backpacking. *Annals of Tourism Research, 33*(1), 109–131. doi:10.1016/j.annals.2005.05.001

Thirumaran, K. (2009). Renewing bonds in an age of Asian travel. Indian tourists in Bali. In T. Winter, P. Teo, & T. C. Chang (Eds.), *Asia on tour. Exploring the rise of Asian tourism* (pp. 127–137). London and New York: Routledge.

Tribe, J. (2006). The truth about tourism. *Annals of Tourism Research, 33*(2), 360–381. doi:10.1016/j.annals.2005.11.001

Tribe, J. (2007). Critical tourism: Rules and resistance. In I. Ateljevic, A. Pritchard, & N. Morgan (Eds.), *The critical turn in Tourism Studies. Innovative Research Methodologies* (pp. 29–39). Oxford: Elsevier.

Tribe, J. (2008). Tourism: A critical business. *Journal of Travel Research, 46*(3), 245–255. doi:10.1177/0047287507304051

UNWTO. (2011). Report of the seminar on tourism ethics for Asia and the Pacific: Responsible tourism and its socio-economic impact on local communities, held in Bali, Indonesia, 11 June 2011. Retrieved from http://ethics.unwto.org/sites/all/files/pdf/report_bali_seminar_short.pdf

Wijesinghe, S., Mura, P., & Bouchon, F. (2017). Tourism knowledge and neocolonialism – A sys-tematic critical review of the literature. *Current Issues in Tourism, 22*(11), 1263–1279. doi:10.1080/13683500.2017.1402871

Wijesinghe, S., Mura, P., & Culala, H. J. (2019). Eurocentrism, capitalism and tourism knowledge. *Tourism Management, 70*(1), 178–187. doi:10.1016/j.tourman.2018.07.016

Wilson, J. (2012). *The Routledge handbook of tourism geographies.* Abingdon: Routledge.

Wilson, E., Small, J., & Harris, C. (2012). Editorial introduction: Beyond the margins? The relevance of critical tourism and hospitality studies. *Journal of Hospitality and Tourism Management*, *19*(1), 48–51. doi:10.1017/jht.2012.2

Winter, T. (2009). Conclusion: Recasting tourism theory towards an Asian future. In T. Winter, P. Teo, & T. C. Chang (Eds.), *Asia on tour. Exploring the rise of Asian tourism* (pp. 315–325). London and New York: Routledge.

Winter, T., Teo, P., & Chang, T. C. (2009a). Introduction: Rethinking tourism in Asia. In T. Winter, P. Teo, & T.C. Chang (Eds.) *Asia on tour. Exploring the rise of Asian tourism* (pp. 1–18). London and New York: Routledge.

Winter, T., Teo, P. & Chang, T. C. (2009b). *Asia on tour. Exploring the rise of Asian tourism*. London and New York: Routledge.

Yamashita, S. (2009). Southeast Asian tourism from a Japanese perspective. In M. Hitchcock, V. King, & M. Parnwell (Eds.) *Tourism in Southeast Asia: Challenges and new directions* (pp. 189–205). Copenhagen: NIAS Press.

Becoming Airbn*beings*: on datafication and the quantified Self in tourism

Claudio Minca ⓘ and Maartje Roelofsen ⓘ

ABSTRACT

Provocatively drawing inspiration from an episode of the Netflix series Black Mirror and China's Social Credit System this article critically examines the politics and practices of datafication, quantification and qualification associated to the Airbnb platform. It first explores some of the ideas and ontological claims that endorse Airbnb's digital infrastructure. Secondly, it looks at how the company's use of data management and metrics has become increasingly instrumental in maintaining control over hosts and guests and obtaining desirable and profitable outcomes. It does so by unpicking various applications and technologies used by Airbnb to monitor, record and measure the behaviour of hosts and guests. Drawing on in-depth interviews with Airbnb hosts and their participation in forum discussions the article discusses how people understand – and resist – Airbnb's 'ranking logic' and the ways in which their Selves and their homes should be rated and ranked and put into circulation as 'value' by the platform. In particular, the article argues that, through the review and rating system incorporated in the platform, both guests and hosts actively contribute to the production of a set of constantly changing hierarchies that represent the driving force of *Airbnb* as a biopolitical social regulator.

摘要

这篇文章从网飞公司（Netflix）出品的电视剧《黑镜》和中国的社会信用体系中获得灵感，批判性地审视了与爱彼迎平台相关的数据化、定量化和资格认证的政治和实践。它首先探索了一些支持爱彼迎数字公共建设的想法和本体论主张。其次，它着眼于该公司如何使用数据管理和指标越来越有助于保持对房东和住客的控制，并获得理想的和有利可图的结果。它通过拆解爱彼迎用来监控、记录和衡量房东和住客行为的各种应用程序和技术来做到这一点。通过对爱彼迎房东的深入采访和他们在论坛上的讨论，这篇文章讨论了人们如何理解和抵制爱彼迎的"排名逻辑"，以及他们自己和他们的房子应该如何被该平台评分和排名，并作为"得分"进行传播。特别是，这篇文章认为，通过整合在平台上的点评和评分系统，住客和房东都积极促成了一套不断变化的等级制度，这些制度代表了爱彼迎作为一个生物政治社会监管者的驱动力。

Black mirror(s)

In a disturbing episode of the Netflix series *Black Mirror* entitled 'Nosedive', the protagonist is a young woman, Lacie Pound, who desperately tries to improve her social ranking. Lacie lives in a dystopian society entirely dominated by a rating system of individuals that produces an endlessly mobile social hierarchy through extremely powerful forms of self-discipline. A lens imbedded in her eyeball allows Lacie to 'see' other people's ratings as she interacts with them. By pointing a smart-phone-like device at others, she may rate them and immediately affect their overall ranking. In 'Nosedive', in fact, individuals rate each other constantly with their devices that have become a sort of corporeal extension of their Self (Figure 1). Lacie is ranked 4.2 at the beginning of the episode, which, on a scale of 1 to 5, should not be a bad ranking at all; 4.2 is not enough, however, for her to get the apartment she wishes to rent and to be included in the social circles marked by the high 4s, where she believes to belong. In order to 'improve' her ranked Self, she thus puts significant effort into learning how to smile, how to move, how to get dressed and especially how to avoid anger outbursts and any manifestation of a difficult temperament.

The 'Nosedive' episode is a fictional representation of a world where every aspect of daily life is mediated and regulated by software-enabled calculative devices. A world in which algorithms affect the entire social realm and implement a powerful regime of self-discipline. The episode clearly takes the pervasive effects of real time rating-and-ranking to an extreme. At the same time, it may be seen as a provocative example of how today's feedback-and-assessment mechanisms underpinning countless digital (tourism) platforms may have significant effects on individuals' self-representations – as ranked 'quantified Selves' (Lupton, 2016). On the one hand, keeping track of the personal digital data that underpin people's 'scores' may be a strategic means of self-promotion. On the other, data are also algorithmically manipulated and represented in certain forms – shaping people's behavior towards specific ends and targeting their vulnerabilities (Pasquale, 2015, p. 38). Moreover, these algorithms may mask and deepen social inequality, as observed in Safiya Noble's study on

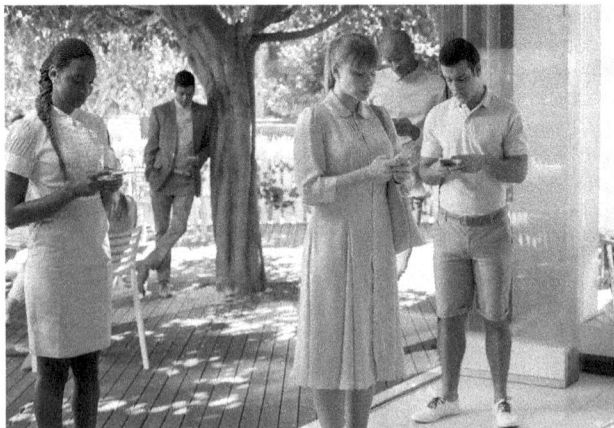

Figure 1. Production still from Nosedive; Lacie Pound (in a pink dress) and others constantly checking their scoring devices (Netflix, n.d.).

racist and sexist biases in Google search results (Noble, 2018). These considerations invite to further reflect on how data on people's routines, behaviors and practices are collected, but also on the ways in which algorithmic management increasingly penetrates our everyday spaces: from how we travel to how we drive, from how we act as 'local residents' to how we open our house to unknown-but-ranked individuals and groups via the platforms of the 'sharing tourist economy'.

Despite being a fictitious depiction of a future dystopian society, the episode has been frequently compared to China's state-driven Social Credit System (SCS) (see Palin, 2018; Griffiths, 2019). Expected to be fully operational in 2020, the SCS is used by various governmental bodies to monitor and assess the behavior and the trustworthiness of individuals and businesses in their compliance with China's laws and regulations (Hoffman, 2018, p. 3; Ohlberg, Ahmed, & Lang, 2017; Creemers, 2017, 2018). The system relies on a central repository of data including financial, criminal and government records, as well as data collected by 'Internet of Things-enabled sensors and personal information that individuals provide to websites and mobile phone applications' (Ohlberg et al., 2017, p. 4). According to the 2014 SCS blueprint, the system is 'part of an openly declared and widely propagated effort to instill civic virtue which is conjoined with propaganda campaigns to raise individuals' consciousness about their actions' (Creemers, 2018, p. 26). Although the SCS was initially centered on financial creditworthiness, its use has rapidly extended to the management of the behavior of individuals to foster trustworthiness and social and political morality (ibid.).

While the SCS does not rely on quantitative scoring, it does however identify miscreants who are inscribed in a public blacklist, in this way limiting their access to specific activities. While Lacie – in one scene of the episode – is denied boarding her plane because of her low ranking, blacklisted individuals in China may face similar measures enforced through the 'Joint Punishment System' (Creemers, 2018), including being barred from travelling on civil aircrafts, staying in luxury hotels, or even going on holiday abroad (Ohlberg et al., 2017, p. 3). Those whose actions reflect particular merit are instead reported on a 'red list' and receive certain privileges, including renting properties deposit-free (ibid.).

Inspired by these two provocative examples, in this article we interrogate how one of the most successful digital platforms in tourism, Airbnb, attempts to monitor and influence the ways in which millions of individuals engage with the practice of hosting in their own homes. We accordingly explore some of the ideas underlying the platform's infrastructure and look at how its metrics and data management have become increasingly instrumental to obtain desirable (and profitable) outcomes. In doing so, we unpick various applications and technologies used by Airbnb to monitor, record and measure the behavior of hosts and guests. We also reflect on the hosts' attempts to align with (and resist) Airbnb's 'ranking logic' by adhering to the rules and the suggestions provided by the platform in order to be highly ranked in its global 'community-of-tomorrow'.

Methodology

In this study, we have primarily relied on a digital ethnographic and autoethnographic approach. We have accordingly 'immersed ourselves' in the digital spaces of the

Airbnb economy, both as participants and as observers of the Airbnb 'community'. Following Pink, Horst, Postill, & Hjorth (2015) reflections on digital ethnography, we argue that digital technologies are increasingly interdependent with the infrastructures of everyday life, something that fundamentally unsettles any dualistic interpretation of the related online and offline environments. Like with other platforms, we consider engagements with the Airbnb platform to be 'based on the real-life dynamics, beliefs, power relationships, and political imaginations that define the everyday life of the groups studied' (Barassi, 2017, p. 410). We are therefore interested in what processes of datafication, rating, ranking and algorithmic management mean to those directly affected by them. Methodologically, this has implied a systematic analysis of our own experiences as hosts, as well as those of other Airbnb hosts. Accordingly, we have first carried out in-depth face-to-face interviews with 18 Airbnb hosts held during our empirical investigations of Airbnb in various European locations as part of a broader project whose main results are published elsewhere (see (Roelofsen, 2018a; 2018b; Roelofsen & Minca, 2018)). In these interviews, we asked our participants about the importance of reviewing and rating, and what effects this form of user-driven auditing culture had on their hosting practices. The interviews were transcribed verbatim and analysed through the qualitative data analysis software MAXQDA. The textual accounts of the interviews were searched via common themes and pseudonyms used to guarantee anonymity.

The second site of digital ethnographic fieldwork was the Airbnb Community Center (ACC) – where Airbnb hosts interact with each other and with Airbnb administrators. We consider the ACC as a form of social media – a web application available to all Airbnb hosts which processes, stores and retrieves user-generated content. In the ACC, Airbnb hosts post blogs or comment on other users' blogs, but also on their experiences with Airbnb and its workings. The ACC is publicly accessible and those who visit it do not need a registered Airbnb account to view the blogs. We paid particular attention to blog posts and comments that showed how digital subjects in the Airbnb economy experienced and made sense of their own data, as well as what tactics they employed to challenge how these data were used by the platform. The ACC therefore is also a site of resistance – a space where Airbnb members think and write politically about Airbnb's use of data and about how these should instead be used. In adopting this approach to the ACC, we have drawn inspiration from studies that challenge any deterministic understandings of the impact of technology on everyday life showing that, whilst technological structures do matter in people's everyday life, they are also (re-)appropriated by those who engage with them (e.g. Barassi, 2017; Ettlinger, 2018; Isin & Ruppert, 2015; Milan & Gutiérrez, 2015; Postill & Pink, 2012).

A third method adopted in this study was digital *auto*ethnography. The use of digital autoethnography has allowed for an intimately lived and embodied experience of the datafication processes underpinning Airbnb's digital infrastructure. In particular, our autoethnographic approach was based on 'selftracking' (Lupton, 2016). Selftracking required us to engage and respond to our behavioral data, reviews and ratings, and to the resulting metrics made available through the Airbnb 'dashboard'. Understanding the ways in which the Airbnb 'dashboard' collated, sorted, categorized, analyzed, profiled, visualized and regulated our own data provided important insights into how Airbnb's algorithms overtly and covertly shaped our chances of becoming 'successful Airbn*beings*'.

Finally, our digital ethnography was supported by a qualitative content analysis: we have thus investigated Airbnb's promotional campaigns, the Airbnb website, affiliated blogs, promotional materials, hospitality guidebooks, together with Airbnb blog posts, discussion boards and policy documents. Studying Airbnb's digital infrastructure and community forum required spending many hours on the internet while clicking and scrolling through hundreds of blogs and comments. We have accordingly come to understand the digital platform of Airbnb as a messy 'fieldwork' site, that is at once observed and constituted through the ethnographer's interactions and narrative (Postill & Pink, 2012).

Tourist studies on Airbnb

Tourism – as a social practice and as discourse – is increasingly digitally mediated. Technologies such as cameras, smart phones, laptops, and navigation devices have by now become ordinary companions of contemporary tourists. They significantly shape how people travel and how people get to know and navigate places (Ash, Kitchin, & Leszczynski, 2016). The so-called 'turistus digitalis' is afforded a central role in 'augmenting and circulating enacted versions of destinations' through their interactions with digital technologies (Munar & Gyimóthy, 2014, p. 251). As such, the digital has the potential to change social and material worlds and (literally) generate new tourism realities. In the past decade, a vast body of scholarly work has explored the digitally enabled 'sharing' or 'collaborative' economies of tourism (Dredge & Gyimóthy, 2015). Both these terms connote a person-to-person economy facilitated by 'digital platforms' (Schor & Attwood-Charles, 2017, p. 2). Whereas the computational meaning of a 'platform' is 'a programmable infrastructure upon which other software can be built and run', in public discourse the term 'platform' is increasingly used to describe companies that offer web 2.0 services and 'afford an opportunity to communicate, interact or sell' (Gillespie, 2017). Platform economies of tourism are centered primarily on mobility, accommodation, food and travel experiences. Airbnb is part of a specific set of digital platforms that facilitate the monetary exchange of residential accommodation (private homes, rooms and beds) and tourist experiences among individuals.

This growing literature on the 'platform' economies of tourism reflects the somewhat traditional and established distinction between two fields of academic enquiry: the business- and management-oriented studies of tourism, mostly quantitative and applied in nature (see, among others, Bridges & Vásquez, 2018; Ert, Fleischer, & Magen, 2016; Teubner, Saade, Kawlitschek, & Weinhardt, 2016; Zervas, Proserpio, & Byers, 2015), and the domain normally identified as 'critical tourism studies' (see Ateljevic, Pritchard, & Morgan, 2007). The present article intends to contribute to existing debates on Airbnb in critical tourism studies. Within these debates, there has been an increased concern for the lived realities of those participating in the tourism platform economy and of those affected by it, spatially, socially, and economically. Critical tourism studies on Airbnb have, for example, shown how the platform often further exacerbates existing processes of uneven socio-spatial development in many cities. Important empirical work has also revealed how the economic benefits produced by Airbnb tend to be unevenly distributed among residents, further adding to existing socio-economic disparities (Arias Sans & Quaglieri Domínguez, 2016; Cócola Gant, 2016; Gil & Sequera, 2018; Gutiérrez, García-Palomares, Romanillos, & Salas-Olmedo, 2017; Mermet, 2017; Roelofsen, 2018b). Other work has discussed the emergence of forms of implicit and explicit racial discrimination on the platform (Cox, 2017; Edelman, Luca, & Svirsky, 2017; Piracha, Sharples, Forrest, & Dunn, 2019), providing further evidence to the claim that Airbnb mostly benefits the already privileged, middle-class, white members of the platform (see Schor, 2017; also, Roelofsen, 2018b). Moreover, the platform has been critiqued for having become a powerful enabler of new business opportunities for big investors, tourist companies, property managers, landlords and other professional actors (Arias Sans & Quaglieri Domínguez, 2016; Gil & Sequera, 2018). Oftentimes ignoring local urban planning regulations, these operators use Airbnb to rent out residential housing to tourists rather than local residents. As such, the housing stock is moved from long-term to short-term rental markets, in this way adding to the already problematic local housing availability of certain cities (Arias Sans & Quaglieri Domínguez, 2016; Roelofsen, 2018a; Wachsmuth & Weisler, 2018). The subsequent rise of housing- and rental prices (e.g. Horn & Merante, 2017; Lee, 2016) has effectively resulted in various forms of direct and indirect displacement of residents from certain neighborhoods (Cócola Gant, 2016). Again, all these processes are seen as an exacerbation of already existing problems in cities but also confirm Airbnb's contribution to rampant manifestations of 'tourism gentrification' (Cócola Gant, 2018).

Critical accounts on the political implications of the platform's technologies are instead relatively scarce. However, a fairly small body of work has recently contributed to the understanding of the power relations entangled in the operations of Airbnb. Bialski (2016), in analyzing Airbnb's digital infrastructure, argues that unravelling the aesthetic design of the platform allows for certain power structures to become visible. For example, hosts are nudged to advertise their homes in order to promote a specific 'way of looking' by uploading photos to the platform that boast 'a certain white-washed aesthetic of homes' (ibid., p. 47). For Bialski, this competition for authenticity and 'coziness' is crucially supported by and entangled in the various digital artefacts and structures that enable the platform to work. Other studies challenge the 'logic' of reviewing and rating and analyze the politics enabled through such technologies. Rather than 'creating trust' between strangers, the market-based reputation that Airbnb users build up is, according to O'Regan & Choe (2017, p. 4), 'often about control, manipulation, and discipline rather than transparency and accountability'. A biopolitical analysis of Airbnb was recently offered by Roelofsen and Minca (2018) where they investigate some of the key technologies and calculative rationalities that drive the making of Airbnb's digital 'global communities', and how specific understandings of the 'spatialities of the home' are central to the quantification/qualification of living spaces generated by the platform. The present article thus speaks directly to this recent literature preoccupied with the political implications of Airbnb's digital technologies. In doing so, it provocatively suggests that, while Airbnb's social disciplining cannot not be compared to the totalitarian force depicted in 'Nosedive', at the same time, *it does influence* people's behavior and contributes to implement modes of social relations that were unknown until its algorithms allowed them to emerge.

People and homes – waiting to be 'unlocked'

'The biggest asset in people's lives is not their home, but their time and potential — and we can unlock that [...] We have these homes that are not used, and we have these talents that are not used. Instead of asking what new infrastructure we need to build, why don't we look at what passions we can unlock? We can unlock so much economic activity, and this will unlock millions of entrepreneurs.' Today's new platforms are unlocking human potential to 'be the people we wanted.'

Brian Chesky, interview excerpt (Friedman, 2017)

The above excerpt of a recent interview with Airbnb's CEO and co-founder Brian Chesky reveals one of the ontological stances that drive the platform's ranking modalities and inform its ideology in dealing with people, home and hospitality. For instance, the platform deploys the concept of 'home' as a material and imaginative site through which hosting and guesting subjects may supposedly express their 'potentials'. A site which, according to Chesky, needs to be 'unlocked'. Airbnb's advertisement campaigns of recent years proclaim that the 'home' allegedly provides a deeper understanding of what it means to be human in today's world – more specifically, what it means to be a *good human*, as clearly stated in Airbnb's Man[sic]kind campaign:

Is man kind? Are we good?

Go see.

Go look through their windows so you can understand their views.

Sit at their tables, so you can share their tastes.

Sleep in their beds, so you may know their dreams.

Go see. And find out just how kind the he-s and she-s of this mankind are.

'Is Mankind?' Airbnb's commercial campaign (Airbnb, 2015)

The above vignette suggests that there is an ontological truth to be discovered by penetrating the daily life of Others in their homes. In the Airbnb economy, the disembodied 'tourist gaze' (Urry & Larsen, 2011) is no longer sufficient in providing the contemporary traveler a mirroring experience with 'hospitable' local residents and their 'culture-in-place' (see Minca, 2011). Airbnb' travelers are instead invited to see, look, sit, eat, and sleep in other people's homes, in order to quench the 'thirst for difference' in a world supposedly plagued by the anxieties of the (post)modern condition. This points at the embodied nature of the practices put in circulation by the platform through datafication; hosts and guests generate personal digital data through their engagement with each other and with the platform, data which keep the Airbnb machinery running. These practices in fact feed into the platform's review and rating system, a system that numerically captures the qualitative 'distinctions' of home and hospitality and extracts 'value' from these same distinctions. The need and desire to participate in this economy of reviewing-and-rating is elicited through a suggested *lack of* – and *need for* an – exhibition of the Self in its most intimate spatialities. Chesky's portrayal of people and places in need to be 'unlocked' also suggests that one can never be public enough, that there is always something more in people's lives to be made available and to produce 'value'. Airbnb thus asserts the (natural, i.e. human) need for a constant flow of people's data to generate new value and a new community of hosts and guests, and that we are all 'lucky' that such data are everywhere around us (see also Sadowski, 2019). Individuals just need to develop their human potential to 'be the people we want[ed]' – as long as they properly produce and consume.

Through the narrative of 'belonging', the production, consumption and circulation of people's most intimate spatialities is thus further ostensibly justified:

> '*Belonging* has always been a fundamental driver of humankind. So, to represent that feeling, we've created a symbol for us as a community. It's an iconic mark for our windows, our doors, and our shared values. It's a symbol that, like us, can belong wherever it happens to be. [...] It's a symbol *for going where the locals go* – the cafe that doesn't bother with a menu, the dance club hidden down a long alleyway, the art galleries that don't show up in the guidebooks. It's a symbol for people who want to welcome into their home new experiences, new cultures, and new conversations. We're proud to introduce the *Bélo: the universal symbol of belonging*'. (Chesky, 2014, italics added)

By stating that 'belonging has always been a fundamental driver of humankind', the *Bélo* campaign disturbingly echoes some of the conservative communitarian ideologies that have capitalized in the past two centuries on a necessary identification between a specific community of people, a specific way of life, and a specific land (see, Esposito, 2010; also, Roelofsen & Minca, 2018). However, since we are in the brave new world of what has been contentiously termed the 'sharing economy' (Ravenelle, 2017), in this community-in-the-making hosts are set up to the task of generating income by providing complete strangers with a sense of 'belonging' in their intimate spatialities. In a similar vein, guests have to prove to be well-behaved and respectful of the house rules while sitting at the hosts' tables, while lying in their beds, while looking through their windows. Overall, what is incentivized here is a powerful form of social regulation based on a (rather reactionary) belief in people's need of belonging and in a related structure of social relationships based on the platform 'guidelines'. In the next section, we thus explore which everyday embodied practices interlace with the platform's digital infrastructure, and how such practices conform to the platform's idea of home and hospitality.

Airbnb's datafied selves

In the Airbnb world of global hospitality, hosts and guests alike are continuously rendered into 'assemblages of digital data' (Lupton, 2016, p. 709). They are *datafied* through their engagement with various data-recording applications incorporated in the platform's digital infrastructure. Airbnb's messaging system, search engine, and review-, rating- and self-tracking applications are some of the key 'machineries' that hosts and guests engage with and that generate data enabling Airbnb's platform economy. These applications are continuously alimented by behavioral data intentionally or unintentionally provided by users. Since the platform is set up as a social network, self-compiled profiles containing personal details on millions of hosts and guests worldwide also constitute a large part of Airbnb's grand datasets. Setting up a detailed profile is in fact the first step necessary to become a 'Airbn*being*' and take part in future Airbnb exchanges. According to Airbnb (2019a), 'when your profile is robust, it helps others feel that you're reliable, authentic, and committed to the spirit of Airbnb'. Profiling oneself as an Airbnb host or guest requires the submission of plenty of personal information such as: email address, full name, date of birth, proof of identity, a clear, front-face profile photo, a verified phone number and credit card details. Optionally, users may also indicate where they went to school, the languages they speak, and their profession. What follows is a process by which hosts and guests textually and visually detail their profiles through biographies, videos and symbols. Airbnb recommends autobiographical descriptions to include: 'things you like', '5 things you can't live without', 'favorite travel destinations, books, movies, shows, music, food', 'what it's like to have you as a guest or host', 'style of traveling/hosting', and a 'life motto'. The hospitality Toolkits for hosts propose even more specific ways to represent one's 'authentic' digital Self, in order to draw in 'guests with similar tastes and interests' and prevent them from having disappointing experiences (Airbnb, 2016).

Upon completion, the publicly viewable profile page includes the member's first name, profile picture, verified ID information, date of registration, preferred language(s), and place and country of residence. Hosts additionally profile their listing(s) through photos and videos and enter a description including pricing, property/room type, number of bedrooms, beds and baths, amenities, location, check-in/out times, cancellation policy, home safety features. They may also post information on fellow inhabitants, the neighborhood, ways of getting around, and the desired level of interaction with guests. The data about the lives of hosts and guests collected through profiling alone is therefore vast and of a sensitive and intimate nature. These personal details, along with other data produced by Airbnb users, are then stored on cloud computing databases.

Other processes of datafication through the Airbnb application often silently operate in the background of the users' computers, smartphones and other digital devices. According to Airbnb's privacy policy (2019b), 'these activities are carried out based on Airbnb's legitimate interest in ensuring compliance with applicable laws and Our Terms, preventing fraud, promoting safety, and improving and ensuring the adequate performance of our services'. Users interacting with the application are in fact subjected to digital surveillance: Airbnb monitors and records their movement and behavior, often without their knowledge or awareness. Andrejevic (2007, p. 304) has referred to this as 'the work of being watched': a form of unpaid labor by which users willingly or unknowingly 'submit themselves to monitoring practices that generate economic value in the form of information commodities'. For example, unless this function is deliberately disabled where the application is installed, the platform collects Geolocation Information, tracking hosts' and guests' location through their IP address or the device's GPS (Airbnb, 2019b). Airbnb also gathers information about the users' interactions, such as pages and content viewed, searches for listings and bookings. Online communication taking place among users, and between users and Airbnb, is also recorded, with the message content being scanned, reviewed and analyzed. Data are collected on the hosts' average response time to booking inquiries and new messages sent out to them by guests, and then factored in the ranking of listings in the search results.

> '[B]ecause making travel plans can be complicated or time-sensitive [...] responses need to be dependable and quick. That's why we calculate response rates and response times and show them on [hosts'] listing pages. We want guests to know what to expect when they reach out to a host. And we ask hosts to stay focused on swift, reliable communication, because we know it will help them confirm more bookings. We want to make sure the meaning of these metrics is clear so that hosts know how to stand out'. (Airbnb, 2014)

Yet other applications incorporated in the platform, such as the review and rating system, gather loads of information on feelings, emotional needs and desires of hosts and guests. Like in the *Nosedive* episode, these expressions are predominantly the result of interactions among individuals. Over the span of more than 10 years, the review-and-rating application has congregated millions of written testimonies of Airbnb experiences all around the world. These reviews and ratings are also the primary source of the users' digital reputation, which in turn has become vastly important in how hosts and guests understand each other, as well as 'in succeeding' in Airbnb's economy. Reviews and ratings report on the supposed 'quality' of hosts' and on the guests' emotional-, caring- and affective capacities in providing each other with an Airbnb experience, as well as on the 'quality' of the related homes and neighborhoods. The insistence on the production of positive feelings is engrained in almost every step of the review process. After each stay, Airbnb asks guests to provide the next guests with an account of what they 'loved' about their host's place through a written testimony. They also encouraged them to rank their hosts' efforts on a scale of 1 to 5 stars along several parameters: Overall Experience, Cleanliness, Accuracy, Communication, Check-in, Value and Location. Additionally, guests are asked to rank the home where they stayed in a range starting from 'Budget' (limited amenities and minimal furnishing) all the way up to 'Upscale' (beautiful space with high-end amenities and decor). Hosts, in a similar vein, are asked to rank their guests' efforts along the parameters of: Communication, Cleanliness, Observance of House Rules, and Overall Experience. The majority of these parameters aim to collect information on the behavior of individual hosts and guests: 'How clearly did the guest communicate their plans, questions, and concerns? How clean was the guest? How responsive and accessible was the host before and during the stay? How observant was the guest of the house rules?'

Hosts and guests are also expected to behave according to Airbnb's 'Community Standards'. These standards 'help guide behavior and codify the values that underpin' the Airbnb community in terms of 'trust and safety' (Airbnb, 2019c). The five 'central pillars' on which the Standards rest include: 'Safety', 'Security', 'Fairness', 'Authenticity', and 'Reliability', each comprising a subset of categories explaining how hosts and guest are to behave around each other and in each other's homes. For example, as far as 'Authenticity' is concerned, Airbnb provides detailed instructions on how not to misrepresent oneself or the spaces within which the 'hosting' takes place. The platform also invites hosts to provide experiences that are 'not merely transactions', as 'Airbnb experiences should be full of delightful moments and surprising adventures' (Airbnb, 2019c). 'Reliability' can be achieved by engaging in timely communication, asking hosts to respond to their guests' messages within 24 hours. The instructions on how to behave are at times (deliberately?) vague, leaving room for reinterpretation and, at the same time, expansion of the arena of self-disciplinary behavior.

Although reviews and ratings are voluntarily submitted to the platform, Airbnb continuously encourages its users via multiple emails and text messages to upload 'thorough reviews' in a timely manner in order to 'aid the decision-making of future guests and hosts' (Airbnb, 2019d). For a host, not being reviewed by a guest after their stay may result in not being promoted to the 'Superhost' status – a status that indicates a 4.8 or higher average overall rating in the past year. Successful Superhosts attract more guests by being 'featured to guests' in search results, in this way increasing their chances of being booked (Airbnb, 2019e). When Superhosts fail to get reviewed according to the 50% benchmark in new assessment rounds, they risk having their Superhost status revoked (Airbnb, 2019f, 2019g). This may lead to less bookings and a lower earning potential. Not to report or not to be reported on is considered adverse behavior in the Airbnb economy and it is directly and indirectly sanctioned.

The deeper politics of reviews, ratings and rankings

As a form of social regulation, the growing importance of each individual's 'digital reputation' is a powerful incentive for the platform's members to act in the 'desired' manner, possibly without Airbnb's direct intervention. Such self-discipline is further enforced through the Airbnb 'dashboard' incorporated in the host's profile page, which provides key metrics of their various expected performances. The dashboard includes an application called 'Progress', which monitors, records, organizes, measures, analyses and presents a variety of data on the hosts' behavior, who are in this way constantly reminded of their 'rated' performances of hospitality over time (Airbnb, 2019h). 'Progress' enables the practice of 'self-tracking', allowing users to reflect on certain patterns in their behavior and accordingly improve their relationships with their guests (see Lupton, 2016). Unlike Airbnb hosts, Airbnb guests are not provided with a tool to self-track their performances, although they are exposed to written reviews after their stay. The guest's metrics, however, are only shared with a specific group of hosts who allow for their listings to be 'instantly booked' without prior communication with the respective guest. If hosts 'ever rate a guest at 3 stars or below', this guest will not be able to instantly book with the same hosts again (Airbnb, 2017). Both hosts and guests, then, are offered (and incentivized to use) digital technologies that render them and others into 'quantified selves' along a set of given parameters and, accordingly, allow them to assess whether or not they wish to engage with each other.

In this process of reciprocal ranking among hosts and guests, qualitative distinctions are translated into quantitative ones, a process that 'actively works to depersonalize and de-particularize the very activities being measured' (Hearn, 2010, p. 428). This, however, does not prevent Airbnb hosts and guests from uploading reviews and ratings not conforming to the level of 'objectivity' or 'sincerity' that the company aspires to obtain. Whilst Airbnb promises that reviews and ratings provide some 'transparency' or 'truth' about individual members and listings (Airbnb, 2019i), the reviewers' sentiments are not unaffected 'by already existing class, gender, race and other social relations' (Hearn, 2010, p. 433). Providing feedback is not 'only ever motivated by an honest desire to do good' (ibid.), since reviews and ratings are oftentimes the result of diverse or even conflicting understandings of 'quality' among hosts and guests. As noted in a recent post by an Airbnb administrator:

> 'Reviews are so important. They not only impact the success of your business, they're also really personal. We know you put a lot of thought and care into your hospitality and that it's frustrating when you receive a review that is uncharacteristically low – be it a mistake, a misunderstanding, or an unfair assessment. [...] We've invested and will continue to invest a lot of thought and effort into how we can make the review system more fair' (ACC, 2018).

Airbnb is thus committed to continue improving its review system and to adopt tools capable of detecting 'outlier reviews' – that is, reviews and ratings that do not accurately represent 'truthful' feedback (ibid.). What also transpires from the above statement is that reviews and ratings are a taken-for-granted aspect in the 'self-governance' of the platform. Truthful and objective *reporting* on each other's behavior is considered of fundamental importance in keeping Airbnb's feeling-intermediary credible and 'risk-free'– behavior that the platform monitors and 'corrects' when necessary. By using predictive analytics and 'machine learning', every Airbnb reservation is 'scored' for risk before being confirmed (Hakim & Keys, 2014), 'instantly evaluating hundreds of signals that help [Airbnb] flag and investigate suspicious activity before it happens' (Airbnb, 2019j). To facilitate this, the platform requires data. In a recent interview with Bloomberg, Nick Shapiro, former CIA's Deputy Chief of Staff and White House counterterrorism and homeland security aide to President Obama and now Global Head of Trust and Risk Management at Airbnb, has explained that:

> 'People need to know that they are going to be safe, they have to feel safe. So, we do a number of things, to use technology to do that. We risk-assess each and every reservation. We run global watchlist checks against all of our users worldwide. We background check hosts and guests in the US. But just being safe isn't enough. We use these technologies to also build connections. There's detailed profiles. There's the messaging system where you can learn more about each other. And there's reviews where you can look at previous history. And on top of that, people need to know that they are not alone. Airbnb is there for them' (Bloomberg Technology, 2017).

What emerges from the quote is that a multitude of data on intimate aspects of the members' everyday life is incorporated by Airbnb into predictive analytics aiming at anticipating their potential behavior. Such behavioral data are of vast importance to enable Airbnb in nudging and coercing social behavior on a large scale. According to Zuboff's reading of surveillance capitalism (Zuboff, 2019, p. 15), in 'this reorientation from knowledge to power, it is no longer enough to automate information flows *about us*; the goal now is to *automate us*' (see also Andrejevic, 2019). What is more, reviews and ratings rely entirely on the free labor of hosts and guests, whose affective participation is voluntary and unpaid (see Terranova, 2000). Whilst often promoted as being fundamental to 'attracting new business', the value created through reviews and ratings crucially contributes to the deeper logic of the platform. Data generated by reviews and ratings in fact feed into algorithms that sift and sort them to optimize future transactions, with the main beneficiary being Airbnb that draws on this massive data to improve its market appeal and nudge the members' social behavior toward the company targets. These algorithms also measure how often hosts cancelled a reservation, the amount of stays they hosted, and their overall ratings by guests. As such, these categories depend on the hosts' dedication and aptness, to inspire a culture of constant 'self-improvement' and invite them to be virtually on-demand every minute of the day.

The review and rating system must thus be continuously alimented by 'embodied data' concerning the behavior of individuals in order to effectively shape future interactions on the platform, while hosts and guests alike are strongly encouraged to provide such information. These 'embodied data' in fact determine how guests, hosts and their homes are ranked in the platform's search index and booking tool, thanks to the software associated with algorithms that identify, classify, structure, and prioritize certain people and certain homes over others. As such, these algorithms are far from being neutral (Kitchin, 2017), since they *do* assess homes and individuals according to specific parameters and 'values' set by Airbnb. As Safiya Noble illustrates in her recent book on Google (2018, p. 2), 'some of the very people who are developing search algorithms and architecture are willing to promote sexist and racist attitudes openly at work and beyond, while we are supposed to believe that these same employees are developing 'neutral' or 'objective' decision-making tools'.

Airbnb search revealed: or, the importance of ranking well

On October 21st, 2017, Airbnb hosts received the platform's monthly newsletter in their inboxes, which included a link to an online post written by Lizzie, Airbnb's 'Online Community Manager'. Lizzie wished to provide more information about the platform's search algorithm, because, according to the stats, one of the most popular topics on the Airbnb-Community fora was: 'how Airbnb Search works?' (ACC, 2017b). Responding to a large cohort of hosts speculating about how the search algorithm classified and ranked their homes, Lizzie's post revealed some of the underlying ideas driving Airbnb search engine. Lizzie explained that the algorithm responds to a specific set of preferences (e.g. dates, location, etc.) that guests enter into the online booking tool. Based on these preferences, the algorithm sifts through existing data on all Airbnb homes and on hosts' past performances to rank those that best 'match' the guest's criteria.

While Lizzie indeed provided some hints, these were never *too specific*:

> 'We have an algorithm that looks at over 100 signals to decide how to order listings in search results. Most of those signals have to do with things that guests care about, like positive reviews and great photos. If you think guests might care about it, it probably factors into your ranking!' (ACC, 2017b)

Similar to the *immunitary* and *control* strategies observed in van Doorn's study (2017) of on-demand platforms such as Uber, Lizzie's comment reveals the strategies employed by Airbnb. The first strategy is Lizzie's appeal to the algorithm as an 'independent' assessor of Airbnb hosts' performances. Designating the algorithm as an objective measure 'shields' Airbnb from dealing directly with those who perform the labor of hosting on the platform. In doing so, the platform rids itself from as much liability as possible (Van Doorn, 2017). Secondly, the post suggests that what guests care about is decisive in the ranking of Airbnb listings. In other words, Airbnb outsources quality-control to the hosts' 'customers' – namely the Airbnb guests.

Lizzie then claimed that hosts do not need a perfect listing or an unbeatable location for ranking well and suggested *how* hosts could possibly improve their position in terms of search results. Lizzie insisted in particular on the importance of activating the Instant Booking feature:

> '[T]ravelers prefer to use Instant Book because they can book quickly, skip the wait time for hosts to respond, and avoid possibly being rejected. Because of the high booking success for hosts and guests, Instant Book gives your listing a boost in searches'. (ibid.)

The main incentive for hosts to use Instant Book derives from not having to franticly maintain what the administrator described as: 'welcoming correspondence and strong response metrics'. Automated messages would do the work of 'responding to the guest in real time', making for the highest possible response rate and a higher rank in search results. A second incentive is the possibility, provided only by Instant Book, to see how other hosts have rated the prospective guests – offering 'more peace of mind' (Airbnb, 2019k). What Lizzie forgot to mention was the considerable advantages for Airbnb in having hosts accept all booking requests without prior consultation. Besides instantly receiving commissions on the booking payment, Airbnb conveniently avoids any difficulty that may arise from human interactions between hosts and guests. Instant Book makes all preliminary and potentially 'unruly' social interactions redundant. An algorithm will do the job!

The responses to Lizzie's post, however, show how many hosts have learned to incorporate but also resist the overall logic of the platform, its capitalization on their home, and its power in attributing value to people and their practices. Janine, a long-time host, declared that when you live with guests in your home, it is vital to have the 'opportunity to choose' who you share your intimate everyday spaces with. Janine added that Instant Book should not be a search factor at all. In fact, Janine's status as a Superhost for over 2 years 'should make a difference!' In another response, Sally contended not to use Instant Book either, because their place 'is not a hotel; it is a home'; since the guests often don't read the House Rules on the listing page, 'a few e-mails back and forth help create a mutually positive experience.' Arguably, the concerns manifested by Janine, Sally and others were not about people being ranked but rather about *how* they were ranked. These concerns clearly emerged from our analysis of ACC posts but also from our interviews. Many hosts in fact complained about the ranking system, since they would have liked to have *more reviews* and *better rankings*, and more personal information on their potential guests.

> 'I think maybe the system could improve reviewing and rating a little bit. They could require more detailed reviews and references. Because now you have a specific set of categories, like cleanliness and tidiness, etc. And then you read a review based on 200 characters and this is it. And then you want to say something personal to your guests. I understand that often you don't have time to write a proper review and describe in detail your experience with your guests or hosts or whatever. But this is essential

especially when you rent a room, the flat is ok but when you rent a room you co-live with this person!' (interview with host Pino, 2015).

Some members felt that the guests should be rated as well. Supported by hundreds of 'likes', one vocal host addressed the 'one-off bad reviews' and practices of black-mailing on the platform. According to Gloria, Airbnb guests 'will be very critical of a property only because they want a big discount and they will threaten their host with a bad review'. One of Gloria's guests recently complained that the taps in the house were 'old fashioned', and should be replaced with modern ones. The guest also 'photographed dust and magnified the pictures and told [the host] that she wanted a refund', a request accepted by Gloria to avoid a bad rating.

All in all, our analysis revealed the implicit power of Airbnb's algorithms in regulat-ing a specific set of social relationships, even when the members expressed dissatisfac-tion and concern about their workings. We have also recorded a number of comments – both on the ACC and in person – that revealed how the availability of personal information on the platform may offer the ground for discriminatory or racist behavior of some members, despite Airbnb's attempts to sanction these practices. In an inter-view, Emma, reflected on the process of selecting guests:

> 'We try to make a selection [before we accept bookings]. We look through guests' profiles and read the comments of previous hosts. We never host a person without comments or references. I decline a lot of requests just because I don't like to get this feeling that these people only come here for the bed and nothing else. So, we are searching for interesting people. Interested to share and exchange something and spent time together'.

Many hosts also admitted having implemented various forms of self-disciplining in order to get a better ranking and become visible via the search tool. Proper feedback was considered by many as paramount to become a proper 'citizen' of the Airbnb glo-bal community, something clearly illustrated by host Madeleine during an interview. Madeleine insisted on reviewing all guests because 'it is really important to share what my impressions were about them […] Feedback is really important to know where you are and what are you doing well or not so well'. Another interviewee, Dave, openly declared having changed his behavior at home to comply with the plat-form's expected standards and get good reviews:

> 'What changes [when guests come over] is that I try to be more calm, quiet. Not to argue a lot with my mother. I would not watch TV louder. I wouldn't have parties. Or, I'll ask if the person is ok with it. And you should clean all the time after you use something. If you go to the toilet or if you use the kitchen and stuff like that'.

Mercedes, not only made a few changes in the home to accommodate potential guests, but also mobilized personal social networks:

> 'I did some renovations in my apartment and I bought many things like sheets, you know. You have to prepare a lot of stuff for Airbnb. I was wondering is it going to work or no? You never know. *You need reviews to get guests and need to have guests to have reviews*. And I asked some friends to write me, not reviews, but a recommendation. And step by step I had some guests.'

Despite the criticism expressed by many members, the review system is commonly considered reliable and truthful, a fundamental tool to build an individual capital of trust that will be reflected in the ranking. Mercedes argued that: 'this is the way to know about some things that the host doesn't mention, like there is no hot water from time to time. So, I think the reviews are very very important'.

A third group of responses showed clear awareness of the possibility that some members may use the ranking system at their own advantage or provide 'false' or unreliable information. For host Ada,

'It's definitely good to have the review system. But as I confessed, I rarely write what I really think. Especially when it's negative. When it's very negative, then I write it. Because when it gets to a point of danger or something like this. I think this is already more security than somewhere else.'

Another host, Frida, argued on the ACC for a more detailed history of how guests reviewed previous hosts, suggesting that the platform should take into account the number of stars that guests have previously given to their hosts. According to Frida, some guests never give 5 stars, something that should be weighed in Airbnb's algorithms.

To return to our initial provocations, the idea that Airbnb acts as a social regulator affecting the behavior of many hosts in their most intimate spaces was clearly supported by the evidence emerging from our interviews and the materials consulted. However, how may this form of social discipline be related in any possible way to the Black Mirror episode or even the Chinese Social Credit System? Notably, in the past years, the Chinese government has exerted increasing pressure on foreign technology companies operating in that country to control their flow of digital information (see Creemers, 2017). Airbnb has swiftly complied with China's regulations and has *proactively* moved to local Chinese servers its user data processed by Airbnb *China* (a separate Airbnb business entity) (Cadell, 2016). In a similar vein, Airbnb has also complied with the government's demand to disclose personal information on individual hosts operating on Airbnb China's platform (Jing & Soo, 2018), while this might extend to data related to guests in the near future (Shen, 2018). Airbnb's Privacy Policy (Airbnb 2019a) tellingly declares: 'Where required under law and you have expressly granted permission, Airbnb China may disclose your information to Chinese government agencies without further notice to you'. Airbnb China' hosts and guests generate data that concern not just their identity, but also their private communications, their geo-locations, their movements, as well as sensitive information about how they engage in with their guests. Details of the most intimate spatialities of people's homes are described on Airbnb with unprecedented depth and specificity – from photos of people's bedrooms to descriptions of hosts' housemates and family members. Airbnb's subjects actively generate data through multiple means, suggesting a 'Panspectric' rather than a 'Panoptic' mode of surveillance and monitoring (Creemers, 2017, p. 96). A key question in this rapidly evolving landscape is what may happen when such databases would become instrumental to the Chinese government in its attempts to stimulate self-disciplining in their citizenry and to impose new 'soft' forms of social regulation via the pervasive use of these digital technologies? Despite the abovementioned Social Credit System is still in its infancy, Creemers (ibid, p. 99) argues that 'it can safely be said that propaganda, public opinion and social management work will be increasingly integrated through technological processes'. The Party-state has in fact clearly shown the intention to deploy these new technologies 'in a manner that renders society legible and predictable' (ibid., p. 88). What is particularly relevant for our argument is that such a strategy aligns particularly well with the disciplining and self-optimizing effects of Airbnb's algorithmic management – which are constructed in ways that induce its 'members' to comply and act in line with the platform's predefined notions of 'Man[sic]kindness'.

Given Airbnb's fierce resistance to sharing its data with New York City (and other cities) – claiming that it would imply a form of 'illegal surveillance' – the company's compliancy with China's controversial laws is contradictory to say the least. China, however, currently is the world's fastest growing travel market and Airbnb has been very vocal in expressing its business interests in that country. In a press release of March 2018, Airbnb Co-Founder and Airbnb China Chairman Nathan Blecharczyk stated that 'China is a critical part to Airbnb's mission of creating a world where anyone can belong anywhere. By 2020, more Airbnb guests will come from China than any other country. We will continue to deepen our commitment with the goal of bringing authentic magical travel experience to Chinese travelers' (Airbnb, 2018).

Concluding thoughts: Airbnb's biopolitics

Imagine a world where (almost) everyone behaves or tries to behave in line with the deeper logic of a series of algorithms. Imagine one day when all homes are part of the Airbnb 'sharing economy' and all residents ranked as hosts. In such a world, every single space of your home would be incorporated, in some form, by algorithms translating individual experiences of your most intimate spatialities into globally advertised hierarchies. In such a world, again, the homes next to yours would be constantly visited by strangers selected by Airbnb.

In response to this provocative scenario, and in line with the main argument of this article, we would thus like to conclude by advancing three theoretical propositions. The first is that Airbnb may be thought of as a powerful biopolitical machinery (on this, see Agamben, 1998, 2002 and related literature; e.g. Campbell & Sitze, 2013; Lemke, 2011; Minca, 2015; on tourism and biopolitics see, among others, Ek & Hultman, 2008; Minca, 2009, 2011; Simpson, 2016). A machinery fed by elements of life, home, care, coziness, local culture – all incorporated by the platform as 'values' and converted into quantitative measures producing a specific kind of hierarchy. Biopolitical machineries, as we know, not only 'contain' elements of life, but also endlessly 'qualify' life. They produce a mapping made of specific representations of life. In the Airbnb world, members are actually mapped out: as individuals, as families, as 'home', but also as travelers, as guests, as providers of care and 'hospitality'. The Airbnb ranking of people and homes via its algorithms in fact comprises the incorporation of elements of real life, real homes, real relationships, into the calculative rationality of the platform. The qualified quantifications of these real-life elements tend to shape what hospitality is for Airbnb and its global community, and how individuals should perform to be highly ranked in a world-made-of-hosts-and-guests.

However, and this is our second theoretical proposition, the Airbnb algorithms are machineries that spin around an empty core. Despite the fact that the ideas of community, home, hospitality, local culture, etc. feeding into the platform's algorithms are linked to real-world contexts – the homes offered are real as are their locations –

when they are translated into the Airbnb ranking they tend to become something else, possibly self-referential metaphors based on Airbnb's calculative rationalities. This does not mean that they do not operate as social regulators, quite the contrary. The Airbnb algorithms – as shown by this article – are in fact part of a biopolitical technology that squeezes value out of a myriad of aspects of everyday life. These aspects are often voluntarily offered by the participants, who willingly put on display a series of intimate and personal elements of their respective lives (and homes) to have them incorporated into the grand metrics of the platform. Many of these hosts actually enjoy being involved in these encounters with the guests, and often interpret their role in ways that resist and somewhat 'twist' the rationale of the platform, as emerged in interviews discussed above. However, despite these subjective (and sometimes even subversive) interpretations of the interplay between guests and hosts, by incorporating 'homes' and 'everyday life' into its metrics the platform somehow tends *to empty them out*, to convert them into elements of datafication and algorithmic management. For this reason, Airbnb, like all biopolitical machinery, needs endless injections of 'new life' (new homes, new intimate encounters, new hosts and guests), new 'stuff' to be incorporated and put into circulation by its broader regulatory system. Indeed, the embodied data offered by millions of hosts and guests to be datified by the platform's algorithms are an important form of capital, without which Airbnb would not be able to operate nor to generate value (see Sadowski, 2019).

What is more, and this our third theoretical proposition, the Airbnb calculative rationale cannot be taken to its most extreme consequences. This is confirmed by the ideal citizen of the Airbnb community, the Superhost, who represents a distilled and embodied abstraction of the deeper logic of its algorithms (see Roelofsen & Minca, 2018). While each hospitable resident of this community should aspire to obtain the condition/status of Superhost, at the same time this is an endlessly mobile condition, since the rules to maintain it are constantly changed by the platform. The Superhost in fact must remain the horizon towards which all hosts move, but that nobody can actually permanently inhabit. As we have learned from the history of all biopolitical regimes, the principle of 'endless improvement' does not produce a perfect(ed) society, because the workings of biopolitics is based on movement, on ever-changing thresholds of inclusion and exclusion (see, again, Agamben, 1998, 2002). This is fundamentally why the Airbnb platform is a biopolitical machinery spinning around an empty core: there is no point of arrival, no community to be realized, no perfect guest or host, since its only possible objective, in the end, is *to reproduce its capitalist Self*.

Disclosure statement

No potential conflict of interest was reported by the author(s).

Funding

This article was funded by Faculty of Arts, Macquarie University. The fieldwork underpinning this study was supported by the University of Graz Doctoral Stipend and the 2015 Rudi Roth Grant.

ORCID

Claudio Minca (iD) https://orcid.org/0000-0001-6619-6614
Maartje Roelofsen (iD) https://orcid.org/0000-0003-0952-0849

References

Agamben, G. (1998). *Homo sacer*. Stanford University Press: Stanford CA.
Agamben, G. (2002). *Remnants of Auschwitz*. London: Zone Books.
Airbnb. (2014). https://blog.atairbnb.com/dependable-communication-builds-community/
Airbnb. (2015). https://blog.atairbnb.com/is-mankind/
Airbnb. (2016). https://www.niido.com/travel-time/
Airbnb. (2017). https://blog.atairbnb.com/guest-star-ratings/
Airbnb. (2018). https://press.airbnb.com/airbnb-deepens-china-commitment-with-the-launch-of-airbnb-plus-and-the-airbnb-host-academy/
Airbnb. (2019a). https://www.airbnb.com.au/help/article/67/why-do-i-need-to-have-an-airbnb-profile-or-profile-photo
Airbnb. (2019b). https://press.airbnb.com/updating-our-terms-of-service-payments-terms-of-service-and-privacy-policy-2/
Airbnb. (2019c). https://www.airbnb.com.au/trust/standards
Airbnb. (2019d). https://www.airbnbcitizen.com/how-airbnb-works/
Airbnb. (2019e). https://www.airbnb.com/superhost
Airbnb. (2019f). https://www.airbnb.com.au/help/article/832/can-i-lose-my-superhost-status
Airbnb. (2019g). https://www.airbnb.com.au/help/article/829/how-do-i-become-a-superhost
Airbnb. (2019h). https://www.airbnb.com.au/help/article/1319/how-do-i-track-my-superhost-status
Airbnb. (2019i). https://www.airbnb.com.au/help/article/4/how-does-airbnb-help-build-trust-between-hosts-and-guests
Airbnb. (2019j). https://www.airbnb.com.au/help/article/2356/what-does-it-mean-when-someone-s-id-has-been-checked
Airbnb. (2019k). https://www.airbnb.com.au/host/instant.
Airbnb Community Center (ACC). (2017a). https://community.withairbnb.com/t5/Hosting/Sorry-another-thread-about-reviews-for-bad-guests/td-p/513905.
Airbnb Community Center (ACC). (2017b). https://community.withairbnb.com/t5/Hosting/Your-top-questions-about-Airbnb-Search/td-p/509644.
Airbnb Community Center (ACC). (2018). https://community.withairbnb.com/t5/Airbnb-Updates/Airbnb-Answers-Protecting-you-from-one-off-bad-reviews/td-p/822623.
Andrejevic, M. (2007). Surveillance in the digital enclosure. *The Communication Review*, *10*(4), 295–317. doi:10.1080/10714420701715365
Andrejevic, M. (2019). Automating Surveillance. *Surveillance & Society*, *17*(1/2), 7–13. doi:10.24908/ss.v17i1/2.12930

Arias Sans, A., & Quaglieri Domínguez, A. (2016). Unravelling Airbnb. Urban perspectives from Barcelona. In P. Russo & G. Richards (Eds.), *Reinventing the local in tourism.* (pp. 209–228). Bristol: Channel View.

Ash, J., Kitchin, R., & Leszczynski, A. (2016). Digital turn, digital geographies?. *Progress in Human Geography, 42*(1), 25–43. doi:10.1177/0309132516664800

Ateljevic, I., Pritchard, A., & Morgan, N. (Eds.). (2007). *The critical turn in tourism studies.* London: Routledge.

Barassi, V. (2017). Ethnography beyond and within digital structures and the study of social media activism. In L. Hjorth, H. Horst, A. Galloway & G. Bell (Eds.), *Routledge Companion to Digital Ethnography* (pp. 406–418). London: Routledge.

Bialski, P. (2016). Authority and authorship: Uncovering the sociotechnical regimes of peer-to-peer tourism. In A.P. Russo & G. Richards (Eds.), *Reinventing the local in tourism* (pp. 35–49). Bristol: Channel View.

Bloomberg Technology. (2017). How Airbnb is using tech to build trust. Retrieved from https://www.youtube.com/watch?v=b5_vknirzSw

Bridges, J., & Vásquez, C. (2018). If nearly all Airbnb reviews are positive, does that make them meaningless?. *Current Issues in Tourism, 21*(18), 2065–2075. doi:10.1080/13683500.2016.1267113

Cadell, C. (2016). Airbnb tells China users personal data to be stored locally. Retrieved from https://www.reuters.com/article/us-airbnb-china-idUSKBN12W3V6

Campbell, T, & Stize, A., Eds. (2013). *Biopolitics: A reader.* Durham: Duke University Press.

Chesky, B. (2014). Belong anywhere. Retrieved from https://medium.com/@bchesky/belong-anywhere-ccf42702d010

Cócola Gant, A. (2016). Holiday rentals: The new gentrification battlefront. *Sociological Research Online, 21*(3), 1–9doi:10.5153/sro.4071

Cócola Gant, A. (2018). Tourism gentrification. In L. Lees & M. Phillips, M. (Eds.), *Handbook of gentrification studies* (pp. 281–293). Cheltenham: Edward Elgar.

Cox, M. (2017). The face of Airbnb, New York City: Airbnb as a racial gentrification tool. Retrieved from http://insideairbnb.com/face-of-airbnb-nyc

Creemers, R. (2017). Cyber China: Upgrading propaganda, public opinion work and social management for the twenty-first century. *Journal of Contemporary China, 26*(103), 85–100. doi:10.1080/10670564.2016.1206281

Creemers, R. (2018). China's Social Credit System: An evolving practice of control. Available at SSRN: https://ssrn.com/abstract=3175792 or http://dx.doi.org/10.2139/ssrn.3175792

Dredge, D., & Gyimóthy, S. (2015). The collaborative economy and tourism: Critical perspectives, questionable claims and silenced voices. *Tourism Recreation Research, 40*(3), 286–302. doi:10.1080/02508281.2015.1086076

Edelman, B., Luca, M., & Svirsky, D. (2017). Racial discrimination in the sharing economy: Evidence from a field experiment. *American Economic Journal: Applied Economics, 9*(2), 1–22. doi:10.1257/app.20160213

Ek, R., & Hultman, J. (2008). Sticky landscapes and smooth experiences: The biopower of tourism mobilities in the Öresund region. *Mobilities, 32*, 223–242. doi:10.1080/17450100802095312

Ert, E., Fleischer, A., & Magen, N. (2016). Trust and reputation in the sharing economy: The role of personal photos in Airbnb. *Tourism Management, 55*, 62–73. doi:10.1016/j.tourman.2016.01.013

Ettlinger, N. (2018). Algorithmic affordances for productive resistance. *Big Data & Society, 5*(1), 1–13. doi:10.1177/2053951718771399

Esposito R (2010) Communitas: The Origin and Destiny of Community. Stanford: Stanford University Press.

Friedman, T.L. (2017). Self-driving people, enabled by Airbnb. Retrieved from https://www.nytimes.com/2017/07/26/opinion/airbnb-experiences-machines-jobs.html

Gil, J., & Sequera, J. (2018). Expansión de la ciudad turística y nuevas resistencias. El caso de Airbnb en Madrid. *Empiria. Revista de Metodología de Ciencias Sociales, 41*, 15–32. doi:10.5944/empiria.41.2018.22602

Gillespie, T. (2017). The platform metaphor, revisited. Retrieved from http://culturedigitally.org/2017/08/platform-metaphor/

Griffiths, M. (2019). How many stars is a smile worth? The social cost of emotional labor. Retrieved from https://www.theguardian.com/media/2019/feb/04/how-many-stars-is-a-smile-worth-the-social-cost-of-emotional-labour

Gutiérrez, J., García-Palomares, J. C., Romanillos, G., & Salas-Olmedo, M. H. (2017). The eruption of Airbnb in tourist cities: Comparing spatial patterns of hotels and peer-to-peer accommodation in Barcelona. *Tourism Management*, *62*, 278–291. doi:10.1016/j.tourman.2017.05.003

Hakim, N., & Keys, A. (2014). Architecting a machine learning system for risk. Retrieved from https://medium.com/airbnb-engineering/architecting-a-machine-learning-system-for-risk-941abbba5a60

Hearn, A. (2010). Structuring feeling: Web 2.0, online ranking and rating, and the digital 'reputation' economy. *Ephemera*, *103*/(4), 421–438.

Hoffman, S. (2018). *Social Credit. Technology-enhanced authoritarian control with global consequences*. Canberra: Australian Strategic Policy Institute.

Horn, K., & Merante, M. (2017). Is home sharing driving up rents? Evidence from Airbnb in Boston. *Journal of Housing Economics*, *38*, 14–24. doi:10.1016/j.jhe.2017.08.002

Isin, E., & Ruppert, E. (2015). *Being digital citizens*. London: Rowman & Littlefield International.

Jing, M., & Soo, Z. (2018). Airbnb complies with China law to hand over guest details as listings double. Retrieved from https://www.scmp.com/tech/article/2139526/airbnb-complies-china-law-hand-over-guest-details-listings-double

Kitchin, R. (2017). Thinking critically about and researching algorithms. *Information, Communication and Society*, *20*(1), 14–29. doi:10.1080/1369118X.2016.1154087

Lee, D. (2016). How Airbnb short-term rentals exacerbate Los Angeles's affordable housing crisis: Analysis and policy recommendations. *Harvard Law and Policy Review*, *10*, 229–254.

Lemke, T. (2011). *Biopolitics*. NYU Press: New York.

Lupton, D. (2016). *The quantified self*. Cambridge, UK: Polity.

Mermet, A. C. (2017). Airbnb and tourism gentrification. Critical insights from the exploratory analysis of the 'Airbnb syndrome' in Reykjavík. In M. Gravari-Barbas & Guinand, S. (Eds.), *Tourism and gentrification in contemporary metropolises* (pp. 52–74). London: Routledge.

Milan, S., & Gutiérrez, M. (2015). Citizens' media meets big data: The emergence of data activism. *Mediaciones*, *11*(14), 120–133. doi:10.26620/uniminuto.mediaciones.11.14.2015.120-133

Minca, C. (2009). The island: Work, tourism and the biopolitical. *Tourist Studies*, *92*, 88–108. doi:10.1177/1468797609360599

Minca, C. (2011). No country for old men. In C. Minca & T. Oakes (Eds.) *Real tourism* (pp. 12–37). London: Routledge.

Minca, C. (2015). The biopolitical imperative. In J.A., Agnew, V., Mamadouh, A.J., Secor, J.P. Sharp, (Eds.) *The Wiley Blackwell companion to political geography*. (pp. 165–186). Wiley-Blackwell: Oxford.

Munar, A. M., & Gyimóthy, S. (2014). Critical digital tourism studies. In A.M. Munar, S. Gyimóthy & L. Cai (Eds.) *Tourism social media*. (pp. 245–262). Bingley: Emerald.

Netflix (n.d.) *Black mirror*. Retrieved from https://www.netflix.com/title/70264888 doi:10.7238/c.n84.1903

Noble, S. U. (2018). *Algorithms of oppression*. New York,NY: NYU Press. Kindle Edition.

O' Regan, M., & Choe, J. (2017). Airbnb and cultural capitalism: Enclosure and control within the sharing economy. *Anatolia*, *28*(2), 163–172. doi:10.1080/13032917.2017.1283634

Ohlberg, M., Ahmed, S., & Lang, B. (2017). *Central planning, local experiments. The complex implementation of China's Social Credit System*. Berlin: Mercator Institute for China Studies.

Palin, M. (2018). China's 'social credit' system is a real-life 'Black Mirror' episode. Retrieved from https://nypost.com/2018/09/19/chinas-social-credit-system-is-a-real-life-black-mirror-nightmare/

Pasquale, F. (2015). The algorithmic self. *Hedgehog Review*, *17*(1), 30–45.

Pink, S., Horst, H., Postill, J., & Hjorth, L. (2015). *Digital Ethnography*. Thousand Oaks, CA: Sage Publishing.

Piracha, A., Sharples, R., Forrest, J., & Dunn, K. (2019). Racism in the sharing economy: Regulatory challenges in a neo-liberal cyber world. *Geoforum, 98*, 144–152. doi:10.1016/j.geoforum.2018.11.007

Postill, J., & Pink, S. (2012). Social media ethnography: The digital researcher in a messy web. *Media International Australia, 145*(1), 123–134.

Ravenelle, A. J. (2017). Sharing economy workers: Selling, not sharing. *Cambridge Journal of Regions, Economy and Society, 10*(2), 281–295. doi:10.1093/cjres/rsw043

Roelofsen, M. (2018a). Performing "home" in the sharing economies of tourism: The Airbnb experience in Sofia, Bulgaria. *Fennia - International Journal of Geography, 196*(1), 24–42. doi:10.11143/fennia.66259

Roelofsen, M. (2018b). Exploring the socio-spatial inequalities of Airbnb in Sofia, Bulgaria. *Erdkunde, 72*(4), 313–327. doi:10.3112/erdkunde.2018.04.04

Roelofsen, M., & Minca, C. (2018). The Superhost. Biopolitics, home and community in the Airbnb dream-world of global hospitality. *Geoforum, 91*, 170–181.

Sadowski, J. (2019). When data is capital: Datafication, accumulation, and extraction. *Big Data & Society, 6*(1), 1–12.

Schor, J. B. (2017). Does the sharing economy increase inequality within the eighty percent? Findings from a qualitative study of platform providers. *Cambridge Journal of Regions, Economy and Society, 10*(2), 263–279. doi:10.1093/cjres/rsw047

Schor, J. B., & Attwood-Charles, W. (2017). The "sharing" economy: Labor, inequality, and social connection on for-profit platforms. *Sociology Compass, 11*(8), e12493–16.

Shen, J. (2018). *Briefing: Zhejiang province to require user data from home-sharing platforms.* Retrieved from https://technode.com/2018/12/18/zhejiang-home-sharing-information/

Simpson, T. (2016). Tourist utopias: Biopolitics and the genealogy of the post-world tourist city. *Current Issues in Tourism, 191*, 27–59. doi:10.1080/13683500.2015.1005579

Terranova, T. (2000). Free labor: Producing culture for the digital economy. *Social Text, 18*(2), 33–58. doi:10.1215/01642472-18-2_63-33

Teubner, T., Saade, N., Kawlitschek, F., & Weinhardt, C. (2016). It's only pixels, badges, and stars: On the economic value of reputation on Airbnb. *Australasian Conference on Information Systems*, Wollongong, Australia.

Urry, J., & Larsen, J. (2011). *The tourist gaze 3.0.* London: Sage.

Van Doorn, N. (2017). Platform labor: On the gendered and racialized exploitation of low-income service work in the 'on-demand' economy. *Information, Communication & Society, 20*(6), 898–914.

Wachsmuth, D., & Weisler, A. (2018). Airbnb and the rent gap: Gentrification through the sharing economy. *Environment and Planning A: Economy and Space, 50*(6), 1147–1170.

Zervas, G., Proserpio, D., & Byers, J. (2015). A first look at online reputation on Airbnb, where every stay is above average. Available at SSRN: https://ssrn.com/abstract=2554500

Zuboff, Z. (2019). *The age of surveillance capitalism (Ebook).* New York: PublicAffairs.

Going on holiday only to come home: making happy families in Singapore

Yinn Shan Cheong ⓘ and Harng Luh Sin ⓘ

ABSTRACT

Family holidays have become a ubiquitous cultural norm – imagined as the epitome of family togetherness and encapsulated through the production of 'happy family' photographs. The pressure to conform to 'naturalized' idealizations of the family is underpinned by pervasive family ideology that remain hegemonic, despite the changing structures of modern family life. But how exactly does the 'happy family' materialise through tourism practices? What is it about the tourism time-space that differentiates it from home, transforming the ordinary family holiday into an extraordinary obligation that desires to be performed? These questions warrant an investigation into the exceptionalism of the family holiday, and we answer them by deconstructing the dilemma between the expectations and lived realities of modern family life, using the typical Singaporean nuclear family as an empirical lens. By amalgamating an 'everyday geographies' conceptual approach with Gilligan's (1982) 'feminist ethics of care' perspective, the paper re-examines the seemingly banal practices of carework on the family holiday. It draws its findings from qualitative data based on focus group discussions conducted with five Singaporean families and an autoethnography of the first author's family holidays. The paper reveals that carework practices on holiday are perceived as a means for families to authenticate nostalgic identities about what a family is – perceived to be lost in modern life. These acts transform the holiday into a home-away-from-home, underpinned by a prioritisation of familial relationship. Hence, this paper conceptually formulates a reinterpretation of 'feminist ethics of care' as a morality whose salience through holiday practices renews family relationships of care with each recurring episode of the family holiday. This theorisation elucidates that the family tourism time-space is exceptional, simultaneously ordinary and extraordinary, existing in-between home and away. Through its (re)production of the 'happy family' identity, going on holiday has become one of the few ways of coming back home to the family.

摘要

家庭度假已经成为一种普遍存在的文化规范——想象成家庭团聚的缩影，并浓缩为拍摄出'幸福家庭'的照片。尽管现代家庭生活结构不断变化，但普遍存在的家庭观念仍然处于霸权地位的，这助长了遵守'自然的'家庭理想的压力。但'幸福家庭'究竟是如何通过旅游实践具体表现出来的呢？旅游的时空与家庭中有什么不同，究竟是如何把普通的家庭度假变成了一种想要履行的特殊义务？这些问题需要对家庭假日的例外主义进行调查，我们通过解构现代家庭生活的期望和现实生活之间的困境来回答这些问题，并以典型的新加坡核心家庭为实证对象。本文将'日常地理'的概念方法与吉利根(1982)的'女权主义关怀伦理'视角相结合，重新审视了家庭假期关怀的看似平庸的做法。该研究基于与5个新加坡家庭进行的焦点小组讨论和第一作者家庭假期的自民族志的定性数据。这篇论文揭示了假期的关怀被认为是一种方法，让家庭对何为家庭进行怀旧的身份检验——认为家庭在现代生活中迷失了。这些行为以家庭关系优先为基础，将这个节日变成了一个家外之家的节日。因此，本文从概念上重新诠释了'女权关怀伦理'，即通过节日实践，以家庭节日每一个情节反复出现，延续了家庭关系关怀的品行。这一理论阐释了家庭旅游时空的特殊性，即既平凡又不平凡，存在于家与外之间。通过对'幸福家庭'身份的重新塑造，度假成为少数几种回家的方式之一。

Introduction

Why do families go on holiday? Rest, relaxation, fun and bonding come to mind, but underneath these seemingly ordinary answers lay broader societal expectations intricately bound to idealised notions of contemporary parenting (Carr, 2011). In today's society filled with work and educational obligations, the family holiday has become both symbolic of an escape from day-to-day routines and responsibilities, and sacralised as an obligation to sustain the performance of the wholesome 'happy family'. Beneath the desire to capture picture-perfect family-togetherness, we argue that family holidays encompass an imagined responsibility imbued with powerful ideological familial norms, and reflect the desire to achieve that elusive completeness of a family. This paper thus considers the prevalence of socially constructed imaginations of the family in shaping travel realities and highlights how the identity of the family and meanings of home are not just made at home, but also through one's mobilities across space. The spaces of tourism therefore permit parenting practices otherwise absent in everyday life.

Family holidays are however often marginalised in tourism research (Schänzel, Smith, & Weaver, 2005) – deemed too private and ordinary to be worthy of critical and scholarly insights (Baerenholdt, Haldrup, Larsen, & Urry, 2004; Schänzel et al., 2005). We argue that neglecting family tourism as an area of study is a critical and missed opportunity. An investigation of family holidays' banal and mundane routines through an everyday perspective can redirect our attention to the role tourism plays in the social reproduction of a family, thereby inspiring new ways to rethink tourism theories. Studying family tourism also presents us with the means to understand the family far beyond simply documenting what families do on holidays – rather, it gives us insights into the contemporary everyday lives of families and how they cope with the 'time bind' (Hochschild, 1997) of conflicting work and family time demands. We posit that the idealised desires of what makes a 'happy family', together with the everyday limitations of doing so, dovetails into how family negotiate and perform family holidays. The research objective of our study is thus to pitch tourism as central to the social understanding of family lives and identities, and how society's expectations of a happy family is built and shaped by tourism. As such, our emphasis is not so much on understanding what families do in their touristic endeavours, but why they do it.

We frame the family holiday as a 'home-making practice' (Blunt & Dowling, 2006), and argue that within the context of Singaporean families' everyday lives, the family holiday is considered not merely as an extension of home – it is home itself, as it allows the family to authenticate an idealised vision of everyday family life fulfilling notions of the complete home. This exists precisely because the 'complete home' is deemed to be missing in everyday family life. The family holiday is hence not an escape from home, it is an escape to an idealised version of home, one in which the 'happy family' becomes a reality, albeit a temporary one. For our respondents, the prioritisation of the family relationship while on holiday is therefore performed through a temporary change in parenting routines[1] – it is within the exceptional, episodic event of the holiday that the family has the opportunity to suspend normal everyday routines and take on roles wherein building relationships are prioritised as signifiers of the 'happy family'. Through studying the family's holiday practices, we observe the reproduction of the family through care practices and understand how family tourism is indeed an act of (re)producing dominant family ideology relevant to its attendant geographies or histories. The family is not merely an empirical unit to study tourism – instead, family tourism provides the lens to examine how the family functions as a hegemonic societal institution in differing social contexts, where Asian familial discourses propagated by the state shape imaginations of ideal family life. Existing literature on family holidays is largely dominated by the normative expectations of the Western family unit. We believe that while the case from Singapore cannot be said to represent all Asian families, it is a useful entry point to look at the under-researched Asian family in tourism (Khoo-Lattimore, Prayag, & Cheah, 2015; Kim, Choi, Argusa, Wang, & Kim, 2010; Wang, Hsieh, Yeh, & Tsai, 2004) and reflect upon the diversified ideas of family and how this influences tourism studies.

A home for the family in tourism research

Modern families' travels now represent one of the largest markets for the tourism industry (Carr, 2006). With the advent of budget travel, family holidays have become ubiquitous, seen as an essential rather than a luxury (Beioley, 2004). However, tourism scholarship is slow to catch up to this burgeoning field. Tourism scholarship on the family has been fragmented, limited to place-based studies of family-friendly tourism destinations (Ong, 2017) or geared towards studying trends in family demographics to cater to the tourism industry (Schänzel & Yeoman, 2015), both of which limit the scope of study by viewing the family only as consumers. There remains a dearth of studies on the experiential dimension of family holidays (Carr, 2011; Shaw, Havitz, & Delemere,2008). In recognising that the family on holiday occupies a liminal space between 'extraordinary places and familial faces' (Haldrup & Larsen, 2003, p. 26), this paper seeks to fill the gap by investigating the experience of ordinary care practices within the family on holiday. We therefore take Obrador's (2012, p. 404) conceptualisation of 'thick sociality and relations of domesticity' that constitute the family group to re-socialise academic perspectives on family tourism – through de-exoticizing tourism (Larsen, 2008), and rejecting the study of tourism as a realm separate from everyday life.

It is hence important to first consider: what defines the family? As recognisable as the term is, every family is different, and its boundaries are impossible to pin down. 'Family' often summons a pervasive, powerful image of the nuclear family, and a prescription of differentiated roles and naturalised categories (Bernardes, 1985) – father as breadwinner, mother as nurturer of children. The family is thus not a naturally preexisting category, but socially constructed and 'a matter of collective definition' (Newman, 2009, p. 5). The hegemonic conception of the ideal family that we stand by today often accentuates unity and happiness – an evolved product of Western familism in the 1970s, which promoted a set of pro-family ideals about how parents and children should live in accordance to family virtues of togetherness (Eichler, 1986). As a cultural symbol, family ideology has thus become deeply embedded as a singular, universalising aspirational entity. Bernardes (1985, p. 288) terms this 'idolistic mystification', wherein family ideology encourages the veneration of a fantastical 'idol', rather than examining family life as it is experienced. In this paper, we emphasise that research should instead focus on the family as a lived process through which meaning can be reproduced or transformed (Barrett, 1980) – where family ideology is not merely as a set of fictive regulating ideals, but as 'a necessary component of everyday social practice' (Bernardes, 1985, p. 277).

Family ideology therefore permeates the spaces we inhabit – particularly, the home space – ideologically saturated as the site for the active construction of the 'happy family' life (Crow & Allan, 1989). More than a material dwelling, home represents a 'relation between material and imaginative realms and processes' (Blunt & Dowling, 2006, p. 22). As such, the spatial scale of the home is understood to be continually recreated through 'everyday home-making practices, which are themselves tied to spatial imaginaries of home' (Blunt & Dowling, 2006, p. 245). To frame the construction of home as a process is to understand it as constantly in a state of becoming. This confers agency to family members who contribute to its creation through everyday practices within the family holiday, and the construction of home becomes imbued with normatising family ideology.

What then is the Singaporean family? Consistently used in political rhetoric as the 'building block of society', Teo (2010) argued that the state glorifies a particular familial form in Singapore – one which straddles the patriarchal conservatism that Western familism epitomised, but also venerates Confucianist virtues of relationality that prioritises the family group over the individual. This heteronormative construct of family is a critical anchor in Singapore's development narrative, sacralised and reinforced continually through state discourse and policies in which this family form is pivotal to economic productivist goals. However, when theorising from Asia, attention is warranted in distilling the nuances of Asian family values (Barr, 2000) without being guilty of a 'post-colonial hangover' that reappropriates Western framings to contextually different spaces. It is thus essential not to immediately assume Confucian or Chinese ideals as starting points for analysing family relations within the Asia context (Martin, 1990) as such sweeping generalisations do more harm in a bifurcating a West vs Rest mentality. Considering this cautionary note, our approach rejects this dichotomising vein of thinking, and instead, we want to highlight the exceptional hegemonic salience of the Singapore family and context of the state's interventionist approach to family policies in Singapore.

The nuclear family is institutionalised into social policies in Singapore. Most salient in the allocation of public housing – family units are accorded various priorities and subsidies in purchasing housing. The state's definition of the family unit is strictly limited to heterosexual married couples with children, and up until recently, unmarried single parents with children were not considered a valid family nucleus eligible for public housing purchase (Oswin, 2010).

The state also reifies the nuclear family as the main provider of intergenerational financial support. Unlike Western social democratic welfare states where pensions contribute to living expenses of the aged population, the care and maintenance of the aged is deemed the responsibility of his or her own child, and legitimised by filial piety as a moral value. Care arrangements, particularly in the case of elder care or childcare, is shaped by state policies such as heavy subsidies for foreign domestic workers (FDW) for families with young children or aged dependents (MOM, 2019). The existence of 3Generation housing and housing proximity grants (given to nuclear families buying a public housing within 1 km of their parents' public housing) further incentivise families to stay closer and facilitate grandparents as a source of childcare (HDB, 2019). These policies have collectively normalised the culture of hiring FDW and dependence on grandparents for childcare, where the normative expectation is that of a dual-income family where parents are both in full-time employment often involving long working hours. FDW working and living within the Singaporean household is thus both seen as a 'solution to the care crisis on the domestic front' (Lam, Yeoh, & Huang, 2006, p. 475) and also often a 'necessity' for care needs when a married couple decides to have a child (see also Yeoh & Huang, 2010).

We thus see a set of contradictory principles that Singaporean families face, on one hand the family is constructed as a self-sufficient unit of care and support, and on the other, it is acknowledged to be never enough. This explains why the family nucleus holds sway in popular imaginations of Singaporean familism, and perhaps the constructed immutable universality of this script creates heightened pressure to conform in order to fit in. The politically motivated aspects of the Singaporean familial narrative provides food for thought, but within the scope of this paper, rather than deconstruct the epistemological reasons underpinning them, we seek to explore the implications of this ideology as manifested in tourism practices, and in our participants' subjectivities of their performed identities.

Understanding the family through studying tourism

The 'performative turn' in critical tourism studies in the 1990s has emphasised a move beyond visual paradigms of seeing (Cloke & Perkins, 1998, p. 189) towards the corporeality of feeling tourist bodies in multi-sensuous experiences (Crouch, 2001). This turn to the emotional emphasises intimate interactions with others, and stresses the importance not just of being there, but being with. Haldrup and Larsen (2003) conceptualisation of the 'family gaze' as a choreographed staging of the 'real family' through photography reinforces the importance of preserving social relations at the core of family holidays. In opposition to the gaze on exotic places (Urry, 1990), the family gaze seeks to capture the 'extraordinary ordinariness' of banal family life. Haldrup and Larsen's astute observations of how photographic memory adheres to idealised conventions of natural family life in accordance to cultural codes, reinforces the notion that family tourism exists not outside everyday life, but along a continuum amidst the banalities of the everyday, something that earlier tourism theories have neglected.

Inspired by Larsen (2008) and Edensor (2007) who frame tourism as a mundane mobility, we therefore use 'everyday geographies' as a conceptual framework to make sense of family tourism's supposed exceptionalism vis-à-vis its intricate ties to meanings of everyday family life. 'The everyday offers itself up as … a contradiction, a paradox: both ordinary and extraordinary…, obvious and enigmatic' (Highmore, 2002, p. 16) where indeed, 'the exceptional can be found within the everyday' (Skelton, 2017, p. 1). Lefebvre (1988, p. 87) distinguishes between two different ideas within the word everyday – *la vie quotidienne*, translated as everyday life, and *le quotidien*, translated as the everyday. Seigworth (2000, p. 245) interprets this succinctly – if everyday life represents the events perceived and experienced by individuals, then the everyday refers to how everyday life is depicted and transformed into a concept via discourse, such as language or photography. The everyday can thus be understood as the 'form that life takes after it is transformed by a system of representation' (Horton & Kraft, 2014, p. 183) – it becomes a constructed entity. In applying this to family life, we draw a corollary between Seigworth's differentiation and Gillis' (1996b, p. xv) conceptualisation of the families we 'live with' in everyday life, set against the families we 'live by', a nostalgic imagination which romanticises family stability and cohesiveness, perpetuated by family ideology. The persistence of such ideals creates standards to which 'the everyday' is to be performed and transforms the way 'everyday life' is practiced during the episodic event of the holiday. This paper thus unravels how the family holiday represents an opportunity for the dialectic between everyday life and the everyday to be negotiated by families.

Research on family tourism is deeply entwined with feminist scholarship within leisure studies. Studies on women's leisure time (Shaw, 1992), perception of leisure (Davidson, 1996) and entitlement to leisure (Henderson & Dialeschki, 1991) often conceptually utilised the 'feminist ethics of care', where women's morality centred on care, relationships and responsibility towards others came to the fore, differentiated from the masculine morality of individual rights (Gilligan, 1982). By framing women's morality and concern with sustaining relationships through care as a human strength (Miller, 1976), feminist scholars have rationalised the 'double shift' (Hochschild, 1989) women were obliged to perform on holiday as a 'labour of love' (Finch & Groves, 1983) and as a source of empowerment. This challenged the relevance of the work/leisure dichotomy to women, but it did not destabilise the gendered feminisation of carework as a women's job. In pronouncing the family holiday as a stressful obligation, Backer and Schänzel (2015) therefore referred solely to women's onus to perform carework responsibilities associated with motherhood (Mottiar & Quinn, 2012), further valorising the mother's role in family tourism scholarship.

Mirroring trends in gender research dominated by concern for women, little is known on fathers' caring roles in family tourism literature (Pritchard, Morgan, Ateljevic, & Harris, 2007; Ryans, 2003), nor the meaning of holidays for fatherhood as an identity (Allen & Daly, 2005) – Shaw et al. (2008) merely feature father's joint parenting voice, while Schänzel and Smith (2011) explain men's contributions as children's entertainers in relation to their facilitation of mother's leisure time. We emphasise in this paper that there is a need to move beyond such binaries and examine the family group as a whole through the lens of caring, and their commitment to each other, which then manifests as the holiday obligation. Indeed, dads can care too. Family holidays allow us to understand fathering in its own context, rather than an adjunct to maternal care (Brotherson, Dollahite, & Hawkins, 2005). This paper thus extends Schänzel and Jenkins' (2017) qualitative documentation of leisure experiences as a responsibility of non-resident fathers, in the negotiation of how fatherhood is experienced. Indeed, modern family ideology does not merely valorise the caring capacities of motherhood, but ascribes importance to 'retrieving the father' (Chopra, 2001) in order to produce a close-knit family unit. This is pivotal in our conceptual framing of the episodic time-space of the family holiday as an opportunity for alternative moralities prioritising relationships to be taken on by the whole family.

The empirical sections highlight instances where these care ethics come to the fore in everyday routines, in order to make sense of the holiday's importance to family relationships, and why it is perceived as essential to an envisioned happy family life. The continuities that family holidays share with everyday life is thus simultaneously mundane and extraordinary, and not a realm of exceptionalism which departs from everyday life. We argue that the time-space of the family holiday is a liminal episode where temporary identities yearn to be represented, but also an important recurring episode over time which sustains a longer-term vision of such family relationships. It is an exceptional realm where the banality of carework takes on a different meaning. No longer perceived as routine chores from which families desire refuge, carework becomes an aspirational construct that families seek to emulate on holiday in becoming the 'happy family'; a nostalgia that families seek to reclaim through the holiday; and a practice through which an authentic self-identity of being a family can be recognised. Family holidays therefore provide that opportunity for a suspension of normal everyday life roles (that prioritises work and schooling in Singapore), to take on caring roles wherein building relationships and connections are prioritised as signifiers of the everyday construct of 'happy family' ideology. The family holiday is where home is truly made.

Whole family methodology

This paper adopts a 'whole family methodology' (Hess & Handel, 1959) ethnographic style, underpinned by a conscious effort to document family life through deep immersion within the context of encounter, rather than focussing only on what is represented. Hess and Handel construct family members as agents in family life, where the family generates its own culture by producing individuals and relationships, drawn selectively from the cultural systems they are embedded in. This idea lays foundational groundwork for our analysis on how families produce their own meanings of the 'happy family' amidst ideologically dominant cultural constructs of family ideology. We therefore chose to base our research on focus group discussions (FGD) with five families and augment this with an autoethnography conducted by the first author. Indeed, as emphasised in Schänzel (2010, p. 556) – 'no single family member is a sufficient source of information for their family as a family constructs life from the multiple perspectives of its members' (Handel, 1997). The inherently private nature of family holidays limits our research access on other families' holidays if we depended solely on focus group discussions and risk obtaining a purely represented approach that can possibly overlook nuances in necessarily embodied experiences embedded in the banal experiences of togetherness in family holidays. The two methods are thus combined strategically to build on each other's blindspots to research the family holiday.

The participating families in this study were dual-parent heterosexual families with at least one dependent below 18-years-old. We acknowledge upfront that this is a specific family typology and our discussions reflect biases because of this selection. However, while it is not our intention to allow this research to perpetuate a normatively defined traditional nuclear family, and we would like to acknowledge that the normative idea(l)s of the modern family are changing, we decided to limit research participants to such family structures as it provides a heuristical convenience and coherence between families' backgrounds when we discuss our findings within the context of Singapore. At the same time, the dual-parent heterosexual family is what is classically represented in official rhetoric promoting 'family life' in Singapore, and can arguably be seen as a model of what makes the 'good life' in Singapore's context. Moreover, including diverse family forms warrants a different set of literature beyond the scope of this paper. Future research into the conformation to or rejection of hegemonic family ideology through holiday practices is certainly encouraged.

FGDs were also chosen over one-to-one interviews as we hoped to gain access into insights produced through interaction as a group (Morgan, 1988), and saw FGDs as a suitable technique for 'collaborative storytelling' (Starkweather, 2011) in eliciting contributing perspectives. This fittingly substitutes our absence on participant families' private holidays by allowing natural observations of inter-personal family exchanges. We acknowledge that doing so meant that individual perspectives might have been omitted at the expense of conforming to a singular group stance. As a mitigation measure, the actual structure of the FGDs included the first author's prompting for each individual member to contest fellow family members' opinions, and this was common instead of a homogenous family position. Our choice of family groups with older children aged 8-17 was made in hope of encouraging their independent voices to emerge despite being interviewed in a group setting. The familiarity between our participants allowed for these debates to take place and contributed to a variety of voices heard.

The first author conducted five in-depth FGDs with her relatives' families across June to July 2017, each lasting between 1 and 1.5 hours (Table 1). Pseudonyms were given to protect the identities of the participants.

All FGDs were conducted in the family home (Larossa, Bennett, & Gelles, 1994), as emplacement of the family in their 'natural habitat' (Kennedy-Eden & Gretzel, 2016, p. 466) increased their propensity to act normal while being observed. As families were selected from the first author's immediate relatives, critical reflection on positionality is

Table 1. Family profiles of focus group discussion participants (Source: Authors).

Family names	Name	Age	Family relation	Date of family holiday(s)
#1 Loh family	Matthew Loh	44	Father	Dec 2016, Macau
	Sarah Ong	44	Mother	& SEA Cruise Ship
	Zack Loh	15	Son	
	Alvin Loh	13	Son	
	Edison Loh	6	Son	
#2 Chang family	Steve Chang	63	Father	June 2017,
	Mathilda Yeow	56	Mother	Tokyo
	Jessica Chang	17	Daughter	
#3 Tan$_1$ family	Keith Tan	47	Father	Dec 2015,
	Marissa Ang	43	Mother	Hong Kong
	Cheryl Tan	17	Daughter	
	Calvin Tan	15	Son	
	Tiffany Tan	11	Daughter	
	Taylor Tan	11	Daughter	
#4 Foo family	Vincent Foo	48	Father	June 2017,
	Diana Tan	38	Mother	Bali & Australia
	Dylan Foo	8	Son	
#5 Tan$_2$ family	Vance Tan	53	Father	Dec 2016,
	Germaine Hong	45	Mother	Phuket
	Eleanor Tan	15	Daughter	
	Dickson Tan	12	Son	

warranted. This decision was made not out of convenience, but as an intended strategy to overcome the insider/outsider binary and destabilise the power inequalities inherent in a research relationship. I[2] sought not to decentre the researcher, but instead weave my experiences into starting points for discussion. In this unique encounter, I become a respondent just as much as my participants, pivotal in creating an open environment for sharing. I paid close attention to participants' self-representations as strategic, partially produced ethnographic material, implicated by my presence within a private space. My position as a relative extended the longevity of shared information beyond the time-frame of the FGD, and this exceptional researcher-researched relationship may impinge on how families sought to 'protect' their 'image'. For example, parents consistently corrected their children's 'misrepresentations' that strayed from ideal representations. This unique dynamic is used to my advantage as it distils a salient self-representation of families' desire to showcase a 'correct' portrayal, which further buttresses our claims on the hegemony of family ideology. Indeed, it must be recognised that families' representations of themselves as 'happy families' through holidays is projected precisely to their own friends and relatives, which is exactly my position to these families we included in the research. This, we believe, worked in extracting a version of families' holidays in more accurate ways beyond simply allowing us access into families. FGDs were then transcribed, categorised and coded according to the themes of the meaning of home and family, the difference between leisure time at home and overseas, the distribution of parenting roles on holiday and the importance of family holidays.

Yet as Lefebvre (1991, p. 15) suggested, 'the familiar is not necessarily known' – an autoethnography was thus done to further allow an integration of 'positionality, involvement and experiences' into the research (Cloke, Crang, & Goodwin, 1999, p. 333), where an everyday perspective was salient. During a planned family holiday, I kept a research diary documenting my thoughts and emotions throughout the day's activities. Using my own family as research subject is strategic as established relations preclude the likelihood of unnatural 'staged' performances. Subjective human experiences are necessarily not emotion or value free (Bondi, 2009). As such, the research diary method is useful in creating 'a gap between [the] everyday self and [the] diary-writing self' (Latham, 2003, p. 2004), which alerts myself to moments of self-representation. In sum, our research techniques intentionally utilise researcher visibility as 'resource rather than contaminant' (DeVault, 1996, p. 42). Through effectively writing the self into research and playing my positionality as a family member to my advantage, we aspire to overcome the 'methodological timidity' (Thrift, 2000, p. 3) that hinders fruitful engagement with positionality. Our subsequent analysis thus incorporates instances of personal reflection and insights drawn from my holiday diary through a self-referential mode of writing. While the FGDs and autoethnography each represent limitations in their own right, we view the two methods in triangulation and hope that in balance they provide the benefits of each method, while mitigating the issues each presents. For example, the autoethnographic approach is sometimes criticised for being overindulgent in the personal, but when used as a supplement to FGDs, it is weaved into a movement between self and other (Ellis & Bochner, 1996) to paint a rich analysis informed by both reflexive modes of thinking and empirical exemplification.

No time to be a family – except on holiday

Unlike traditional tourism research that sees the holiday in a linear perspective of the before, during, and after stages, what stood out in this research was the continual reference by respondents to holidays both past and upcoming, suggesting a more cyclical process at play. The 'happy family' is (re)produced with every recurring episodic event of the holiday because it offers the opportunity to be a home to the family – something that our respondents find lacking. An ambiguous and not-yet-planned-for holiday in the future is often expected, and past experiences of holidays in different destinations and times were typically conflated in discussions. Importantly, our respondents' expectations of being in this continuous loop of being either on holiday or in a simultaneous state of being after one holiday and before another holiday, is largely tied to their view that the home space in Singapore does not provide for the idealised family time they envisioned. This conforms to works that suggest that the modern family is overworked (Daly, 1996) to the extent that family time has taken on an 'industrial tone' to accommodate the pressures of work (Hochschild, 1997). The exceptional carving out of time even for holiday is a manifestation of timetabling that remains 'tethered' to the clock-bound nature of work (De Grazia, 1962, p. 326) – leisure becomes rationed and regimented (Rybczynski, 1991). In the following excerpt, the breathlessness and weariness in the father's tone of voice displays his exhaustion, as he recounts the various activities that preoccupy them on the regular weekends in Singapore. The family holiday, offering uninterrupted time together, is then perceived in contrast as a space of exception:

> I mean we do spend the whole day together in Singapore as well, but in the holiday context, its different *lah*! Weekend we don't work and there is no school, but it's not uninterrupted *lah*! We have church, Sunday school, and when he is doing his CCA[3],

Chinese classes then it's also dropping him off, and there is time apart! But when you are on holiday, its 24/7 together! So it does build the bond *lah*, because we get the family together. (*Vance*)

This is especially the case for the first author's family, as my father's work abroad means the family rarely spends time together, apart from the scheduled annual holidays. While this may seem exceptional, the discordance between expectations and realities of modern family life are more real than imagined. When asked if the family convenes to plan holidays, our respondent laughed bitterly about the dismal home situation:

We hardly have time to sit down together, we just [discuss] along the way! (*Mathilda*)

Regardless, families continue to pursue 'prolonged family time' (*Sharon*) against a 'socially constraining reality' (Daly, 2001, p. 284). We interpret this as a sign that hegemonic prescriptive beliefs about family time remains central in facilitating the 'happy family' and are resistant to change. It is not that the home can be performed in the holiday; rather, it craves to be performed, because 'everyday life' does not provide the opportunity for this idealised vision of home. In order to perform home, ironically, the family must get out of the home; in order to achieve idealised notions of 'the everyday', the family needed to break out of temporal routines of 'everyday life'. The family holiday therefore represents a nostalgic yearning for uninterrupted family time that 'anchored families to a secure past' (Daly, 2001, p. 288) within the context of highly scheduled lives.

Our findings further reveal that the spatial element of being away from familiarity is of essence to a successful holiday. When asked about domestic staycations, respondents lamented the lack of quality family time, citing distractions from locally based commitments as the aggravating factor. This is particularly evident in Vance's complaints of how work responsibilities continue to pervade should one choose not to leave the country:

If you are so close to your work, how do you really enjoy? And your phone is always ringing like as if you're not going on holiday. You cannot focus, that's why I would prefer, [go] out of this country, because you have a clean break from it all, to minimise disturbances. (*Vance*)

The spatial distancing from home thus takes on an exceptional meaning in family tourism. More than an escape from work commitments (Chen & Petrick, 2013), the holiday is a space-time wherein commitment to the family, the decision to focus on relationships in the here and now, and the enjoyment of moments of togetherness and sociality is (finally) prioritised. The holiday becomes a realm where the family allocates time to each other, thereby facilitating the abovementioned avoidance of 'disturbances' and reinforcing the perception that this is only achievable away from the lurking familiarity of the home space of 'everyday life'. Beyond emphasising a temporal dimension of the family holiday, many respondents also indicated the importance of the spatial dimension. Place is important insofar as it allows the family 'to be away from home'. This does mean that the actual places visited and activities engaged in on holiday is downplayed – families are seemingly less preoccupied with *where* the next holiday destination is, as opposed to *when* they can spend time together away from Singapore as a family.

The discussion thus far buttresses the notion that family holidays not only obfuscate the binary of home and away, but are clearly a liminal zone of exception in which spatial-temporal dimensions factor into its extraordinary position between lived realities and a nostalgia for untainted family time. However, while an academic view of the family holiday may interpret this as a better understanding of the lived mobilities of modern family life, we cannot tune out our own personal voice, nor the views of our respondents, that perhaps the family holiday is not perceived as different after all. In bringing to light the layman association families have of holidays as home, we suggest that home is ideologically bound to the bodies of significant family members.

> To me, over the years, travelling together, we actually watch them grow, from carrying them, to now, I don't think I can lift them up anymore, so all this time the changes, the different stage of their lives, you will see that more when you are overseas... When you are overseas together as a family, you are just doing the family thing! That's what I treasure la, a lot of these things through the holiday. (*Vance*)

Vance's notion of togetherness on family holidays as the 'family thing' speaks volumes. Even though our participants constantly contrasted the holiday experience with work schedules, domestic staycations, parenting arrangements and interactions of 'everyday life' back home, the holiday was perceived as no different from home, perhaps even, as an amplification of banal family life made visible through the holiday. As the following excerpt from the Tan family illustrates, it is simply another avenue for home to be lived, so much so that place of travel matters less than the company of family:

Interviewer: 'At which point in the trip did you feel most at home?'

Marissa: 'I mean, how you want me to differentiate between home and family?'

Cheryl: 'Family is part of home, and your friends can be your home, and your significant other can be your home too, because I personally believe that home is not just a place. It doesn't really matter what we are doing, it is who we are doing it with. I feel for me there was no particular point in time where I felt at home, because it was just everything'.

Marissa: 'The whole family is together, still a home'.

Calvin: 'We could go anywhere, as long as it's the six of us together'.

The micro-scale embodiment of tourism's mundaneness through the family group reveals that carework is a 'home-making practice' situated at the scale of the body, particularly, interpersonal relationships between bodies of co-present family members. Thus, to the family, despite the tangible exceptionalism of the tourism space, there is no imagined distinction between holiday space and home space. Instead, it is the corporeality of being together that symbolises the home-away-from-home; it is not the act of travelling but the sustained relationships that is prioritised in family travel. Families are not seeking to escape from home. Through tourism, they are seeking the lost mundaneness of 'everyday life' associated with time spent together, in reclaiming a family life imagined to be endangered.

Quoting Gillis (1996a, p. 6), 'in this new understanding of temporality, time is no longer a passive measure of social change but itself an active agent'. Modern living has taken mundaneness away from everyday life, such that the very ordinariness of family life together on holiday becomes a spectacle that warrants documentation; the performance of carework is not a banal routine but an unfamiliar affair; and the everyday interaction of the family on holiday becomes unexpectedly remarkably exciting. This challenges what we imagine as the exotic unknown in tourism – it challenges the very idea of the family that is familiar. But at the same time, the ideological coherence of the family remains strong. The family that we 'live by' continues to inform practices, which explains why families see no dissonance between the home space and the holiday space – it is the people who make the difference, it is the family which makes the home complete, regardless of where.

Reproducing the 'happy family' on holiday

It would be naïve to theorise that the mere time spent together on holiday is what makes a family. How then do families become a family? Using the 'feminist ethics of care' as a lens to understand what families do on holiday, we now turn our focus not to the bonding activities most associated with family holidays, but to the everyday routines of family life that permeate the holiday space. Contrary to imaginations of tourism as an intentional respite from the demands of routines, we argue that the family holiday is not an escape from the mundane 'life's work' (Katz, Marston, & Mitchell, 2004) of parenting. Rather, the opposite holds – the parents we interviewed reflected a desire to perform the very banal chores that elude them in 'everyday life'. The holiday exists as a space which allows for a validation of their imagined authentic selves as parents, in a time where expectations of 'the everyday' as a happy loving family do not live up to the realities of 'everyday life' – particularly when carework is outsourced to FDW and grandparents to make way for the demands of work in Singapore. The holiday thus serves as a mediation between the time stress families face in balancing work and family life, a means to cope with the 'time bind' (Hochschild, 1997).

The ideology of 'intensive mothering' (Hays, 1996) expects mothers to remain as primary caretakers of children, even amidst demands of work. This version of contemporary motherhood reflects the expectations mothers hold for themselves within our research, and working Singaporean mothers are thus caught in a double bind when the 'supermom' ideal fails to materialise in the home spaces of 'everyday life'. In order to cope with work demands, care is often outsourced to grandparents, childcare centres and FDW (Huang & Yeoh, 1998). While this is logistically necessary, it often leaves mothers feeling inadequate in fulfilling their mothering role, perpetuated by popular media representations that make accusations of working mothers neglecting their children at home, or not doing their jobs well as they are 'distracted' with familial commitments (Abetz & Moore, 2018). These further their perceived insufficiency in complying with 'hegemonic motherhood' (Newman & Henderson, 2014). The family holiday thus serves as an important avenue to reconcile their absence in 'everyday life', to take on motherhood aligned closer to the expectations of family ideology, perceived to be the banal household chores – often carried out by the domestic helper – that elude them in everyday life. The following two excerpts illustrate this:

During holiday, I'm the helper[4]! *laughs* I guess mainly because when we are not having a holiday, I would be working, not much time to, you know, take care of the physical needs of the children. So I think this is the time I have to look into their needs. (*Sarah*)

One important point to add is, since Dylan was two, we started staying with our parents, his grandparents. So holiday gives us time as a family unit, not with extended family. We are more involved, there is no, 'hey can you just watch him for five minutes' kind of help that we can get, so ... he gets the full attention from us... *smiles* It just gives me more time to be the maid[4]! (*Diana*)

Modern families are more time-stretched than before (Hochschild, 1997) – dual-income households make up a 53.8% majority of Singaporean families (SingStat, 2015). The 'commodification of intimate life' through the 'outsourced self' (Hochschild, 2012) where a mother can 'pay not to feel' (Stephens, 2015, p. 210) by compensating 'emotional labour' with substitute stand-ins, such as the FDW, is fairly common in Singapore. However, this arrangement is not unproblematic. Caregiving arrangements may have changed, but the standards that families 'live by' remain unchanged. They 'play an important role in preserving continuity with the myths and values of the past' (Daly, 2001, p. 293), and are further reinforced by state discourses on the ideal family nucleus. The holiday thus provides a rare opportunity for mothers to perform a longed-for nostalgic identity, envisioned as a compensation for their lack of involvement at home.

This performance of carework is a means that mothers seek out 'intra-personal existential authenticity' (Wang, 1999) where the temporary space of family holidays activates what they consider to be their 'authentic self' as an ideal mother. The two excerpts illustrate a mixture of emotions – guilt and pride. It is through performance of the 'emotional work' (Hochschild, 1989) of social reproduction, that both mothers take pride in fulfilling a 'natural' everyday responsibility but simultaneously lament a guilt-ridden 'feeling of loss' (Giddens, 1990, p. 98) through their inability to realise this goal outside the holiday. The 'feminist ethics of care' explains this internal struggle that mothers face between a yearned for accomplishment of familial responsibilities in line with discourses of 'the everyday' family, and the insufficiency of the self in 'everyday life' which motivates a prioritisation of imagined 'real' mothering roles through the holiday.

To solely focus on mothering however, would be to reproduce the relegation of fathers from the domain of carework. Instead, we posit that family tourism presents uncharted possibilities in re-examining the caring roles fathers take on. Indeed, the liminal space of the family holiday allows the opportunity for flexibility of a 'joint-parenting synergy' (*Germaine*), and resists the compartmentalisation of parental roles, in contrast to the home space. As illustrated in the following, the traditional division of labour is blurred during the holiday:

Mathilda: 'Back home, the roles and responsibility are a bit more distinct, because back home things in the house and Jessica (daughter) is mostly taken care of by me, whereas on holiday the roles are ... a bit more blurred, like ... '

Steve: 'Like we play it by ear, because you don't know what to expect'.

Contemporary fatherhood promotes the nurturing, 'involved' father (Dermott & Miller, 2015) but has yet to reconcile how fathering should be practised in the home space (Dermott, 2008), leading men to feel like a burden in struggling to define a new role for themselves at home (Varley & Blasco, 2001, p. 130). But on holiday, because it is a break from routine in the home space, this segmentation of roles can become more fluid, hence allowing alternative possibilities to be lived out. We propose that care as a 'practice-based reasoning' (Ruddick, 1995, p. 56), rather than an intuitive, natural state unique to women that Gilligan implies, means that the space of family holidays holds potential for fathers to redefine fatherhood in new spaces, and our respondents have indeed used the opportunity to take on these context-specific self-defined caring roles, as *Matthew*'s quote illustrates:

> My carework will be driving, to make sure we know how to get there. I know what the activities are for the whole day, so I'm the family's planning guy. (*Matthew*)

The 'ethics of care' is therefore not unique to the way women prioritise their family's needs, but can be extended to fathering practices too. In their efforts to rekindle familial relationships through care, fathers can also be 'maternal thinkers' (Ruddick, 1995). This is however, not without awkwardness. Fathers are aware of the rejection they may face, as illustrated by this father's self-deprecating humour:

> Usually when it comes to the last day, the packing part, that one I'm off, whatever comes I cannot do! *chuckles* Our luggage, if it's packed by me, I will have to buy another one. (*Vance*)

The family tourism space hence reveals insights into 'fathering and faltering' (Aitken, 2000) as an emotional practice that is negotiated in different spaces of 'everyday life', where incidences such as the aforementioned reinforce contradictions in expectations of fatherhood as an idea and in practice. In the first author's observed family holiday, my father's attempts to display care are materialised in his efforts to prepare breakfast for us children, only to have his efforts 'overridden' by my mother's 'takeover'. Thus, while the holiday opens up alternative possibilities to perform romanticised visions of 'the everyday', it remains constrained by 'an insidious patriarchal logic' (Aitken, 1998) which pervades social relations and routines of 'everyday life'.

Indeed, the holiday space is an avenue for all family members to build intra-family relationships. In deconstructing family time, Daly (2001, p. 290) contrasts the ideal of harmonious, positive experiences with the reality of tension and conflict in families' 'everyday life'. Undoubtedly, just like the obligations of carework, these do not disappear on holiday. Yet the meanings attributed to these experiences change when practised away from the banal setting of home – they become opportunities for rekindling miscommunication and bridging understanding, for the sake of nurturing a stronger family relationship. Quoting Tronto (1987, p. 658), 'the perspective of care requires that conflict be worked out without damage to the continuing relationships ... accommodating the needs of the self and of others ... balancing competition and cooperation'. The notion of the holiday as a unique opportunity window for building tolerance is displayed in the following excerpt from the Tan family:

> *Germaine*: 'We learn to adapt to each other, so when we travel together, we sort of grow together'.

> *Eleanor*: 'On a day-to-day basis we don't really spend a lot of time together obviously, so we really have to interact with each other, you can already understand each other better'.

Germaine: 'I think when you spend so much time together, you learn both the good and the bad. There will always be friction, so that friction also gives that opportunity to build tolerance'.

This resonates well with the first author's family holiday experiences, where I constantly feel apprehensive of a 'family meltdown' breaking out through disagreement and arguments, and no doubt when it happens, I am awash in the reality that the perfect family we portray through our photographs is but an illusion. Yet, in the heat of the moment, I am thankful for the open communication we embrace, as I attempt to mediate the argument and seek consensus. Looking back, these moments display how the tourism space becomes an important means of achieving compromise and building relationships, and how family holidays are intricately bound to the identities we co-construct as a family. While it is not to say that we will be less of a family if we did not have the chance to holiday together, it is important to highlight the integral role tourism provided in shaping the family we have today.

The idealised 'happy family' is therefore not a pre-existing entity but a work-in-progress; it requires effort to sustain its relations. The holiday's exceptional living arrangements of a family cramped under-one-roof in a shared space, is perhaps what compels this opportunity for cherished interaction, opposed to 'everyday life'. But more significantly, the holiday builds the family relationship because it is understood to be an ongoing cyclical event, rather than a one-off episode. Holidays are part and parcel of family life, and because of this, it makes little sense to talk about a single holiday trip, but rather, we theorise that holidays should be seen as a constant in the continuum of life experiences that accompany the family as it transits through the life stages. It matters less what is done and where we went to, but more who we went with.

Concluding remarks

One could say that family was put into cultural production, representing itself to itself in a series of daily, weekly, and annual performances that substituted for the working relationships that had previously constituted the everyday experience of family life. (Gillis, 1996a, p. 13)

Modern families' lives are different from that of a nostalgic past, yet it is the latter which fuels our hegemonic tenets of family ideology. Is the family holiday one of the ways the family 'represents itself to itself' then? This question begs a serious re-examination of the institution of the family in reality. The 'happy family' is not so clear cut – one might even question if the episodic event of the family holiday, where time is frozen and routines put on pause, is merely a façade over which instability lurks. Nevertheless, what we have displayed is how families use the holiday to negotiate the dilemma of the ideal family in the context of Singapore. More importantly, we have illustrated how family life is renewed, reconstructed, with every holiday, because it is the holiday which makes possible the notion of a complete home, and a 'happy family'. The family is constantly remade everyday, and this paper has illustrated how the holiday is an important 'home-making practice', because of the opportunity it presents for living banal family life that has become a rarity, and for nurturing authentic relationships through care practices. Perhaps, this is why families look forward to planning the next holiday, not to escape home, but instead, as the title of this paper implies, because they desire to 'come home' to the family.

This paper clears the path for the development of a 'human centred geography' (Eyles, 1989), wherein personal everyday experiences enrich academic perspectives, rather than exist apart from it. It has thus endowed tourism studies with a novel theoretical angle to deconstruct the banal family holiday, by interpreting care practices as a bridge between 'everyday life' and the social construction of family ideology which undergirds our aspirational understanding of 'the everyday'. As Lefebvre (1987, p. 9) articulates: 'why wouldn't the concept of everydayness reveal the extraordinary in the ordinary?' This paper illustrated the liminality of family tourism as at once exceptional as it is perceived as ordinary, existing in-between home and away, through its reproduction of the 'happy family'. It is a call to action for tourism scholars to re-examine seemingly mundane practices as unusual and exotic, and signals a much-needed departure from originally conceived tourist typologies.

The discussions in this paper tell us not only of the family holiday, but offer a glimpse into the powerful hold that family ideology has on everyday family life within the context of overly scheduled lives. Looking at family tourism often gives us an insight into life beyond the holiday: It is a mirror of everyday family life and reflects what we lack or desire. While our findings are based in Singapore, the phenomena of busy scheduled lives and outsourced emotional labour characterises the plight of the squeezed middle class in many industrialised nations. Within the Asian context, the traditional conservatism of familism still looms large in public consensus as a normative standard for many countries in the region, in direct conflict with the changing structure of the family. It thus commands a powerful rethink on both the work-life balance families seek to achieve outside of the holiday and the hegemony of family ideology as outdated and oppressive. The lens of family holidays thus serves as one possible springboard for further debates on the dilemmas of strained relationships and impossible expectations present within the contractions of modern family life. This thus shifts attention from seeing the discordant nature of family time as a personal trouble requiring a private solution, such as the holiday, to a pervasive systemic dilemma that challenges all families (Morgan, 1975).

Fundamental to this research is the assumption that family holidays are familiar components of family life. Even as we deconstruct family ideology, we acknowledge how we perpetuate a particular privileged vision of the heterosexual, dual-parent 'happy family' with the ability to travel together. This limitation necessitates a reflexive acknowledgement that family holidays are still luxury goods out of reach for the less privileged who fall below the ranks of the middle-class, and that even family tourism spaces are exclusionary to families who defy societally sanctioned norms of heterosexuality, obstructing their perceived accessibility to desired leisure practices that this research takes for granted. Thus, further studies on how non-traditional families negotiate such ideological family norms are warranted, filling an important gap in family literature. Lastly, while we have illustrated that care transcends the space of the home through familial bodies and emerges through the family holiday, that is not to narrowly limit this care ethic to the holiday. Future studies should look into how this care ethic surfaces in other aspects of mundane family life, as this will reveal insights into how family ideology is put into practice and negotiated in everyday life. Laconically, this paper is not merely about the family holiday, but instead, about the (re)making of the modern family through the family holiday.

Notes

1. It is important to note that throughout this paper, there is a seeming emphasis on parenting routines conducted during family holidays, rather than touristic activities by families and destination details. While the later remains important travel motivations, the emphasis on the former are a reflection of what our respondents shared when asked questions like why they travelled as a family. Our focus on parenting routines is thereby an endeavour to accurately represent their views in this paper.
2. Referring to the first author in the rest of this section.
3. Co-Curricular Activities. It is also common for Singaporean children to be enrolled in several enrichment classes in the weekends – these typically include music classes, sports, and dance.
4. Referring to the FDW.

Disclosure statement

No potential conflict of interest was reported by the authors.

ORCID

Yinn Shan Cheong ⓘ http://orcid.org/0000-0002-3665-0617
Harng Luh Sin ⓘ http://orcid.org/0000-0001-7850-993X

References

Abetz, J., & Moore, J. (2018). "Welcome to the mommy wars, ladies": Making sense of the ideology of combative mothering in mommy blogs. *Communication, Culture and Critique, 11*(2), 265–281. doi:10.1093/ccc/tcy008

Aitken, S. (1998). *Family fantasies and community space*. New Brunswick, NJ: Rutgers University Press.

Aitken, S. (2000). Fathering and faltering: "Sorry but you don't have the necessary accoutrements". *Environment and Planning A: Economy and Space, 32*(4), 581–598. doi:10.1068/a3236

Allen, S. M., & Daly, K. (2005). Fathers and the navigation of family space and time. In W. Marsiglio, K. Roy, & G. L. Fox (Eds.), *Situated fathering: A focus on physical and social spaces* (pp. 49–70). London: Routledge.

Backer, E., & Schänzel, H. (2015). Family holidays - Vacation or obli-cation? *Tourism Recreation Research, 38*(2), 159–173. doi:10.1080/02508281.2013.11081742

Baerenholdt, J. O., Haldrup, M., Larsen, J., & Urry, J. (2004). *Performing tourist places.* Aldershot: Ashgate.

Barr, M. (2000). Lee Kuan Yew and the "Asian values" debate. *Asian Studies Review, 24*(3), 309–334. doi:10.1080/10357820008713278

Barrett, M. (1980). *Women's oppression today: Problems in Marxist feminist analysis.* London: Verso.

Beioley, S. (2004). Meet the family – Family holidays in the UK. *January.* Retrieved from www.insights.org.uk/articleitem.aspx?title=Meet+the+Family+-+Family+Holidays+in+the+UK

Bernardes, J. (1985). 'Family ideology': Identification and exploration. *The Sociological Review, 33*(2), 275–297. doi:10.1111/j.1467-954X.1985.tb00806.x

Blunt, A., & Dowling, R. (2006). *Home: Key ideas in geography.* London: Routledge.

Bondi, L. (2009). Emotional knowing. In R. Kitchin & N. Thrift (Eds.), *International encyclopedia of human geography.* Cambridge: Elsevier.

Brotherson, S. E., Dollahite, D. C., & Hawkins, A. J. (2005). Generative fathering and the dynamics of connection between fathers and their children. *Fathering, 3*(1), 1–28. doi:10.3149/fth.0301.1

Carr, N. (2006). A comparison of adolescents and parents holiday motivations and desires. *Tourism and Hospitality Research, 6*(2), 129–142. doi:10.1057/palgrave.thr.6040051

Carr, N. (2011). *Children's and families' holiday experiences.* London: Routledge

Chen, C.-C., & Petrick, J. F. (2013). Health and wellness benefits of travel experiences: A literature review. *Journal of Travel Research, 52*(6), 709–719. doi:10.1177/0047287513496477

Chopra, R. (2001). Retrieving the father: Gender studies, "father love" and the discourse of mothering. *Women's Studies International Forum, 24*(3-4), 445–455. doi:10.1016/S0277-5395(01)00168-6

Cloke, P., Crang, P., & Goodwin, M. (1999). *Introducing human geographies.* London: Arnold.

Cloke, P., & Perkins, H. C. (1998). "Cracking the canyon with the awesome foursome": Presentations of adventure tourism in New Zealand. *Environment and Planning D: Society and Space, 16*(2), 185–218. doi:10.1068/d160185

Crouch, D. (2001). Spatialities and the feeling of doing. *Social & Cultural Geography, 2*(1), 61–75. doi:10.1080/14649360122124

Crow, G., & Allan, G. (1989). Chapter 1: Introduction. In G. Crow & G. Allen (Eds.), *Home and family: Creating the domestic sphere* (pp. 1–13). London: Palgrave Macmillan.

Daly, K. J. (1996). *Families and time: Keeping pace in a hurried culture.* Thousand Oaks, CA: Sage.

Daly, K. J. (2001). Deconstructing family time: From ideology to lived experience. *Journal of Marriage and Family, 63*(2), 283–294. doi:10.1111/j.1741-3737.2001.00283.x

Davidson, P. (1996). The holiday and work experiences of women with young children. *Leisure Studies, 15*(2), 89–103. doi:10.1080/026143696375648

De Grazia, S. (1962). *Of work, time and leisure.* New York, NY: The Twentieth Century Fund.

Dermott, E. (2008). Paradoxes of contemporary fatherhood. In E. Dermott (Ed.), *Intimate fatherhood: A sociological analysis* (pp. 7–24). London: Routledge.

Dermott, E., & Miller, T. (2015). More than the sum of its parts? Contemporary fatherhood policy, practice and discourse. *Families. Families, Relationships and Societies, 4*(2), 183–195. doi:10.1332/204674315X14212269138324

DeVault, M. L. (1996). Talking back to sociology: Distinctive contributions of feminist methodology. *Annual Review of Sociology, 56*, 2–29. doi:10.1146/annurev.soc.22.1.29

Edensor, T. (2007). Mundane mobilities, performance and spaces of tourism. *Social & Cultural Geography, 8*(2), 199–215. doi:10.1080/14649360701360089

Eichler, M. (1986). *The pro-family movement: Are they for or against families?* Ottawa, ON: Canadian Research Institute for the Advancement of Women.

Ellis, C., & Bochner, A. (1996). *Composing ethnography: Alternative forms of qualitative writing.* Walnut Creek, CA: AltaMira.

Eyles, J. (1989). The geography of everyday life. In D. Gregory & R. Walford (Eds.), *Horizons in human geography.* Basingstoke: Macmillan.

Finch, J., & Groves, D. (1983). *A labour of love: Women, work and caring.* London: Routledge.

Giddens, A. (1990). *The consequences of modernity.* Cambridge: Polity Press.

Gilligan, C. (1982). *In a different voice: Psychological theory and women's development.* London: Harvard University Press.

Gillis, J. (1996a). Making time for family: The invention of family time(s) and the reinvention of family history. *Journal of Family History, 21*(1), 4–21. doi:10.1177/036319909602100102

Gillis, J. (1996b). *A world of their own making: Myth, ritual and the quest for family values.* New York, NY: Basic Books

Haldrup, M., & Larsen, J. (2003). The family gaze. *Tourist Studies, 3*(1), 23–45. doi:10.1177/1468797603040529

Handel, G. (1997). Family worlds and qualitative family research. *Marriage & Family Review, 24*(3-4), 335–348. doi:10.1300/J002v24n03_06

Hays, S. (1996). *The cultural contradictions of motherhood.* New Haven, CT: Yale University Press

HDB. (2019). Living With/Near Parents or Child. Singapore. Retrieved from https://www.hdb.gov.sg/cs/infoweb/residential/buying-a-flat/resale/living-with-near-parents-or-married-child

Henderson, K. A., & Dialeschki, M. D. (1991). A sense of entitlement to leisure as constraint and empowerment for women. *Leisure Sciences, 13*(1), 51–65.

Hess, R. D., & Handel, G. (1959). *Family worlds.* Chicago: The Unversity of Chicago Press.

Highmore, B. (2002). *Everyday life and cultural theory: An introduction.* London: Routledge

Hochschild, A. (1989). *The second shift: Working parents and the revolution at home.* New York: Berkeley Viking Penguin.

Hochschild, A. (1997). *The time bind: When work becomes home and home becomes work.* Metropolitan Books.

Hochschild, A. (2012). *The outsourced self: Intimate life in market times.* New York, NY: Routledge.

Horton, J., & Kraft, P. (2014). Everyday geographies. In J. Horton & P. Kraft (Eds.), *Cultural geographies: An introduction* (pp.181–199). London: Routledge.

Huang, S., & Yeoh, B. (1998). Negotiating public space: Strategies and styles of migrant female domestic workers in Singapore. *Urban Studies, 35*(3), 583–602. doi:10.1080/0042098984925

Katz, C., Marston, S., & Mitchell, K. (2004). *Life's work: Geographies of social reproduction.* Hoboken, NJ: Wiley-Blackwell.

Kennedy-Eden, H., & Gretzel, U. (2016). Modern vacations – modern families: New meanings and structures of family vacations. *Annals of Leisure Research, 19*(4), 461–478. doi:10.1080/11745398.2016.1178152

Khoo-Lattimore, C., Prayag, G., & Cheah, B. L. (2015). Kids on board: Exploring the choice process and vacation needs of Asian parents with young children in resort hotels. *Journal of Hospitality Marketing & Management, 24*(5), 511–531. doi:10.1080/19368623.2014.914862

Kim, S. S., Choi, S., Argusa, J., Wang, K. C., & Kim, Y. (2010). The role of family decision makers in festival tourism. *International Journal of Hospitality Management, 2*, 308–318. doi:10.1016/j.ijhm.2009.10.004

Lam, T., Yeoh, B. S. A., & Huang, S. (2006). Global householding' in a city-state: Emerging trends from Singapore. *International Development Planning Review, 28*(4), 475–497. doi:10.3828/idpr.28.4.3

Larossa, R., Bennett, L. A., & Gelles, R. J. (1994). Ethical dilemmas in qualitative family research. In G. Handel & G. G. Whitchurch (Eds.), *The psychosocial interior of the family* (4th ed., pp. 109–126). New York, NY: Aldine de Gruyter.

Larsen, J. (2008). De-exoticizing tourist travel: Everyday life and sociality on the move. *Leisure Studies, 27*(1), 21–34. doi:10.1080/02614360701198030

Latham, A. (2003). Research, performance, and doing human geography: Some reflections on the diary-photograph, diary-interview method. *Environment and Planning A: Economy and Space, 35*(11), 1993–2017. doi:10.1068/a3587

Lefebvre, H. (1987). The everyday and everydayness. *Yale French Studies, 73*, 7–11. doi:10.2307/2930193

Lefebvre, H. (1988). Toward a Leftist cultural politics: Remarks occasioned by the centenary of Marx's death. In C. Nelson & L. Grossberg (Eds.), *Marxism and the interpretation of culture* (pp. 75–88). Urbana: University of Illinois Press

Lefebvre, H. (1991). *Critique of everyday life.* London: Verso.

Martin, L. G. (1990). Changing intergenerational family relations in East Asia. *The Annals of the American Academy of Political and Social Science, 510*(1), 102–144. doi:10.1177/0002716290510001008

Miller, J. B. (1976). *Towards a new psychology of women.* Boston: Beacon Press.

MOM. (2019). Levy concession for a foreign domestic worker. Singapore. Retrieved from https://www.mom.gov.sg/passes-and-permits/work-permit-for-foreign-domestic-worker/foreign-domestic-worker-levy/levy-concession

Morgan, D. (1975). *Social theory and the family.* London: Routledge & Kegan Paul.

Morgan, D. (1988). *Focus groups as qualitative research.* Newbury Park, CA: Sage.

Mottiar, Z., & Quinn, D. (2012). Is a self-catering holiday with the family really a holiday for mothers? Examining the balance of household responsibilities while on holiday from a female perspective. *Hospitality & Society, 2*(2), 197–214. doi:10.1386/hosp.2.2.197_1

Newman, D. (2009). *Families: A sociological perspective.* New York, NY: McGraw-Hill.

Newman, H., & Henderson, A. (2014). The modern mystique: Institutional mediation of hegemonic motherhood. *Sociological Inquiry, 84*(3), 472–491. doi:10.1111/soin.12037

Obrador, P. (2012). The place of the family in tourism research: Domesticity and thick sociality by the pool. *Annals of Tourism Research, 39*(1), 401–420. doi:10.1016/j.annals.2011.07.006

Ong, C. E. (2017). 'Cuteifying' spaces and staging marine animals for Chinese middle class consumption. *Tourism Geographies, 19*(2), 188–207. doi:10.1080/14616688.2016.1196237

Oswin, N. (2010). The modern model family at home in Singapore: A queer geography. *Transactions of the Institute of British Geographers, 35*(2), 256–268. doi:10.1111/j.1475-5661.2009.00379.x

Pritchard, A., Morgan, N., Ateljevic, I., & Harris, C. (2007). Editor's introduction: Tourism, gender, embodiment and experience. In A. Pritchard, N. Morgan, I. Ateljevic, & C. Harris (Eds.), *Tourism and gender: Embodiment, sensuality and experience* (pp. 1–12). Walliford: CABI.

Ruddick, S. (1995). *Maternal thinking: Towards a politics of peace.* Boston, MA: Beacon Press.

Ryans, C. (2003). *A new wave-or beached fathers! Gender issues in academic tourism literature-where is the dad?* Paper Presented at the CAUTHE Conference Proceedings Coffs Harbour, Australia.

Rybczynski, W. (1991). *Waiting for the weekend.* New York, NY: Viking.

Schänzel, H. A. (2010). Whole family research: Towards a methodology in tourism for emcompassing generation, gender, and group dynamic perspectives. *Tourism Analysis, 15*(5), 555–569. doi:10.3727/108354210X12889831783314

Schänzel, H. A., & Jenkins, J. (2017). Non-resident fathers' holidays alone with their children: Experiences, meanings and fatherhood. *World Leisure Journal, 59*(2), 156–173. doi:10.1080/16078055.2016.1216887

Schänzel, H. A., & Smith, K. A. (2011). The absence of fatherhood: Achieving true gender scholarship in family tourism research. *Annals of Leisure Research, 14*(2-3), 143–154. doi:10.1080/11745398.2011.615712

Schänzel, H. A., Smith, K. A., & Weaver, A. (2005). Family holidays: A research review and application to New Zealand. *Annals of Leisure Research, 8*(2-3), 105–123. doi:10.1080/11745398.2005.10600965

Schänzel, H. A., & Yeoman, I. (2015). Trends in family tourism. *Journal of Tourism Futures, 1*(2), 141–147. doi:10.1108/JTF-12-2014-0006

Seigworth, G. J. (2000). Banality for cultural studies. *Cultural Studies, 14*(2), 227–268. doi:10.1080/095023800334878

Shaw, S. (1992). Dereifying family leisure: An examiniation of women's and men's everyday experiences and perception of family time. *Leisure Sciences, 14*(4), 271–286. doi:10.1080/01490409209513174

Shaw, S. M., Havitz, M. E., & Delemere, F. M. (2008). I decided to invest in my kids' memories: Family vacations, memories, and the social construction of the family. *Tourism Culture & Communication, 8*(1), 139–154. doi:10.3727/109830408783900361

SingStat. (2015). General household survey 2015 - Key Findings. Singapore. *SingStat.* Retrieved from https://www.singstat.gov.sg/docs/default-source/default-document-library/publications/publications_and_papers/GHS/ghs2015/findings.pdf

Skelton, T. (2017). Everyday geographies. In D. Richardson, N. Castree, M. Goodchild, A. Kobayashi, W. D. Liu, & R. Marston (Eds.), *The international encyclopedia of geography* (pp. 1–3). Oxford: Wiley-Blackwell.

Starkweather, S. (2011). Telling family stories. *Royal Geographic Society, 44*(3), 281–295.

Stephens, J. (2015). Reconfiguring care and family in the era of the 'outsourced self'. *Journal of Family Studies, 21*(3), 208–217. doi:10.1080/13229400.2015.1058847

Teo, Y. (2010). Shaping the Singapore family, producing the state and society. *Economy and Society, 39*(3), 337–359. doi:10.1080/03085147.2010.486215

Thrift, N. (2000). Dead or alive? In I. Cook, D. Crough, S. Naylor, & J. Ryan (Eds.), *Cultural turns/geographical turns: Perspectives on cultural geography* (pp. 1–6). Essex: Prentice-Hall.

Tronto, J. (1987). Beyond gender difference to a theory of care. *Signs: Journal of Women in Culture and Society, 12*(4), 644–663. doi:10.1086/494360

Urry, J. (1990). *The tourist gaze.* London: Sage.

Varley, A., & Blasco, M. (2001). Exiled to the home: Masculinity and aging in urban Mexico. In C. Jackson (Ed.), *Men at work: Labour, masculinities, development* (pp.115–138). Portland, OR: Frank Cass.

Wang, K. C., Hsieh, A. T., Yeh, Y. C., & Tsai, C. W. (2004). Who is the decision- maker: The parents or the child in group package tours? *Tourism Management, 25*(2), 183–194. doi:10.1016/S0261-5177(03)00093-1

Wang, N. (1999). Rethinking authenticity in tourism experience. *Annals of Tourism Research, 26*(2), 349–370. doi:10.1016/S0160-7383(98)00103-0

Yeoh, B. S. A., & Huang, S. (2010). Transnational domestic workers and the negotiation of mobility and work practices in Singapore's home-spaces. *Mobilities, 5*(2), 219–236. doi:10.1080/17450101003665036

Linkages between tourist resorts, local food production and the sustainable development goals

Regina Scheyvens (iD) and Gabriel Laeis

ABSTRACT

The United Nations' Sustainable Development Goals (SDGs), ratified in 2015, are set to guide global development through to 2030. As an industry, tourism receives considerable attention in development discussions and in planning for development in the Global South, particularly in small island developing states (SIDS). A number of authors see particular prospects in the area of food producer linkages with the tourism industry, based on the notion that it should be possible to enhance both local food systems and tourism industries in SIDS by putting more local food on the menu. In this article, we test the assumption that building strong linkages between tourism and food production systems, especially agriculture and fisheries, is possible, and desirable, and would lead to more sustainable development. We do this by drawing on data from research in hotel kitchens and on farms in the South Pacific country of Fiji. When reflecting on the Fijian data in light of SDGs 2, 12 and 14, it is apparent that there are considerable constraints to developing food-tourism linkages in a way that will deliver more sustainable development in future. While 'island night' menus do use a lot of local produce, and small numbers of guests enjoy consuming Fijian foods at fine dining restaurants, incorporating local produce into mainstream resort menus had been difficult for chefs. Overall, the way large-scale multinational resorts are structured in terms of their food requirements does not make them ideal partners for food producers in developing countries, particularly SIDS.

摘要

2015年联合国批准的可持续发展目标(SDGs)将指导全球发展到2030年。旅游业作为一个产业, 在全球南方, 特别是在小岛屿发展中国家的发展讨论和发展规划中受到相当的重视。一些作者认为, 在食品生产者与旅游业的联系方面有特别好的前景, 因为他们认为, 应该能够通过将更多的地方食品列入菜单, 加强小岛屿发展中国家的地方食品系统和旅游业的联系。在本文中, 我们检验了以下假设, 即在旅游业和食品生产系统之间, 特别是农业和渔业之间建立强有力的联系是可能的, 也是可取的, 并将导致更可持续的发展。我们的研究数据来自南太平洋国家斐济的酒店厨房和农场。当根据可持续发展目标2、12和14对斐济的数据进行反思时, 很明显, 在以一种将在未来实现更可持续发展的方式发展食品与旅游业的联系方面存在着相当大的限制。虽然"岛屿之夜"的菜单确实使用了大量当地农产品, 而且少数客人喜欢在高级餐厅享用斐济食品, 但将当地农产品纳入主流度假村菜单对厨师来说一直很困难。总的来说，大型跨国度假村根据其食品需求进行组织的方式并不使它们成为发展中国家，特别是小岛屿发展中国家食品生产者的理想伙伴。

Introduction

The United Nations' Sustainable Development Goals (SDGs) are potentially transformative. In this article, we consider whether tourism businesses can deliver on the SDGs particularly with regard to food-related goals. As an industry, tourism receives considerable attention in planning for development in the Global South (Telfer & Sharpley, 2008). This is especially the case in small island developing states (SIDS) as they have sustained growth in tourist arrivals above that of all other tourist destinations (Nowak & Sahli, 2007, pp. 49–50). Many SIDS are heavily dependent on tourism as a major source of foreign exchange earnings; in fact, the top 10 countries in terms of having a high share of their GDP from tourism are all SIDS (Schubert, Brida, & Risso, 2011). Simply put, tourism-dependent small islands 'are good economic performers' (Scheyvens & Momsen, 2008, p. 498). Interestingly, tourism is also an industry singled out for attention in the targets of several of the SDGs (UNWTO, 2015), almost as if there is an implicit assumption that this industry is more socially and environmentally benign than many others, while still offering good potential to deliver economic development.

A number of authors see particular prospects in the tourism industry's linkages with food producers, based on the notion that it should be possible to enhance both the agriculture and tourism industries in SIDS by putting more local food on the menu (Berno, 2015; Rogerson, 2012; Timms & Neill, 2011; Torres & Momsen, 2004). In the case of the South Pacific, many countries are seen as fertile, idyllic environments abundant in tropical fruit and vegetables (Manner & Thaman, 2013), thus a closer marriage between tourism and agriculture would seem like a natural fit (Berno, 2011, 2015). To add to this, they are based in the world's largest ocean which can supply a variety of seafood options.

Translating this potential into stronger linkages, and associated benefits, is not an easy process. To start with, the agriculture and fisheries sectors are facing their own challenges in SIDS. In the Pacific, traditional small-holder agriculture and fisheries is still pivotal to the food security, income and livelihoods of more than 8 million people (Campbell, 2009; Morgan, 2013). However, food production per capita has decreased across the region (Campbell, 2015; Connell, 2015) due to the promotion of export-driven agriculture, urbanization, and a decline in agrobiodiversity and ecosystem functions, along with a growing preference for imported foods and Westernized diets (Thaman, 2008; Thow et al., 2010). Concurrently there has been a financing and investment crisis in agriculture, leading to lack of extension services and technical support for farmers as well as lack of improved tools and plants, and difficulty in accessing credit (Sisifa, Taylor, McGregor, Fink, & Dawson, 2016). A similar issue affects fishing, yet donor funding to both sectors has declined. Research by Blasiak and Wabnitz (2018) has revealed that between 2010 and 2015, aid agencies reduced their allocations to the fisheries sector by over 30%. Moreover, younger generations tend to not see agriculture (or fisheries) as an interesting livelihood option: there has been a decline in the prestige of such work (Connell, 2015). Meanwhile, despite SIDS having some of the largest Exclusive Economic Zones for fishing per land area around the globe, the livelihoods of artisanal fishers are being compromised due to the impacts of industrial fishing (James, Tidd, & Kaitu, 2018), and climate change-induced reductions in fish stocks (Le Cornu, Doerr, Finkbeiner, Gourlie, & Crowder, 2018).

Economic leakages from the tourism industry—money leaving the economy for some form of imported resource—are reported to be among the highest in SIDS, reaching up to 80% (Anderson, 2013, p. 72). A recent study from Fiji on fresh produce found that tourist hotels and resorts purchase 52% of imported fruits and vegetables (International Finance Corporation, 2018, p. 3). This, however, does not include other foods, such as meat, dairy products or dry goods.

Given the above challenges, this research aims to test the assumption that building strong linkages between tourism and food production systems, especially agriculture and fisheries, is possible, and desirable, and would lead to more sustainable development in SIDS. We do this by drawing on data from research in hotel kitchens and on farms in Fiji (Figure 1), asking the question: are tourist resorts in SIDS likely to contribute to achieving food-related SDGs through their food purchasing and menu choices? The 'food-related SDGs' we consider are as follows: SDGs 2 (End hunger, achieve food security and improved nutrition and promote sustainable agriculture), 12 (Ensure sustainable consumption and production patterns) and 14 (Conserve and sustainably use the oceans, seas and marine resources for sustainable development). The ensuing discussion identifies areas of promise as well as revealing that there are considerable

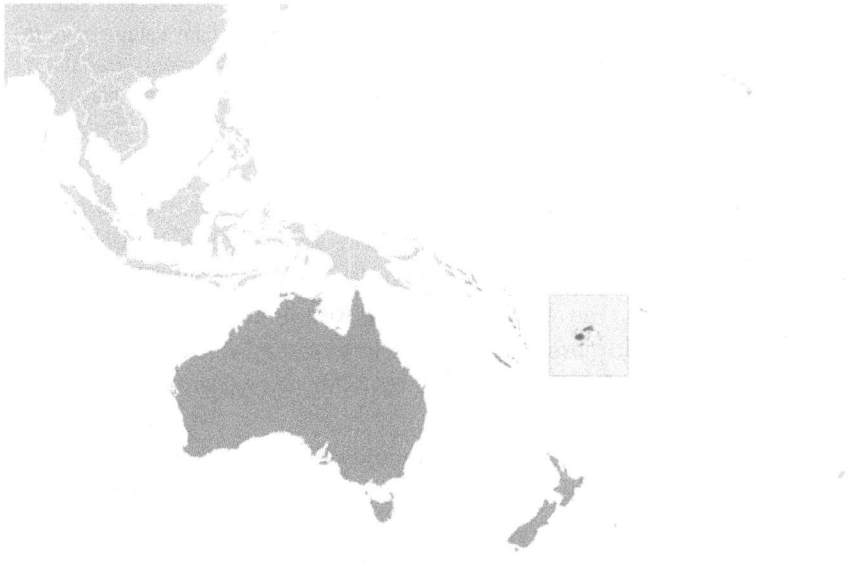

Figure 1. Location of Fiji in the South Pacific.
Source: CIA (2019).

constraints to tourism businesses being able to deliver more sustainable development in the future.

SDGs 2, 12 and 14: linking food systems and tourism

As noted, we have taken a purposive approach to identifying specific SDGs and targets that provide particular insights regarding food and tourism linkages. In doing so we carefully examined all of the SDGs and their targets then focused on those in which food production and consumption was explicitly mentioned. We also took guidance from Cañada et al. (2017), which provide the only independent analysis to date of how each of the SDGs relates to tourism. In Table 1, we outline questions that could be asked of tourism businesses in relation to the most relevant targets within the SDGs to prompt them to achieve these SDGs through actions to support food producers and the environments they rely upon. While it is beyond the scope of this article to answer all of these questions, we hope they will provide guidance for other researchers, as well as for businesses seeking to adopt more sustainable food practices.

SDG 2 end hunger, achieve food security and improved nutrition and promote sustainable agriculture

From a development perspective, SDG 2 is a particularly important goal as promoting sustainable agriculture has the potential to contribute to ending hunger and enhancing food security within SIDS as well as increasing the earning potential of farmers. This is recognized by the UNWTO:

> Tourism can spur agricultural productivity by promoting the production, use and sale of local produce in tourist destinations and its full integration in the tourism value chain. In addition, agro-tourism, a growing tourism segment, can complement traditional agricultural activities. The resulting rise of income in local communities can lead to a more resilient agriculture while enhancing the value of the tourism experience (UNWTO, 2015).

Moreover, agricultural systems directly impact if and how we 'protect, restore and promote sustainable use of terrestrial ecosystems', as required by SDG15 (UN, 2016). It is estimated that about 11% of the earth's land surface is under agricultural production (Bruinsma, 2003, p. 127). The way food is produced influences the rates of soil degradation, deforestation and desertification, the loss of biodiversity and the health of biospheres, to name but a few factors (FAO, 2017; McIntyre, Herren, Wakhungu, & Watson, 2009).

We suggest in Table 1 that more could be done by hotels and restaurants to support SDG 2 by examining how they source food, and prioritizing local suppliers. They could also develop strong relationships with agencies which support fresh produce production, to enhance local purchasing of food, and aim to source produce grown in a sustainable agro-ecological manner. Another question we raise in Table 1 concerns whether hotels and restaurants pay employees a living wage so they can feed themselves well, and provide nutritious meals to employees.

Table 1. Linking food-related sustainable development goals and selected targets with questions that could be asked of tourism businesses.

Sustainable development goal	Selected targets that relate closely to tourism-food system linkages	Questions that could be asked of tourism businesses
2. End hunger, achieve food security and improved nutrition and promote sustainable agriculture	2.3 By 2030, double the agricultural productivity and incomes of small scale food producers, in particular women, Indigenous peoples, family farmers, pastoralists and fishers, including through secure and equal access to land and other productive resources and inputs. Knowledge, financial services, markets and opportunities for value addition and non-farm employment. 2.a: Increase investment, including through enhanced international cooperation, in rural infrastructure, agricultural research and extension services, technology development and plan and livestock gene banks in order to enhance agricultural productive capacity in developing countries, in particular least developed countries	• How are hotels, cruise ships and restaurants sourcing food? Are they purchasing from local suppliers wherever possible, especially small-scale food producers? • Do hotels and restaurants partner with farmers, government or development agencies to enhance agricultural production for the tourism sector? • Are they sourcing produce grown in a sustainable manner?
12. Ensure sustainable consumption and production patterns	12.3 By 2030 halve per capita global food waste at the retail and consumer levels and reduce food losses along production and supply chains, including post-harvest losses. 12.6 Encourage companies, especially large and transnational companies, to adopt sustainable practices and to integrate sustainability information into their reporting cycle. 12.b Develop and implement tools to monitor sustainable development impacts for sustainable tourism that creates jobs and promotes local culture and products	• Are restaurants minimizing food waste? • Are hotels maximizing use of local products (including promoting local cuisine) where possible, rather than prioritizing high-end imported brands? • Are hotels raising awareness of guests about consuming sustainable products, including 'eating local and seasonal'? • Are hotels monitoring and reporting systematically on their sustainability practices?
14. Conserve and sustainably use the oceans, seas and marine resources for sustainable development	14.1 By 2025, prevent and significantly reduce marine pollution of all kinds, in particular from land-based activities, including marine debris and nutrient pollution 14.2 By 2020, sustainably manage and protect marine and coastal ecosystems to avoid significant adverse impacts, including by strengthening their resilience, and take action for their restoration in order to achieve healthy and productive oceans 14.7 By 2030, increase the economic benefits to small island developing States and least developed countries from the sustainable use of marine resources including through sustainable management of fisheries, aquaculture and tourism 14.b Provide access for small-scale artisanal fishers to marine resources and markets	• Are hotels and marine vessels effectively treating and directing their wastes so they do not pollute marine environments and contaminate seafood? • Are hotels protecting coastal environments and repairing damage caused by past construction? • Are hotels educating guests about protecting the environment (including coral reefs)? • Are hotels supporting the livelihood activities of neighboring communities dependent on marine environments for food? • Are hotels supporting local small-scale fishers by purchasing seafood from them?

Source: Authors (SDGs and Targets based on various sections of UN (2016).

SDG 12 ensure sustainable consumption and production patterns

From a tourism industry perspective, food is a fundamental requirement in order to meet the needs of guests: it is not an 'add on' like many forms of corporate social responsibility by hotels, such as donations to schools or community projects. Attention to procurement strategies could therefore make a significant contribution to SDG 2—as well as SDG 15 to a certain degree—if locally produced, sustainably grown fruit, vegetables and seafood were priorities (Scheyvens & Hughes, 2015). Shortening the food supply chain would also reduce use of energy for transport and refrigeration, save food miles (Gössling, Garrod, Aall, Hille, & Peeters, 2011) (contributing to SDG 13 on combatting climate change) as well as enhancing local development prospects (SDG 1 on eliminating poverty).

Relevant questions raised in Table 1 include whether restaurants of hotels, cruise ships and the like are maximizing use of local products and encouraging their guests to use local products, including 'eating local'. Another question challenges restaurants to minimize their food wastes. In addition, we believe it is essential that hotels are asked whether they are systematically monitoring and reporting on their sustainability practices. Note that Ling (2016, p. 97) is concerned at the weak working of the associated target which uses language which 'encourages' sustainable practices and monitoring, rather than requiring this.

SDG 14 conserve and sustainably use the oceans, seas and marine resources for sustainable development

Tourism relies heavily upon, and has a great impact on, coastal areas and marine environments around the world. This is pertinent because concerns about coastal tourism are particularly high in the case in SIDS which derives the majority of their foreign exchange earnings from tourism (Pratt, 2015). In the Cook Islands, where there were 161,362 visitors in 2017 to a resident population of only 11,700 (South Pacific Tourism Organisation, 2018), the impacts on the natural environment are clear. For example, a 2015 news report claimed that 'The golden egg is cracked', referring to sewage from tourist establishments and other sources polluting Rarotonga's Muri lagoon, a major tourist drawcard (Newsnow, 2015, December 30) .

In addition to waste disposal and pollution from tourism in SIDS, the impacts of tourism-related construction on marine environments are a critical issue that demands attention. In the past, ecosystems have been destroyed when mangroves have been cleared and large construction projects have taken place to create idyllic resorts with pristine white, sandy beaches that tourists desire (Domroes, 2001, p. 127). In the light of a surge of cruise tourism around the globe, environmental impacts of vessels taking tourists through the Caribbean Sea, Indian Ocean and Pacific Ocean also need to be considered.

The UNWTO endorses this concern regarding SDG 14, noting that:

> Coastal and maritime tourism, tourism's biggest segments, particularly for Small Island Developing States' (SIDS), rely on healthy marine ecosystems. Tourism development must be a part of integrated Coastal Zone Management in order to help conserve and preserve fragile ecosystems and serve as a vehicle to promote the blue economy ... (UNWTO, 2015).

SDG 14 is also connected to SDG 15 in so far, as that agricultural land use is a determining factor in coral reef health (Hoffmann, 2002). We suggest in Table 1 that hotels should not only protect coastal environments, but also repair past damage (e.g. from removal of mangroves, or construction which compromised ecosystems). We also think it essential that hotels and cruise ships have effective systems for minimizing and treating wastes, and ensure that they are not contaminating seafood supplies that are important for local subsistence or which offer a local livelihood opportunity. Another question explores whether hotels are willing to support local small-scale fishers.

Food, agriculture, fisheries and tourism

Local communities, food producers, tourists, and society at large support the idea of close linkages between local food producers and tourism enterprises in developing countries (Department for International Development, 1999). Some studies note how tourism-agriculture linkages can contribute to rural development (e.g., Asiedu & Gbedema, 2011; Butler & Rogerson, 2016; Torres & Momsen, 2004). From a community and producer perspective it has been asserted that increasing the demand for local food produce stimulates the market for farmers and other food producers (Pretty, 2001; Pretty, Ball, Lang, & Morison, 2005). This argument builds on the fact that 70% of the poor residing in rural areas are dependent on agriculture (World Bank, 2016). A substantial body of research points towards the benefit of supporting small-scale farmers in developing countries to promote more equitable, empowered, food-secure and resilient communities (International Federation of Organic Agriculture Movements, 2011; McIntyre et al., 2009). The menus of restaurants can in turn showcase and promote local products and producers (Herzog & Murray, 2013).

From the tourist's perspective local food provides more than just sustenance: it is seen as a significant part of the tourism experience, presenting an opportunity to consume a supposedly authentic aspect of the host's culture (Hall & Sharples, 2003). Figures on tourist spending patterns emphasise the significance of food in tourism in general. Tourists spend between 26 and 28% of their travel budget on food and beverages, and dining in restaurants is one of the most frequently seen activities (Hall & Sharples, 2003, pp. 3–5). In Fiji, it is estimated that tourists spend around 20% of their travel funds on food (Sustainable Tourism Development Consortium, 2007).

Interest in local food mostly stems out of a growing concern of Western consumers about the sustainability of their food choices. Local food consumption in international tourism, however, is a question of the neophilia of consumers (Fischler, 1988). Some tourists are more adventurous in their efforts to discover unknown foods (neophilic) in their holiday destinations, whereas others prefer to eat what they are already familiar with (neophobic) (Sengel et al., 2015). The promotions of local cuisines (Okumus, Kock, Scantlebury, & Okumus, 2013) and agritourism (Phillip, Hunter, & Blackstock, 2010) have tried to create tourism products that foster the use of local food, deepen the tourists' understanding of their destinations' food environments and help local farmers to partake in the tourism industry. However, other studies suggest that only certain consumers are interested in local food. Many tourists ask for what they are familiar with, which might not support demand for foreign local foods. Kim and Eves (2012) found that it took a heightened interest in cultural experience, the willingness to escape routines as well as a desire to learn about and build personal relationships with host communities in order for tourists to seek out local food.

Unsurprisingly, the suggested synergies between tourists, tourism industry and local food producers rarely materialize (Timms & Neill, 2011; Torres & Momsen, 2011). Researchers have found a plethora of factors that may inhibit such linkages. Issues have so far been grouped into four categories: 'supply/production', 'demand', 'marketing/intermediary' and 'role of national government'.

Supply- and production-related constraints mainly focus on the lacking quality, quantity and consistency of supply from food producers. This is a result of farmers in developing countries often having little access to capital, technology and training (Rogerson, 2012). Poor economies of scale, lack of rural infrastructure and conflicting terms of payment between producers and buyers (e.g. resorts and restaurants) aggravate the situation (Rogerson, 2012; Torres & Momsen, 2004). On the other hand, demand is often seasonal, and it is very much focussed on what tourists like to eat (Torres & Momsen, 2004). Tourist preferences can be at odds with local production systems and the type of foods a particular environment supports. Furthermore, hotels under international management often prefer to import food, because they find it more reliable in terms of quantity and quality (Torres & Momsen, 2004).

Issues in marketing mostly pertain to questionable practices of food intermediaries (Torres & Momsen, 2004). This can range from exploitation of farmers to the establishment of monopolies and even corruption. Less severe, but still challenging, can be the simple absence of local intermediaries (Rogerson, 2012), a sense of mistrust between farmers, suppliers and hotels, as well as inadequate transportation, processing and storage facilities.

Lastly, governments tend to not actively plan tourism-agricultural linkages and simply assume that they will come into existence (Pillay & Rogerson, 2013; Torres & Momsen, 2004). Therefore, such linkages are rarely supported, for instance by training for chefs and local farmers to foster mutual understanding. Governments may also have a preference for export-oriented agriculture to increase their foreign exchange earnings (Rogerson, 2012). Added to this are concerns around the general under-financing and lack of investment in agriculture in SIDS in recent decades, as noted in the introduction to this article (Sisifa et al., 2016).

Small island states have not been a focus of most studies to date, with the exception of some interesting research on Caribbean island states (Rhiney, 2011; Richardson-Ngwenya & Momsen, 2011; Timms & Neill, 2011) and the South Pacific (Berno, 2011, 2015; Scheyvens & Russell, 2012; Veit, 2007). This research overall confirmed the constraints mentioned in the four groups described above. Some unique issues highlighted for small island states were, firstly, their limited natural resource base. Small islands may be attractive for tourists, but are often not able to support freshwater- and land-intensive agriculture (Berno, 2011), yet larger islands, such as Samoa and Fiji, certainly are (Berno, 2015; Veit, 2007). Secondly, Scheyvens and Russell (2012) noted the importance of both scale and land ownership in Fiji as factors in where food is sourced from: where 'small-scale tourism businesses are located on customary land, the procurement of local goods is usually quite high, if available' (p. 423).

While there is a lot we can learn from these existing studies, literature on tourism-agriculture linkages is limited somewhat by its economistic perspective. If we are interested in the prospects of tourism-agricultural linkages contributing to more sustainable forms of development, a broader focus is needed. We need more studies that adopt an environmental perspective. One rationale for utilising local food is that this reduces the greenhouse gas emissions attributed to the consumption of food transported over long distances (Gössling et al., 2011) and helps to fight climate change. Secondly, there should be consideration of socio-cultural circumstances (Østrup Backe, 2013; Sims, 2009; Timothy & Ron, 2013). For example, Berno's work (2015) draws attention to the poor perception of Samoan cuisine by international visitors and the lack of confidence of chefs to prepare their local food for tourists: such factors directly impede tourism-agricultural linkages. Similarly, Oliver, Berno, and Ram (2010) present anecdotal evidence that through a history of colonialisation chefs in SIDS became dismissive of their own culinary heritage and favoured Western cuisine. Therefore, their traditional island cuisine has not been developed into contemporary versions that connect local agriculture and current consumer desires.

There appear to be limited studies which examine the possibilities for tourism to be planned in conjunction with supporting fishing livelihoods, for example through offering culinary tourism, apart from those focusing on wealthier countries (see, for example, G. Kim, Duffy, Jodice, & Norman, 2017). In fact, a number of studies in the Global South focus instead on the tensions between these two industries, citing competing claims of fishers and those developing infrastructure for tourism (see, for example, Hugé, Van Puyvelde, Munga, Dahdouh-Guebas, & Koedam, 2018). Others decry the negative impacts of tourism on livelihoods of those relying on reef fishing (Eastwood, Clary, & Melnick, 2017; Garcia Rodrigues & Villasante, 2016). Even the establishment of marine protected areas, which can theoretically be set up in ways that serve the interests of both industries, can impede the livelihoods of fishers (Said, MacMillan, Schembri, & Tzanopoulos, 2017).

There is competing evidence over whether selling fish to tourist enterprises supports local livelihoods. For example, a study across 25 Pacific island countries and territories identified that there has been a decline in the small-scale commercial fisheries catch; nevertheless it was suggested that even small catches are economically significant to local fishers if they can sell to tourist enterprises (Zeller, Harper, Zylich, & Pauly, 2015). Scheyvens and Russell (2012), for instance, observed that small-scale hospitality enterprises on customary land in Fiji purchased fish valued at around US$200 per week locally, supporting the livelihoods of several fishing people. In contrast, Garcia Rodrigues and Villasante (2016) found that the presence of a tourism industry is likely to drive up the price for fish and potentially make it unaffordable for locals. Moreover, restaurateurs profited much more from fish and seafood sales than did traders or fishers.

Overall, various economic factors, such as demand and supply, marketing processes, agro-ecological circumstances and governmental policies possess key roles in negotiating linkages between food producers and tourism businesses. A critical appreciation of the farming and fishing development context is vital to understanding the underlying issues of ailing food producer-tourism linkages (Torres, 2003). As works by Berno (2015) and Oliver et al. (2010) suggest, there may also be cultural factors at play that so far have not received much scholarly attention. To follow, the methodological approach to this study is detailed, focusing around collection of primary data in Fiji.

Methodology

In this study, we draw from the first author's experiences working on a research project on 'Sharing the riches of tourism' in Fiji (2009–2010), and a later project examining the role of resorts in promoting community development in Fiji (2013–2015). Primarily, however, we rely on four months of fieldwork in 2017 by the second author in Fiji, which set out to examine how the restaurant menus of large-scale upmarket tourist resorts in Fiji affect the ability of small-scale farmers to participate in the tourism economy. Methods included participant observation, semi-structured interviews and review of documents. A case study approach was taken, employing ethnographic methods in an extended fieldtrip to Fiji, April to July 2017. The second author thus spent one month doing participant observation as a cook in the kitchen of a large-scale, internationally managed luxury resort. Participant observation allowed for deep immersion into social and physical aspects of food networks, helping to unravel the agency of all actors. The case study resort was purposefully sampled for being: large-scale (>100 rooms), high-class (5 stars), internationally branded and managed, within a major tourism area of Fiji (Coral Coast on Viti Levu) and close to a food producing area (Sigatoka river valley, known as the 'salad bowl of Fiji'). Having had professional exposure in the hospitality industry previously, the second author was able to become to a large extent an 'insider' (O'Reilly, 2012). This enabled him to create rapport with about 70 members of the kitchen team, understand their worldviews and avoid Hawthorne effects (Arendt et al., 2012) in observations, casual conversations and interviews. This author also did document analysis of menu cards of four of the case study resort's restaurants and compared them to what was actually sold on a daily basis. Guests' perspectives on their food experience were integrated by reviewing about 300 guest feedback cards from the four restaurants, using an inductive thematic analysis (Braun & Clarke, 2013, p. 175). Finally, the resort's food purchasing spreadsheet for a representative month was analysed with descriptive statistics to see which goods were purchased, from where and at what prices.

Working in the kitchen enabled the second author to make direct contact with food suppliers. This provided access to three local, independent fruit and vegetable intermediaries who were interviewed. Other interviewees included one international aid agency, six members of a farmers' cooperative, and the executive chefs from six resorts similar to the case study resort, plus one boutique resort and one other restaurant. Key informants from the Fijian Hotel and Tourism Association (FHTA), the Ministry of Industry, Trade and Tourism (MITT), Ministry of Agriculture (MOA), and the University of the South Pacific (USP) were interviewed to triangulate findings from the agricultural and hospitality sectors. Finally, 54 professional cookery students attending a technical college in Fiji were surveyed to determine what they were learning about 'appropriate' food for tourists.

Overall, the second author conducted 38 semi-structured interviews. All interviews were transcribed and together with a reflexive field journal, analysed systematically via manual coding (Taylor-Powell & Renner, 2003, pp. 2–5). Emerging categories were bundled under leading categories or further refined into subcategories, if necessary. These codes and categories build the foundation for qualitative analysis and the presentation of findings in the following section.

Findings from hotel kitchens and farms in Fiji

The following section will elaborate on the main case study resort in greater detail, as well as highlighting similarities and differences across all resorts and restaurants visited. At average occupancy, the main case study resort hosted about 500 guests per day. About 80% came from Australia, another 10% were New Zealanders and the remaining 10% were mostly Europeans, U.S. Americans and Asians. This guest structure is similar to foreign visitor arrivals statistics in Fiji (Fiji Bureau of Statistics, 2018b).

'Hotel food' practices

Overall, Western 'hotel food' dominated the menus. One executive chef even noted that 'everybody is doing the same thing', referring to the food offerings of large upmarket resorts along Fiji's Coral Coast. For example, at the main case study resort, the breakfast buffet offered standard Western breakfast fare, including cereal, toast, eggs, bacon and hash browns, with a few localised highlights, such as fresh tropical fruit, pineapple jam and freshly grated coconut. For lunches, guests had the choice of eating at two pool bars and a steak and pizza restaurant. Fish and chips, chicken nuggets, sandwiches, burgers and wraps were the standard fare. For dinner, patrons had the choice of buffet dinners, a pool bar serving international and Asian cuisines, the steak and pizza restaurant and a fine-dining restaurant specialising in South Pacific-international fusion food. In all large-scale beach-side resorts, the pool bars received the most customers for lunch, while during dinner the seafood, steak and pizza restaurants were most popular.

Within this culinary setup, food sourced from within Fiji was utilised in three different ways.

As a substitute: A dish familiar to guests is made partly by using local produce as a substitute for Western staples. 'Moca'[1] leaves (a local Amaranth species[2]) substitute spinach, or coconut cream stands in for cow's cream, for example. While this might work with some products and recipes, there are limitations. For instance, efforts to substitute potato-based foods with the abundantly available root crops 'cassava' (Manihot esculenta) or 'dalo' (Colocasia esculenta) have usually failed for two reasons. Firstly, the majority of guests do not accept it; secondly, they are not available as a convenience product (i.e. ready-to-use, deep frozen) and therefore would need to be prepared manually, which ties up too much of the kitchen work force. The exception to this rule was the boutique resort. Here, the chef managed a small, but regularly changing menu based almost entirely on local produce.

As a local dish the local way: Some local recipes seemed to resonate well with the guests' palates in the large resorts. For example, Fijian 'kokoda' is a popular dish and based only on local produce: fresh fish, coconut milk and lime juice. Some local Indo-Fijian curries count in this category, too, even though spices need to be imported. However, other local recipes were not accepted. Guests would often just 'try once', as many chefs agreed. The large-scale beach-side resorts offered local island nights in their buffet restaurants once or twice a week. The dishes certainly imitated a number of genuinely local recipes, and drew on locally produced food, but these dishes were nevertheless described by local Fijian staff as not very authentic. 'The lolo [Fijian for coconut milk] comes out of the tin. It's not like at home!' staff pointed out, for instance. However, all large-scale resort chefs emphasised that they had to have some form of 'island night show' on offer. It was a 'gimmick' that tourists demanded, one chef noted. Overall, local dishes were promoted in a tokenistic way to please tourists' desire to see their idea of local food on offer.

As a local dish the Western way, the third and least applied option fuses local recipes with concepts well known to tourists to create dishes that appear familiar, but also offer unknown aspects. This strategy was mostly applied in the large-scale resorts' fine-dining restaurants. In the main case study resort, three chefs served around 15 customers per night. The menu offered, for instance, a 'mangrove crab wonton bisque' and 'ricotta and green papaya cannelloni' (menu main case study resort, 2017).

This cooking style requires an understanding of and training in local as well as international cuisines, as executive chefs pointed out. Local cooks with little experience and exposure could not conceive of how to use their local produce other than in the way they were used to from home. A young iTaukei (indigenous Fijian) cook wondered why anyone would make a salad out of local root crops, similar to a potato salad. In her worldview, that was not how root crops were supposed to be used. In this respect, Fiji has seen a number of initiatives to help chefs gain an understanding of how to locally-based, yet contemporary dishes that appeal to tourists. For example, Oliver, Berno, and Ram (2013) and Oliver et al. (2010) published cookbooks drawing largely on local produce, but including Western food concepts, too. Chefs Colin Chung and Robert Oliver held trainings with Fijian resort chefs, explaining how to best use local produce (Chefs for Development, 2016; DEPTFO News, 2016, December 11) . The president of the FHTA noted, however, that through these trainings he realised that in many resorts the needs for basic cooking and hygiene skills were even more of a priority than for how to incorporate more local produce. He concluded that despite all efforts to promote contemporary Fijian cuisine to chefs and guests, there was little evidence of it on restaurant menus (Dixon Seeto, personal communication, June 2017).

Observations in all large-scale resorts revealed that the use of local produce, in the sense of more traditional island produce, was rare. In the main case study resort, traditional foods made up about 2% of the locally produced food served to tourists. Plants such as those mentioned above as well as edible fern species (e.g. *Diplazium esculentum*), the spinach-like 'bele' (*Hibiscus manihot*), breadfruit (*Artocarpus altilis*) or different cooking bananas (Musa AAB Group), can be grown in ways that are agro-ecologically sustainable scale, especially in the form of 'agroforests, humanized woodlands, and polycultural gardens of traditional Pacific agricultural systems' (Manner & Thaman, 2013, p. 350). Moreover, because they are part of a traditional diet, they are mostly grown by iTaukei, small-scale and semi-subsistence farmers. However, with the exception of the boutique resort, chefs have found few ways of incorporating these foods in their menus.

In terms of creative engagement with food in the large-scale resort, the options appeared limited. Reportedly, restaurant and buffet menus were reworked about once a year. In the remaining time, they stayed mostly rigid. Short-term specials were uncommon. This was surprising as specials might pose an opportunity to use seasonal local produce. The issue is that this strategy requires a skilled and flexible workforce. To this end a senior resort manager noted:

They [cooks] don't deal with change very well. You want to shift something from here to there, it's a big undertaking! It's not a dynamic workforce. ... You walk into the kitchen and say 'guys, we've got this tuna now and I want you to cook it!' That's like (pauses) the wheels will come off! (Personal communication, April 2017)

Six of the eight chefs interviewed as well as all managerial personnel were concerned about the lack of skill level of their younger chefs. Good chefs were hard to find, everyone agreed. That is why the only instance in which a product that is suddenly available could be successfully integrated into menus was when it easily replaced an already existing product.

The large-scale resorts' most popular dishes were usually in the range of burgers, sandwiches, fish and chips, French fries, pizzas, steaks and salads. A review of about 300 guest feedback cards from the main case study resort's four main restaurants revealed that most patrons liked this kind of food. As far as the buffet dinners were concerned, the local island nights were less popular then other buffets. A senior manager believed that 'the way [the resort's] menus are set up right now is pretty much where the guests want to see them based on feedback' (personal communication, April 2017). In this respect a strongly emerging category was a phrase used often by kitchen staff: 'You can't go without...' What followed were Western dishes or produce—imported high-quality steaks, burgers, bacon, or imported apples and oranges. This category was closely related to that of 'Hotel food'. It was notable that all executive chefs of the large-scale resorts believed that Western cuisine was in most cases the only feasible choice for their menus.

Reasons for limited use of local produce

Fiji's population is made up of two main ethnic backgrounds: Indigenous Fijians (known as iTaukei) (57%) and Indo-Fijian (known as Fijian) (37%) (Fiji Bureau of Statistics, 2018a). The latter are descendants of Indian labourers brought to Fiji during British colonial rule. Therefore Fiji's cuisine is made up of iTaukei as well as Indian foods. In the large-scale resorts, Indo-Fijians made up the bulk of chefs. But regardless of their ethnic background, everyone expressed pride in their own culinary background, but also professed that this was not what should be served to guests. One Sous Chef even believed that 'French cuisine is always about being perfect. Obviously! Everybody knows. French is perfect.'

The focus on Western food very much reflected in the resort's purchasing budgets. Overall, the resort imported about 65%. To a large extent, in terms of the cost of the food, this share consisted of high-quality beef cuts, pork meat (together 22 percentage points) and dry goods (16 percentage points; e.g. sauces, tinned goods and convenience products). About 60% of fruit and vegetables were sourced locally, but this made up only 6% of the resort's entire monthly food cost (procurement spreadsheet for an average month of 2017). Other executive chefs of large-scale resorts in the area confirmed these findings.

When it came to local food sourcing, a majority of interviewees referred to frustrating past experiences, denoted by the often-used phrase 'We tried, but...' This referred to trying to integrate more local food into menu items, which was subsequently rejected by guests, or discouraging experiences with local suppliers and farmers, who did not deliver the right quality, quantity or consistency. As Chef Jack recalled:

> So I called the guy and asked 'what's going on? I ordered 40 kg for Wednesday and 40 kg for Saturday'. And he said 'oh, no, I didn't go fishing!' And I went 'why didn't you go fishing?' He said 'oh, you paid me enough on Wednesday!'

ITaukei farmers and fishers engaged mostly in selling their produce or catch when they needed money; for instance, at the beginning of the school year when facing school fee and uniform expenses. ITaukei economic activity was also aligned to their cultural and communal needs and obligations.

Food safety considerations also dissuaded some chefs from purchasing directly from farmers and fishers, or from local markets. This was a concern expressed by Chef Peter, especially with respect to seafood: 'I don't know where it's coming from. And you have to be very careful, especially as an international chain. I don't want to end up with food poisoning for 120 guests'.

Fruit and vegetable farmers from the Indo-Fijian community, especially in the Sigatoka river valley, were criticised for their unwillingness to plant crops that the resorts needed. Despite Fiji's suitable climate, many chefs noted that it was almost impossible to source different kinds of tomatoes, basil, onions, garlic, strawberries, green asparagus or fennel, for example. Most executive chefs and suppliers knew farmers who had tried in the past to grow these items, but none established long-term operations, a reliable source and required quantities. Chefs and purchasing officers were, concurrently, reluctant to source from a range of small-scale farmers and therefore resorted to imported produce.

Given the size of their operations and managerial pressure on cost and human resources, the executive chefs and purchasing officers of the six large-scale resorts favoured reliability above all else. A predictable supply meant that they could at all times commit to a fixed menu, train their staff in the use of certain products, provide continuous quality to their guests and keep the cost for ingredients at a fixed rate.

Farmers, meanwhile, found numerous challenges in directly accessing the resort market. Yet all farmers believed that tourist resorts were a valuable marketing opportunity because they paid higher prices than local markets. Indirect access through intermediaries was an option, but resulted in farmers compromising up to 50% of revenue. Direct access was impeded by the following factors (in no specific order). Firstly, the quality, consistency and volume requirements were hard to meet, given the farmers' comparatively basic farming capacity in terms of training, irrigation systems, mechanisation and farm size. For example, most farmers never received formal agricultural education; shade-houses that were destroyed by recent tropical cyclones had not been rebuilt; and around 44% of all Fijian farms are subsistence farms of less than one hectare (Department of Agriculture, 2009, p. 33; Ministry of Agriculture, 2016, p. 20). Secondly, many farmers were unsure of how to approach resorts and what produce they were interested in. On this issue, some farmers, as well as chefs, believed that purchasing officers were corrupt. Thirdly, the resorts' purchasing strategy proved difficult for many farmers. Most challenging was the fact that all resorts used a tender system requiring farmers to apply to be included in weekly tender calls. Even if they were respected in the tender process, farmers did not necessarily receive business for each product they had listed. Small-scale farmers, especially, felt this gave them little planning security.

Foreign aid agencies, such as the Taiwan Technical Mission (TTM) in collaboration with the MOA have established farmer groups that were able to generate higher efficiency rates, volumes and bargaining power (Taiwan ICDF, 2018). Interviews with a farmer group supported by the TTM and MOA revealed that members welcomed the TTM's training, financial support and marketing assistance. Through forming a cooperative-like structure and receiving agricultural training they successfully addressed the above issues. However, all farmers of this group were established Indo-Fijian farmers that could draw on financial resources, had enough time to attend meetings and trainings and had secure access to land. A former iTaukei member of the farmer group mentioned that it was difficult to attend all meetings and become an active member of a farmer group due to communal obligations, the need to guard their farm or inadequate transportation options. Indo-Fijian farmers that did not participate in such groups often favoured export-oriented farming. Export companies mostly dealt with only a few products. They provided farmers with fixed contracts and therefore planning security.

With respect to seafood, all chefs emphasised how difficult it was for them to source fresh local fish. On the one hand, they believed that fishers catching large pelagic species, for instance tuna or mahi-mahi, mostly sold their catch to exporters because of the higher prices they could fetch when selling to Japan, China or North America. Little of it was sold to the Fijian market, unless it was frozen and therefore of a lower quality. On the other hand, buying from local small-scale fishers was challenging because of quality, quantity and reliability issues. Chefs noted especially the limited understanding of hygienic post-harvest handling. WWF Pacific addressed the latter issue by launching a Sustainable Seafood Project which aimed to connect small-scale iTaukei fisher villages in the norther Macuata province with resorts and hotels in Suva, Nadi and the Coral Coast. The project took a two-fold educational approach. Fishers were trained about sustainable fishing practices and post-harvest handling; chefs were trained about preparation of local reef fish species and basic cooking skills (Esterhazy, 2017, November 3) . As far as the three interviewed chefs who had been involved in this project were concerned, the project did result in a few shipments of high-grade fish, but it did not turn into a reliable source. The chefs were therefore reluctant to further cooperate and the project was discontinued.

Discussion

When reflecting on the information above from hotel managers, chefs, procurement officers, farmers, government officials and donors in Fiji, it is apparent that there are considerable constraints facing those endeavouring to develop stronger links between local food systems and tourism. A number of the questions raised around the SDG targets in Table 1 focused on the purchasing decisions of hotels and restaurants, and whether they prioritised local products and producers. However this potential was limited by the fact that the large-scale multinational resorts reported on above served predominantly Western food. A strongly emerging theme was that of the importance and desirability of Fijian local food as a tourist experience (Sims, 2009; Timothy & Ron,

2013), but this only went as far as tokenistic 'island night shows'. This is further exem- plified by the economic insignificance of the initiatives to integrate locally grown, trad- itional produce and recipes. If at all, they materialized in the form of fine-dining restaurants serving fusion cuisine for a few affluent guests.

If we think of local food procurement as a form of corporate social responsibility, then resorts are only interested in the business case and safeguarding market share (Chakravorti, Macmillan, & Siesfeld, 2014). The customer is king, after all, and custom- ers frequenting large-scale multinational resorts on Fiji's Coral Coast are, according to the customer satisfaction cards they fill out, apparently not interested in exposure to genuine Fijian food. The majority asks for familiar dishes from within their comfort zone (Quan & Wang, 2004) and only a few are interested, educated and affluent enough (Y. G. Kim & Eves, 2012) to try some form of fusion cuisine. The implication of such marginalisation of local food and produce is cultural erosion through tourism, as feared by Pleumarom and Ling (2016). It perpetuates a system in which Western food is seen as desirable by chefs and local food is not, confirming findings from Berno (2015) and Oliver et al. (2010). In this respect, employing ethnographic methods was valuable to reveal the views and daily routines of resort employees, such as chefs. They are at the centre stage of negotiating guests' demands, available produce and local cuisine. Decades ago colonialisation brought messages of 'the greater efficacy of Western foods' (Pollock, 1992, p. 235) to South Pacific islands; today, it is mass tourism that favours the foreign over the local. It does not seem surprising that resorts cater- ing to this type of tourism import the majority of their food. In this respect, it also seems unlikely that mass tourism is able to support SDG 12 in terms of sustainable food production and consumption. Moreover, we argue that this means large-scale tourism in Fiji is generally unsupportive of small-scale and iTaukei farmers and fishers, thus limiting its contributions to SDG 2 and 14.

Local producers who would like to sell their products to these resorts, meanwhile, need to follow standards—standards set by efficient, profit-oriented multinationals seeking to satisfy the demands of mostly Australian mass tourists. Generally, food needs to be procurable in a reliable quality and quantity. In sum, the impediments to agriculture-tourism linkages from a demand, supply, marketing and policy perspective laid out above (summarised by, for example, Rogerson, 2012; Timms & Neill, 2011; Torres & Momsen, 2004, 2011) resonate well with the Fijian case study presented here.

What has not been addressed so far in research on tourism-food producer linkages, however, is that local producers may neither be willing nor feel the need to adjust to the multinational resort industry. Through ethnographic enquiry and interviews a strong theme emerged of a mismatch between the fundamental business orientations of resorts on the one hand and farmers and fisher people on the other. ITaukei often own land and engage in economic activity according to their own cultural needs and customs. These may or may not resonate with the resorts' needs. Indo-Fijian farmers, in contrast, may find it easier to deal with food exporters, who generally provide greater planning security than resorts do. As a result, links between resorts and food producers remain weak and food import figures high.

Furthermore, aligning the agricultural policy of SIDS with the resort industry's demands for beef and dairy, which could be lucrative because of the large proportion of food budgets spent on these items, would come at a considerable environmental cost. Large-scale cattle farming in fragile island ecosystems is likely to be detrimental to terrestrial and marine ecosystems (Hoffmann, 2002; Steinfeld, Gerber, Wassenaar, Castel, & De Haan, 2006) and therefore not supportive of SDGs 14 and 15.

In this respect, it is important to note that there is often a lack of coherence which gets in the way of the achievement of SDGs (see Scheyvens, Banks, & Hughes, 2016). For example, a locally sited multinational hotel might implement corporate policies on food safety that mean they forgo the produce from local markets, despite the social, economic and environmental benefits that can come from buying locally. In addition, while more local purchasing makes sense on many levels, within a hotel there are often debates between chefs versus procurement officers versus managers, about whether to prioritise local produce: there is not a coherence of views even within a single hotel property.

Conclusion

In this article, we have examined the assumption that building strong linkages between tourism and food production systems, especially agriculture and fisheries, is possible, and desirable, and would lead to more sustainable development. We found that the way large-scale multinational resorts are structured in terms of their food requirements does not make them ideal partners for food producers in SIDS. This is partly due to producer capacity in some SIDS but, as this article has shown, there are many other factors at play. Unfortunately, the type of tourism that is facilitated by such resorts is unlikely to provide significant support for the food-related aspects of SDGs 2, 12 and 14. Even efforts of chefs to incorporate local produce as a substitute for imported components of 'hotel food' have proven to be largely unsuccessful to date.

Reflecting on the case study-based research from a resort in Fiji, which was largely corroborated by executive chefs from eight other establishments who were interviewed, it seems that a number of the questions we have raised around SDGs 2, 12 and 14 in Table 1 have not actually been considered by scholars. This is no different from similar studies of tourism-agriculture linkages cited in our literature review. Questions overlooked include those around minimising food waste, educating guests about eating locally and seasonally, reporting systematically on sustainability practices around food, and ensuring that no pollution of land or sea resources occurs because of their practices. We urge researchers interested in food systems and tourism to consider such questions in future as they directly challenge hotels, restaurants and marine vessels to be more sustainable in their practices.

While linkages between local food systems and tourism may be weak in SIDS, this should not stop governments and donors from addressing the underinvestment in agriculture and fisheries that has occurred in SIDS over recent decades, as noted in the introduction to this article. This needs to be urgently remedied in the interests of food security (SDG 2), sustainable production and consumption of food (SDG 12) and sustainable fisheries (SDG 14), in line with meeting domestic nutritional needs as well as, potentially, earning revenue from tourism.

Notes

1. Local Fijian (i.e. Bauen) terms are highlighted in single quotation marks.
2. All local and scientific names based on Jansen, Parkinson, and Robertson (1990).

Disclosure statement

No potential conflict of interest was reported by the authors.

ORCID

Regina Scheyvens http://orcid.org/0000-0002-4227-4910

References

Anderson, W. (2013). Leakages in the tourism systems: Case of Zanzibar. *Tourism Review, 68*(1), 62–76. doi:10.1108/16605371311310084

Arendt, S. W., Roberts, K. R., Strohbehn, C., Ellis, J., Paez, P., & Meyer, J. (2012). Use of qualitative research in foodservice organizations: A review of challenges, strategies, and applications. *International Journal of Contemporary Hospitality Management, 24*(6), 820–837. doi:10.1108/09596111211247182

Asiedu, A. B., & Gbedema, T. K. (2011). The nexus between agriculture and tourism in Ghana: A case of unexploited development potential. In R. M. Torres & J. H. Momsen (Eds.), *Tourism and agriculture: New geographies of consumption, production and rural restructuring* (pp. 28–46). London, U.K.: Routledge.

Berno, T. (2011). Sustainability on a plate: Linking agriculture and food in the Fiji Islands tourism industry. In R. M. Torres & J. H. Momsen (Eds.), *Tourism and agriculture: New geographies of production and rural restructuring* (pp. 87–103). London, UK: Routledge.

Berno, T. (2015). Tourism, food traditions and supporting communities in Samoa: The mea'ai project. In P. Sloan, W. Legrand, & C. Hindley (Eds.), *The Routledge Handbook of Sustainable Food and Gastronomy* (pp. 338–347). London, UK: Routledge.

Blasiak, R., & Wabnitz, C. C. (2018). Aligning fisheries aid with international development targets and goals. *Marine Policy, 88*, 86–92. doi:10.1016/j.marpol.2017.11.018

Braun, V., & Clarke, V. (2013). *Successful qualitative research*. London, UK: Sage.

Bruinsma, J. (Ed.) (2003). World agriculture: Towards 2015/2030 - An FAO perspective. London, UK: FAO and Earthscan. Retrieved from http://www.fao.org/docrep/pdf/005/y4252e/y4252e.pdf.

Butler, G., & Rogerson, C. M. (2016). Inclusive local tourism development in South Africa: Evidence from Dullstroom. *Local Economy: The Journal of the Local Economy Policy Unit*, *31*(1-2), 264–281. doi:10.1177/0269094215623732

Campbell, J. R. (2009). Islandness: Vulnerability and resilience in Oceania. *Shima*, *3*(1), 85–97.

Campbell, J. R. (2015). Development, global change and traditional food security in Pacific Island countries. *Regional Environmental Change*, *15*(7), 1313–1324. doi:10.1007/s10113-014-0697-6

Cañada, E., Karschat, K., Jäger, L., Kamp, C., Man, F. d., Mangalasseri, S., … Monshausen, C. T. (2017). (Eds.). Transforming tourism: Tourism in the 2030 Agenda. Alba sud, Spain, Arbeitskreis für Tourismus & Entwicklung, Switzerland, ECPAT Deutschland, Germany, Fresh Eyes - People to People Travel, United Kingdom; Kabani, India; Kate e.V., Germany; Naturefriends International, Austria; Retour Foundation, the Netherlands; Tourism Watch - Brot für die Welt, Germany: Retrieved from http://www.transforming-tourism.org/fileadmin/baukaesten/sdg/downloads/sdg-complete.pdf.

Chakravorti, B., Macmillan, G., & Siesfeld, T. (2014). Growth for good or good for growth? How sustainable and inclusive activities are changing business and why companies aren't changing enough: Citi Foundation, Fletcher School at Tufts University and Monitor Institute. Retrieved from https://www.citigroup.com/citi/foundation/pdf/1221365_Citi_Foundation_Sustainable_Inclusive_Business_Study_Web.pdf.

Chefs for Development. (2016). Culinary training in the Pacific. Retrieved from http://chefs4dev.org/index.php/cuisine/culinary-education/

CIA. (2019). The World Factbook: Australia - Oceania: Fiji. Location Map. Retrieved from https://www.cia.gov/library/publications/the-world-factbook/geos/fj.html

Connell, J. (2015). Food security in the island Pacific: Is Micronesia as far as ever? *Regional Environmental Change*, *15*(7), 1299–1311. doi:10.1007/s10113-014-0696-7

Department for International Development. (1999). *Sustainable tourism and poverty elimination study.* Report for Department for International Development by Deloitte & Touche, IIED and ODI.

Department of Agriculture. (2009). Fiji national agricultural census 2009. Suva, Fiji: Department of Agriculture, Economic Planning and Statistics Division. Retrieved from http://www.fao.org/fileadmin/templates/ess/ess_test_folder/World_Census_Agriculture/Country_info_2010/Reports/Reports_3/FJI_ENG_REP_2009.pdf.

DEPTFO News. (2016, December 11). Yasawa tourism marketing cooperative chefs training held at Botaira Resort. *Fiji Sun online.* Retrieved from http://fijisun.com.fj/2016/12/11/yasawa-tourism-marketing-cooperative-chefs-training-held-at-botaira-resort/

Domroes, M. (2001). Conceptualising state-controlled resort islands for an environment-friendly development of tourism: The Maldivian experience. *Singapore Journal of Tropical Geography*, *22*(2), 122–137. doi:10.1111/1467-9493.00098

Eastwood, E. K., Clary, D. G., & Melnick, D. J. (2017). Coral reef health and management on the verge of a tourism boom: A case study from Miches, Dominican Republic. *Ocean & Coastal Management*, *138*, 192–204. doi:10.1016/j.ocecoaman.2017.01.023

Esterhazy, L. (2017, November 3). Fiji - it's all about quality [Blog post]. Retrieved from https://www.wwf.org.nz/media_centre/blogs/?15361/Fiji-its-all-about-quality

FAO. (2017). *The state of food and agriculture.* Rome: Author.

Fiji Bureau of Statistics. (2018a). Population and demography: 2007 population census. Retrieved from https://www.statsfiji.gov.fj/statistics/social-statistics/population-and-demographic-indicators

Fiji Bureau of Statistics. (2018b). Visitor arrivals statistics. Retrieved from http://www.statsfiji.gov.fj/statistics/tourism-and-migration-statistics/visitor-arrivals-statistics

Fischler, C. (1988). Food, self and identity. *Social Science Information*, *27*(2), 275–293. doi:10.1177/053901888027002005

Garcia Rodrigues, J., & Villasante, S. (2016). Disentangling seafood value chains: Tourism and the local market driving small-scale fisheries. *Marine Policy*, *74*, 33–42. doi:10.1016/j.marpol.2016.09.006

Gössling, S., Garrod, B., Aall, C., Hille, J., & Peeters, P. (2011). Food management in tourism: Reducing tourism's carbon 'foodprint'. *Tourism Management*, *32*(3), 534–543. doi:10.1016/j.tourman.2010.04.006

Hall, C. M., & Sharples, L. (2003). The consumption of experiences or the experience of consump-
tion? An introduction to the tourism of taste. In C. M. Hall, L. Sharples, R. K. Mitchell, N.
Macionis, & B. Cambourne (Eds.), *Food tourism around the world* (pp. 1–24). Oxford, UK:
Butterworth-Heinemann.

Herzog, C., & Murray, I. P. (2013). Is 'local' just a hot menu trend? Exploring restaurant patrons'
menu choices when encountering local food options. In C. M. Hall & S. Gössling (Eds.),
Sustainable culinary systems - local foods, innovation, tourism and hospitality (pp. 122–134).
London & New York: Routledge.

Hoffmann, T. C. (2002). Coral reef health and effects of socio-economic factors in Fiji and Cook
Islands. *Marine Pollution Bulletin*, *44*(11), 1281–1293.   doi:10.1016/S0025-326X(02)00260-6
doi:10.1016/S0025-326X(02)00260-6

Hugé, J., Van Puyvelde, K., Munga, C., Dahdouh-Guebas, F., & Koedam, N. (2018). Exploring
coastal development scenarios for Zanzibar: A local microcosm-inspired Delphi survey. *Ocean
& Coastal Management*, *158*, 83–92. doi:10.1016/j.ocecoaman.2018.03.005

International Federation of Organic Agriculture Movements. (2011). The role of smallholders in
organic agriculture. (IFOAM Position Paper). Bonn, Germany: IFOAM Head Office.

International Finance Corporation. (2018). From the farm to the tourist's table: A study of fresh
produce demand from Fiji's hotels and resorts. Retrieved from https://www.ifc.org/wps/wcm/
connect/dab246c4-7bee-4960-b3a6-25eb8d0932b9/
From+the+Farm+to+the+Tourists+Table+Final+Report.pdf?MOD=AJPERES

James, P. A., Tidd, A., & Kaitu, L. P. (2018). The impact of industrial tuna fishing on small-scale
fishers and economies in the Pacific. *Marine Policy*, *95*, 189–198. doi:10.1016/j.marpol.2018.03.
021

Jansen, A. A. J., Parkinson, S., & Robertson, A. F. S. (Eds.). (1990). *Food and nutrition in Fiji, a historical review: Food production, composition and intake* (Vol. 2). Suva, Fiji: Dept. of Nutrition
and Dietetics, Fiji School of Medicine: Institute of Pacific Studies of University of the South
Pacific.

Kim, G., Duffy, L. N., Jodice, L. W., & Norman, W. C. (2017). Coastal tourist interest in value-added,
aquaculture-based, culinary tourism opportunities. *Coastal Management*, *45*(4), 310–329. doi:
10.1080/08920753.2017.1327345

Kim, Y. G., & Eves, A. (2012). Construction and validation of a scale to measure tourist motivation
to consume local food. *Tourism Management*, *33*(6), 1458–1467. doi:10.1016/j.tourman.2012.
01.015

Le Cornu, E., Doerr, A. N., Finkbeiner, E. M., Gourlie, D., & Crowder, L. B. (2018). Spatial management in small-scale fisheries: A potential approach for climate change adaptation in Pacific
Islands. *Marine Policy*, *88*, 350–358. doi:10.1016/j.marpol.2017.09.030

Ling, C. Y. (2016). SDG 12 Ensure sustainable consumption and production patterns. In *Reflection
Group on the 2030 Agenda for Sustainable Development* (Ed.), Spotlight on Sustainable
Development (pp. 94–100): Social Watch, Global Policy Forum, Development Alternatives with
Women for a New Era, Third World Network and Arab NGO. Retrieved from https://www.
2030spotlight.org/sites/default/files/contentpix/spotlight/pdfs/Agenda-2030_engl_160713_
WEB.pdf.

Manner, I. H., & Thaman, R. R. (2013). Agriculture. In M. Rapaport (Ed.), *The Pacific islands:
Environment and society* (2nd ed., pp. 341–354). Honolulu: University of Hawai'i Press.

McIntyre, B. D., Herren, H. R., Wakhungu, J., & Watson, R. T. (Eds.). (2009). *International assessment
of agricultural knowledge, science and technology for development (IAASTD): Synthesis report*.
Washington DC: Island Press.

Ministry of Agriculture. (2016). *In-depth country assessment of the national system of agricultural
and rural statistics in Fiji*. Suva, Fiji: Author. Retrieved from http://pafpnet.spc.int/attachments/
article/744/Fiji%20IDCA%20Report%202016.pdf.

Morgan, W. (2013). *Growing island exports: High value crops and the future of agriculture in the
Pacific*. [Asia and the Pacific Policy Studies Research Paper, No. 05/2013]. Retrieved from
Australian National University: https://www.researchgate.net/profile/Wesley_Morgan5/publica-
tion/272302097_Growing_Island_Exports_High_Value_Crops_and_the_Future_of_Agriculture_

in_the_Pacific/links/5af3c693aca2720af9c470eb/Growing-Island-Exports-High-Value-Crops-and-the-Future-of-Agriculture-in-the-Pacific.pdf.

Newsnow. (2015, December 30). 'The golden egg is cracked' - Rarotonga tourism industry threatened by lagoon sewage-seep. *TV New Zealand*. Retrieved from https://www.tvnz.co.nz/one-news/world/the-golden-egg-is-cracked-rarotonga-tourism-industry-threatened-by-lagoon-sewage-seep?variant=tb_v_2

Nowak, J.-J., & Sahli, M. (2007). Coastal tourism and 'Dutch disease' in a small island economy. *Tourism Economics*, *13*(1), 49–65. doi:10.5367/000000007779784452

Okumus, F., Kock, G., Scantlebury, M. M. G., & Okumus, B. (2013). Using local cuisines when promoting small Caribbean island destinations. *Journal of Travel & Tourism Marketing*, *30*(4), 410–429. doi:10.1080/10548408.2013.784161

Oliver, R., Berno, T., & Ram, S. (2010). *Me'a Kai: The food and flavours of the South Pacific*. Auckland, New Zealand: Random House.

Oliver, R., Berno, T., & Ram, S. (2013). *Mea'ai Samoa: Recipes and stories from the heart of Polynesia*. Auckland, New Zealand: Random House.

O'Reilly, K. (2012). *Ethnographic methods* (2nd ed.). London, UK: Routledge.

Østrup Backe, J. (2013). Culinary networks and rural tourism development - constructing the local through everyday practices. In C. M. Hall & S. Gössling (Eds.), *Sustainable culinary systems - local foods, innovation, tourism and hospitality* (pp. 47–63). London, UK: Routledge.

Phillip, S., Hunter, C., & Blackstock, K. (2010). A typology for defining agritourism. *Tourism Management*, *31*(6), 754–758. doi:10.1016/j.tourman.2009.08.001

Pillay, M., & Rogerson, C. M. (2013). Agriculture-tourism linkages and pro-poor impacts: The accommodation sector of urban coastal KwaZulu-Natal, South Africa. *Applied Geography*, *36*, 49–58. doi:10.1016/j.apgeog.2012.06.005

Pleumarom, A., & Ling, C. Y. (2016). Tinkering with "sustainable or eco tourism" hides the real face of tourism. In Reflection Group on the 2030 Agenda for Sustainable Development (Ed.), *Spotlight on Sustainable Development* (pp. 96): Social Watch, Global Policy Forum, Development Alternatives with Women for a New Era, Third World Network and Arab NGO. Retrieved from https://www.2030spotlight.org/sites/default/files/contentpix/spotlight/pdfs/Agenda-2030_engl_160713_WEB.pdf.

Pollock, N. J. (1992). *These roots remain: Food habits in islands of the Central and Eastern Pacific since Western contact*. Laie, HI: The Institute for Polynesian Studies.

Pratt, S. (2015). The economic impact of tourism in SIDS. *Annals of Tourism Research*, *52*, 148–160. doi:10.1016/j.annals.2015.03.005

Pretty, J. N. (2001). *Some benefits and drawbacks of local food systems. [Briefing Note for TVU/Sustain AgriFood Network]*, Essex, UK: University of Essex.

Pretty, J. N., Ball, A. S., Lang, T., & Morison, J. I. L. (2005). Farm costs and food miles: An assessment of the full cost of the UK weekly food basket. *Food Policy*, *30*(1), 1–19. doi:10.1016/j.foodpol.2005.02.001

Quan, S., & Wang, N. (2004). Towards a structural model of the tourist experience: An illustration from food experiences in tourism. *Tourism Management*, *25*(3), 297–305. doi:10.1016/S0261-5177(03)00130-4

Rhiney, K. (2011). Agritourism linkages in Jamaica: Case study of the Negril all-inclusive hotel subsector. In R. M. Torres & J. H. Momsen (Eds.), *Tourism and agriculture: New geographies of production and rural restructuring* (pp. 117–138). London, UK: Routledge.

Richardson-Ngwenya, P., & Momsen, J. H. (2011). Tourism and agriculture in Barbados: Changing relationships. In R. M. Torres & J. H. Momsen (Eds.), *Tourism and agriculture: New geographies of production and rural restructuring* (pp. 139–148). London, UK: Routledge.

Rogerson, C. M. (2012). Strengthening agriculture-tourism linkages in the developing world: Opportunities, barriers and current initiatives. *African Journal of Agricultural Research*, *7*(4), 616–623. doi:10.5897/AJARX11.046

Said, A., MacMillan, D., Schembri, M., & Tzanopoulos, J. (2017). Fishing in a congested sea: What do marine protected areas imply for the future of the Maltese artisanal fleet? *Applied Geography*, *87*, 245–255. doi:10.1016/j.apgeog.2017.08.013

Scheyvens, R., Banks, G., & Hughes, E. (2016). The private sector and the SDGs: The need to move beyond 'business as usual. *Sustainable Development, 24*(6), 371–382. doi:10.1002/sd. 1623

Scheyvens, R., & Hughes, E. (2015). Tourism and CSR in the Pacific. In S. Pratt & D. Harrison (Eds.), *Tourism in Pacific Islands* (pp. 134–147). New York, NY: Routledge.

Scheyvens, R., & Momsen, J. H. (2008). Tourism in small island states: From vulnerability to strengths. *Journal of Sustainable Tourism, 16*(5), 491–510. doi:10.2167/jost821.0

Scheyvens, R., & Russell, M. (2012). Tourism and poverty alleviation in Fiji: Comparing the impacts of small- and large-scale tourism enterprises. *Journal of Sustainable Tourism, 20*(3), 417–436. doi:10.1080/09669582.2011.629049

Schubert, S. F., Brida, J. G., & Risso, W. A. (2011). The impacts of international tourism demand on economic growth of small economies dependent on tourism. *Tourism Management, 32*(2), 377–385. doi:10.1016/j.tourman.2010.03.007

Sengel, T., Karagoz, A., Cetin, G., Dincer, F. I., Ertugral, S. M., & Balık, M. (2015). Tourists' approach to local food. *Procedia - Social and Behavioral Sciences, 195*, 429–437. doi:10.1016/j.sbspro. 2015.06.485

Sims, R. (2009). Food, place and authenticity: Local food and the sustainable tourism experience. *Journal of Sustainable Tourism, 17*(3), 321–336. doi:10.1080/09669580802359293

Sisifa, A., Taylor, M., McGregor, A., Fink, A., & Dawson, B. (2016). Pacific communities, agriculture and climate change. In M. Taylor, A. McGregor, & B. Dawson (Eds.), *Vulnerability of Pacific Island agriculture and forestry to climate change* (pp. 5–45). Noumea, New Caledonia: Pacific Community (SPC).

South Pacific Tourism Organisation. (2018). Annual review of visitor arrivals in Pacific island countries 2017. Suva, Fiji: Author. Retrieved from https://www.corporate.southpacificislands. travel/wp-content/uploads/2017/02/2017-AnnualTourist-Arrivals-Review-F.pdf.

Steinfeld, H., Gerber, P., Wassenaar, T. D., Castel, V., & De Haan, C. (2006). *Livestock's long shadow: Environmental issues and options.* Rome: Fao.

Sustainable Tourism Development Consortium. (2007). *Fiji tourism development plant 2007-2016.* Suva, Fiji: Author.

Taiwan ICDF. (2018). Vegetable Production, Marketing Extension and Capacity Building Project. Retrieved from http://www.icdf.org.tw/fp.asp?fpage=cp&xItem=29299&ctNode=30043&mp=2

Taylor-Powell, E., & Renner, M. (2003). *Analyzing qualitative data.* Wisconsin, WI: University of Wisconsin.

Telfer, D. J., & Sharpley, R. (2008). *Tourism and development in the developing world.* London, U.K.: Routledge.

Thaman, R. R. (2008). Pacific Island agrobiodiversity and ethnobiodiversity: A foundation for sustainable Pacific Island life. *Biodiversity, 9*(1-2), 102–110. doi:10.1080/14888386.2008.9712895

Thow, A. M., Swinburn, B., Colagiuri, S., Diligolevu, M., Quested, C., Vivili, P., & Leeder, S. (2010). Trade and food policy: Case studies from three Pacific Island countries. *Food Policy, 35*(6), 556–564. doi:10.1016/j.foodpol.2010.06.005

Timms, B. F., & Neill, S. (2011). Cracks in the pavement: Conventional constraints and contemporary solutions for linking agriculture and tourism in the Caribbean. In R. Torres & J. H. Momsen (Eds.), *Tourism and agriculture: New geographies of production and rural restructuring* (pp. 104–116). London, UK: Routledge.

Timothy, D. J., & Ron, A. S. (2013). Heritage cuisines, regional identity and sustainable tourism. In C. M. Hall & S. Gössling (Eds.), *Sustainable culinary systems: Local foods, innovation, tourism and hospitality* (pp. 275–290). London, UK: Routledge.

Torres, R. M. (2003). Linkages between tourism and agriculture in Mexico. *Annals of Tourism Research, 30*(3), 546–566. doi:10.1016/S0160-7383(02)00103-2

Torres, R. M., & Momsen, J. H. (2004). Challenges and potential for linking tourism and agriculture to achieve pro-poor tourism objectives. *Progress in Development Studies, 4*(4), 294–318. doi:10.1191/1464993404ps092oa

Torres, R. M., & Momsen, J. H. (2011). Introduction. In R. M. Torres & J. H. Momsen (Eds.), *Tourism and agriculture: New geographies of consumption, production and rural restructuring* (pp. 1–10). London, UK: Routledge.

UN. (2016). Sustainable development goals. Retrieved from https://sustainabledevelopment.un.org/?menu=1300

UNWTO. (2015). Tourism and the sustainable development goals. Madrid, Spain: Author. Retrieved from https://www.e-unwto.org/doi/pdf/10.18111/9789284417254.

Veit, R. (2007). Tourism, food imports, and the potential of import-substitution policies in Fiji. Suva, *Fiji: Fiji AgTrade - Ministry of Agriculture, Fisheries & Forests.*

World Bank. (2016). Agriculture and rural development. Retrieved from http://data.worldbank.org/topic/agriculture-and-rural-development

Zeller, D., Harper, S., Zylich, K., & Pauly, D. (2015). Synthesis of underreported small-scale fisheries catch in Pacific Island waters. *Coral Reefs, 34*(1), 25–39. doi:10.1007/s00338-014-1219-1

Food safety and tourism in Singapore: between microbial Russian roulette and Michelin stars

Nicole Tarulevicz (ID) and Can Seng Ooi (ID)

ABSTRACT

Drawing on multiple culinary traditions, foodways, and networks of trade, food is both good and important in Singapore. Brand Singapore relies on food culture to market itself to the world, but also to its citizens. Hawker food, that is, street foods, are at the core of that marketing, becoming a by-word for Singaporean culinary culture. Cheap and delicious food was used to shift Singapore from a stop-over to a destination. But this also reinforces ideas about high and low culture, embodied in what a recent travel blog described as the "golden rule": "When you're travelling in Asia, whether you're in Sri Lanka or Thailand, in Singapore or Vietnam, Malaysia or China, cheap food is the best food." What makes Singapore distinctive in the framing of 'cheap Asian food' is that it is considered much safer, travelers can try new things without engaging in the "microbial Russian roulette of street food" elsewhere. At the same time, regulations and systems that keep people safe can be perceived by tourists to make Singapore, and by extension its culture, too clean, safe, and hygienic. As Singapore emerges as a global food destination with Michelin stared restaurants and a destination-fine-dining culture, the Singapore Tourism Board continues to recreate the Oriental mystique of the destination by cloaking the modern manifestations of Singapore with stories of its Asian and colonial heritage. In focusing on food safety, this paper highlights the tension between high and low food culture, between safe and unsafe, between street food and fine dining, but it also considers how they are being negotiate in Singapore. Taste, its arbiters, makers, and guardians, are raced and hierarchical. Singapore's food culture provides an example of these orthodoxies are both reinforced and challenged.

摘要

新加坡利用多种烹饪传统、饮食方式和贸易网络, 使得美食在新加坡既品质佳又意义非凡。新加坡品牌依靠饮食文化向世界推销自己, 也向其公民推销自己。小贩食品, 也就是街头食品, 是这种营销的核心, 成为新加坡烹饪文化的代名词。廉价而美味的食物把新加坡从一个中途停留的地方变成一个目的地。但这也强化了人们对高雅文化和低俗文化的看法。最近的一个旅游博客描述美食的"黄金法则"为:"当你在亚洲旅行时, 无论是在斯里兰卡还是泰国, 无论是在在新加坡还是越南、马来西亚或中国, 便宜的食物是最好的食物。"新加坡之所以在"廉价亚洲食品"的定义上与众不同,是因为新加坡美食被认为安全得多,游客可以尝试新事物,而不用担心其他地方发生"微生物感染的俄罗斯轮盘赌"(不小心感染食品的微生物中毒)。与此同时,保护人们安全的规章制度可以被游客感知到,从而使新加坡及其文化变得非常干净、安全和卫生。随着新加坡逐渐成为全球美食胜地,新加坡拥有米其林(Michelin)的豪华餐厅和以美食为目的地的高端饮食文化,新加坡旅游局(Singapore Tourism Board)继续以其亚洲和殖民遗产的故事,掩盖新加坡的现代表现,重现目的地的东方神秘。本文在关注食品安全的同时,也强调了高端与低端餐饮、安全与不安全餐饮、街头小吃与精品美食之间的紧张关系,同时也考虑了它们新加坡如何进行协商。口味,它的仲裁者,创造者和守护者,是竞争和划分等级的。新加坡的饮食文化为那些即受到强化又受到挑战的正统观念提供了一个实例。

Introduction

The late food writer Jonathan Gold, writing about Singapore's street food in the American food magazine *Savuer* told readers that "While eating street food in Bangkok or Djakarta can be a game of microbial Russian roulette, the hawker center stalls in Singapore operate under stringent health controls, and thus are unlikely to make you sick" (2007). Gold both celebrated (it is clean and safe) and derided the regulation: "This being Singapore, of course, actual street food was outlawed long ago, its vendors taken off the city's byways and confined to government regulated 'hawker centers' or food courts—garlic-steeped monuments to multiculturalism and terrific Asian food, built into the first floors of apartment complexes and market buildings all over the city" (2007). The tension between reassuring cleanliness and an imagined exotic and Orientalist food culture threads through both international tourism materials and local ones. This paper traces the history of food safety and of tourism in Singapore to show the ways they have become interconnected, especially in discursively on print culture and visual media. Cheap and delicious food was used to shift Singapore from a stop-over to a destination, but this also reinforces ideas about high and low culture. As Singapore emerges as a global food destination with Michelin starred restaurants and a destination-fine-dining culture, the Singapore Tourism Board (STB) continues to recreate the Oriental mystique of the destination by cloaking the modern manifestations of Singapore with stories of its Asian and colonial heritage. In focusing on food safety, this paper highlights the tension between high and low food culture, between safe and unsafe, between street food and fine dining, but it also considers how they are being negotiated in Singapore at the local level, and in promoting Singapore as a food destination. Krishnendu Ray (2016) prompts us to remember that taste, its arbiters, makers and guardians, are raced and hierarchical. Singapore's food culture provides an example of how these orthodoxies are both reinforced and challenged.

Food is increasingly used to market destinations. Or to be more accurate, specific food cultures, identities, and styles are used to market destinations (Lin, Pearson, & Cai, 2011, p. 43). Positive food experience, especially around food hygiene has been clearly linked to intention to return (Lertputtarak, 2012, p. 116). Tourism, as Rickly-

Boyd (2013, p. 684) observed is a "place-based endevor" and perceptions about place matter. There is a growing scholarly interest in the connection between food safety and tourism (Ooi & Tarulevicz, 2019). As Maclaurin (2004, p. 250) notes, the association of specific destinations with food-borne illness and perceived food safety concerns affects tourists destination choices. Their personal experience with food poisoning and perceptions of destination development levels likewise had an affect (Maclaurin, 2004, p. 253). As the case of Singapore illustrates, travelers have long been preoccupied with the safety of the food in places they visit, and the historical lessons around balancing convenience, cleanliness, and demand, are still relevant.

History of food safety in Singapore: managing the game of the microbial Russian roulette

Often described as the 'Emporium of the East,' Singapore leveraged its geographic advantages, what Wong (1978) described as its "nodal position at the tip of the Malay Peninsula, in the Archipelago, flanked by the Pacific and Indian Oceans … combined with such natural advantages as a splendid harbour and sheltered anchorage," with its tax free status, to come to pre-eminence. With that growth came public health challenges. The history of food safety in Singapore began with attempts to improve sanitary conditions. Singapore's limited water supply became quickly polluted by European settlement (Hon, 1990) and tropical conditions made the management of waste critical. Attempts at this were enacted through legislation passed in 1860 which focused on preventing the congregation of food vendors (hawkers) to improve the flow of traffic and to prevent the establishment of spontaneous night-markets.

Calaresu and van den Heuvel (2016) remind us that selling food on the street has a long global history. The economic and social centrality of street food globally is evident. In colonial Singapore, street vendors, or hawkers, provided a vital service in Singapore, offering inexpensive food to feed workers, many of whom did not have access to cooking facilities, or to the labor of family members to cook for them (Tarulevicz, 2018, p. 292). Hawkers played an essential role in building the economy of Singapore, selling basic meals, cooked and uncooked foods, snacks and beverages, and building small businesses (Warren, 1986, 153).

Hawkers are celebrated in contemporary Singapore as preserving a national cuisine and culture, they have been viewed historically as both a necessity and a problem (Tarulevicz, 2018). Mendiola García (2017) using Mexico as her example, stresses that it is the informal nature of street vending that often causes regulatory challenges. In Singapore, hawkers were seen as an unruly industry to administer, unclean and perceived to be prone to breaking regulations in pursuit of profit. The connection between cleanliness and food safety was a constant concern for regulators, in both the colonial era and subsequently. Hawking was a public health issue because of the spread of diseases such as cholera and typhoid. Contaminated water was especially problematic. It might be used in drinks, but it could also be the source of water-borne diseases when left stagnate by hawkers, or on the ground after they had moved on. Diseases could be spread through the bodies of hawkers, through inadequately cleaned hands and utensils. It was feared that hawkers were carriers of gastroenteritis, enteric fever (typhoid), dysentery, cholera, and parasitic infections such as hookworm and roundworm.

Food waste drew insects and rodents. The tropical conditions of Singapore increased the rate of decay and speed of the spread of contamination. Itinerancy itself was also a problem (Tarulevicz, 2015, p. 6), making hawkers, by their itinerancy, a vector for the spread of disease through their own movement and through the movement of their waste across multiple areas. Night markets often appeared spontaneously, making tasks such as street cleaning more difficult. Town-cleaning laborers tended to avoid markets, or if they did not, they often clashed with the hawkers. Congregating was a continual source of tension. Convenience and cleanliness were not always successfully balanced (Tarulevicz, 2018).

Thousands of letters to the editor have been written to Singaporean newspapers about hawkers, their habits, risks and food safety practices, or their lack of. Many of these letters were critical of hawkers, and of the colonial authority for not regulating hawking better. They were authored by residents and visitors alike. A "Bewildered Traveller" (The comforts of Singapore, 1876) for example, suggested that if the authorities could not manage it, then hotels must take up the responsibility of "keeping all those annoying and impudent hawkers away," so that "for the credit of Singapore something might be done to keep visitors from being perpetually worried." "Peaceful" (Making Night Hideous, 1910) was concerned about noise: "from daylight till dark the yells of street vendors and hawkers of various sorts jar on the ear and nerves almost incessantly." Other letter writers had seasonally driven concerns, especially around tropical fruits and the management of waste. Serangoonite's letter (Rubbish Strewn Roads, 1911) provides an exemplar:

> ... the filth that has been allowed to collect is something awesome. The ditches along each side of the road are almost choked and durian and mangosteen skins, where not scattered over the highway, are piled up in heaps alongside to rot. The stench from the decaying refuse coupled with the plague of flies, and mosquitoes that haunt the district cannot do otherwise than enhance the possibility of the surrounding inhabitants contacting diseases ...

The connection between waste and disease and civic services is explicitly made in this, and many other letters.

At the May 1924 meeting of the Singapore Municipal Commission the Health Officer, Dr. P. S. Hunter, proposed the abolition of street hawkers, and their replacement with licensed eating shops. In an editorial the *Straits Times* (Municipal Matters, 1924) noted that while the usual complaint was that people in Singapore did not pay sufficient attention to municipal matters, in the case of the hawker matter the rate of letter writing conveyed "a loud public clamour." Letters to the editor matter, they reflect the political conditions of colonialism and the issues that garnered public attention. A situation that the *Straits Times* was aware of, and noted in their editorial: " ... in a town such as Singapore – with no Municipal elections and practically no public meetings – the letter to the editor is the only way of expressing opinion." They were also acutely aware of the importance of the topic, stating: "The subject of food hawkers and what to do with them has caused more stir than most, and letters continue to come in" (Municipal Matters, 1924).

In contrast to the tone of previous letters, those published in 1924 tended to defend hawking. Letters described the importance of hawking to individual hawkers and their families, to preventing widespread poverty and most frequently, to the broader community. As one letter writer put it ("Your food will cost," 1924) "the suggested abolition of food-hawkers will have a strong repercuss [sic] on the whole community." He described the office clerks in his building who relied on cheap hawker meals and what would happen under abolition: "If they have to go to a shop for their food it will cost them more. Weshall have to increase their wages accordingly." The impact he feared would be even wider: "Our rickshaw pullers also get their lunch from street hawkers. Their food will cost them more. Their fares will cost us more." He predicted a 20–33 percent increase in food costs and further economic and social effects, such as a rise in rents with the greater demand for shops, which in turn would create a reduction in residential housing stock and a consequent increase in rents. All of this he argued would "strike a blow at the public health which will more than counteract any improvement caused therein by a possible but problematical improvement in the quality of public food." This kind of critique was not exceptional, rather it reflected widespread and nuanced critiques of a proposed policy.

By August of 1924 opposition was formalizing. The Chinese Chamber of Commerce held a large public meeting, which unanimously voted for a statement to be forwarded to the Colonial Secretary and Municipal Commissioners against the recommendations of the Medical Officer. There was extensive media coverage of the meeting and of the subsequent change of direction. Dr. Hunter issued a second report, which was partially reproduced in a range of newspapers, often several thousands of words of it (Health Officer Defines His Proposal, 1924). What is clear, from the report, and the media coverage of it, is that Dr. Hunter's objections to hawkers and their food goes beyond matters of public health and was part of broader attempts of ordering and cleaning the city and the citizenry. If we use Foucault's (1976) concept of biopolitics in the case of attempts at hawker abolition and regulation we can understand them as mechanism that control and discipline the population, and visitors. As will be discussed later, food safety development and tourism development, as in other social and economic engineering projects, in Singapore stem from a comprehensive program to discipline, govern and modernize the city-state (Gulrud & Ooi, 2015; Kong, 2007). Cooked food hawking, Dr. Hunter told readers of Singapore's newspapers: "... is undoubtedly the calling pursued by many who are unfit for hard work, but it is equally the calling chosen by those who don't want to work" (Hawkers in Singapore, 1924).

It was not only concerned citizens and officials that wrote to the press. G.H. Kiat, local bookseller, and one of the "rising young men in the Straits Chinese community" ("The hawkers case," 1924) wrote a multi-columned letter, using his own name, to argue against the way the issues had been reported and while he supported the amendments to Dr. Hunters' report, he was offended by the assumptions in it; specifically that the community meetings did not reflect public opinion and that the needs of the poor were not being taken seriously (Kiat, 1924). Kiat (1924) quoted a participant in the Chinese Chamber of Commerce public meeting who sums up the views of many letter writers: "What we want is not abolition. It is better control, better regulation and better supervision." Residents and visitors got none of these out of the recommendations of 1924, and in 1932 a committee was formed to report of the "Hawker Question." The report enumerated the problems of hawking, but was also aware of its necessity and of the critical role it provide in feeding many, concluding that abolition was not "practicable" ("Report of the Committee," 1932).

The transition from unregulated to regulated street vending is not an exclusively Singaporean story. In fact, the transformation of street vending is gathering increasing scholarly attention. Work on street vending in Rio de Janeiro by Acerbia (2017) highlights the tensions between economic and political freedoms. And it is no coincidence that the major transformation of Singapore's hawkers—from unregulated street vending to high ordered covered spaces—coincided with Singapore's transition to a postcolonial nation state, in which hawkers gained both economic and political rights in return for being regulated (Tarulevicz, 2018). The 1950 Hawker Inquiry Commission lay the foundations for these key changes, and succeeded in providing, a quarter of a century later, what the speaker at the Chinese Chamber of Commerce asked for: "better control, better regulation and better supervision."

To be specific, the Hawker Inquiry Commission destroyed and preserved hawking in Singapore. The commission acknowledged the centrality of hawking to the Singaporean way of life, and, as geographer Lily Kong (2007) suggested, the decision to regulate rather than prohibit hawkers was a major shift in colonial administrative thinking. By laying the foundation for moving hawkers into controllable spaces, the commission actually preserved it, albeit in a modified form. As infrastructure, the transformation from itinerancy to hawker centers required a bureaucracy (most notably an army of inspectors), a legal apparatus (empowering police and inspectors), and a shift in popular knowledge. Hawkers and their trade, once seen as dirty and providers of unsanitary foods, were cleaned up, their cooking equipment inspected and standardized. The places where they plied their trade were eradicated or repurposed and replaced with "ordered" spaces—hawker centers. In turn, these spaces were increasingly policed, made cleaner and made orderly at every turn. It was not just hawkers who were being cleaned up, Singapore itself was undergoing rapid transformation of the built environment. Tourism too was changing, in order to understand these changes we trace the history of tourism in Singapore.

History of tourism in Singapore

In 1881, celebrated futurist writer Jules Verne observed (Verne, 2017, p. 18): "Singapore was simply one large warehouse, to which Madras sent cotton cloth; Calcutta, opium; Sumatra, pepper; Java, arrack and spices; Manilla, sugar and arrack; all forthwith dispatched to Europe, China, Siam, &c." He lamented: "Of public buildings there appeared to be none. There were no stores, no careening-wharves, no building yards, no barracks, and the visitors noticed but one small church for native converts" (Verne, 2017, p. 18). As a former British colony and trading post, the island has been receiving travelers over the centuries. In contrast to Verne's observations, gleaning new skyscrapers now mark modern Singapore but the city remains an entrepot. *Tales of old Singapore* present an Orient that was both uncouth and exotic (Manley, 2017; Ooi, 2005b; Wise, 1985).

Figure 1. Chinese children and street hawkers, pre-1914. Source: E. A. Brown Collection, National Library of Singapore.

American author and critic, George Hamlin Fitch, for example, described a scene in Singapore in 1913 (Fitch, 2017, p. 155). He contrasted exotic food culture: "The most conspicuous places of business on these streets were the large restaurants, where hundreds of Chinese were eating their chow at small tables. The din was terrific, and the lights flashing on the naked yellow skins, wet with perspiration, made a strange spectacle," with erotic culture: "Next to these eating houses in number were handsomely decorated places in which Chinese women plied the most ancient trade known to history." Street food dominated, as in this pre-WWI image of hawkers (Figure 1).

In another description, British diplomat R.H. Bruce Lockhart described eating *satai* (more commonly spelled as "satay" today) in 1936. He began by refracting it through the familiar, a European food tradition:

> Here, occupying various pitches like the men who sell hot Frankfurter sausages in the vegetable market in Prague, were the Malay "Satai" cooks ... I got out and went over to an empty stall kept by a plump Malay gentleman, who obviously enjoyed his own cooking [...]

"Satai, Tuan?" he asked and, when I nodded, he began the ritual. With a small straw-plaited fan he began to stir the charcoal to a bright flame. While it burnt he chopped up onion and cucumber on a plate. When the fire was at the proper glow, he produced from his box pieces of chicken meat fixed on little sticks in much the same way the Russians prepare "shashlik". Then, dipping the sticks of meat into a can of fat, he grilled them for several minutes on his fire. He bowed. The "Satai" were ready. (Lockhart, 2017, p. 153).

While shirtless perspiring men eating in the streets are rare today in Singapore, satay remains a popular local dish, and the iconography is persistent. It is still prepared in a similar manner but in a more hygienic environment. Health and hygiene standards in Singapore are now high, but they were issues when Stamford Raffles colonized Singapore in 1819. American journalist Emily Hahn tells the apocryphal story of pest management to illustrate:

> It is alleged that "Singapura" is one of the Indian tongues means "Lion City" but the only animals on the island were rats, who made up in number what they lacked in size. So many were they that, far from fearing the cats that came ashore off the ships, they killed them by ganging up, until Raffles offered a bounty of one wang – i.e., two cents ha'penny for every dead rat brought to him. Thousands were killed in the following days, until there was not one rat left. (Hahn, 2017, p. 20)

The chaotic scenes of street markets, dirty roads and unruly behavior have been replaced by clean spaces and orderly crowds in Singapore. The imagination of Singapore as an exotic Oriental destination is blended with images of a global financial city. The creation of modern Singapore is tied closely to its tourism development, which in turn is tied to nation building.

Modern tourism development in Singapore started with the founding of the Singapore Tourist Promotion Board (STPB), now STB, in 1964, before Singapore's full independence. At that time, the island received 91 000 visitors (Singapore Tourism Board, 2019). By 2005, it received nearly 9 million (Singapore Tourism Board, 2008). The number has since increased to more than 17 million in 2018 (Singapore Tourism Board, 2018). The island-state has a population of 6 million, and its physical size is just about $700\,km^2$. By comparison, the over-popular island destination of Venice is about $415\,km^2$, has a population of a quarter million and receives 20 million visitors a year. Tourism contributes about four percent to Singapore's Gross Domestic Product (Singapore Tourism Board, 2019).

In the three decades of STPB's existence, between 1960s and 1990s, the emphasis in Singapore tourism development was on building attractions, including the zoo, bird park and the tourist island of Sentosa. The strategy changed in 1996, together with the change of name to STB. STB devised a strategy to become a tourism hub. The hub concept concentrates on making Singapore into a node for international travels within Asia and between continents (Lee, 2004; National Tourism Plan Committees, 1996; Ooi, 2004; Teo & Chang, 2000). The island does not just connect international travelers but should entice them to get out of the airport and stay the night. Singapore wants to be a destination that is not just attractive but also relevant to international travel and in the global economy (Ooi, 2002). The tourism hub strategy has evolved with the city-state's attempts at making itself a central node in trade and international business in the global economy.

To understand the official tourism development plans in Singapore, it is essential to accentuate a primary principle in the plans. Tourism is always weaved into the various aspects of Singaporean society, beyond reaping the economic benefits from that industry. The development and "touristification" strategies of Singapore are in locked-step. With the island's limited space, interaction between residents and visitors is inevitable, as they visit, play, eat and wonder in many of the same spaces. Singapore has to be comfortable for both residents and visitors, and also enticing for residents to stay, foreign talents to migrate to, and visitors to return.

An example of how tourism and national social development of Singapore are in sync is the "Garden City" program which started in 1967 (Barnard, 2014; Gulrud & Ooi, 2015). The visioning of Singapore as a green city is part of a wider identity making process that requires the taming of the environment (Barnard, 2014). Besides planting trees, lessening the physical impact of industrialization and to beautify the city, the Singapore Botanic Gardens was expanded to elevate the garden city idea (Barnard & Heng, 2014, p. 291). The Botanic Gardens is a tourist draw, and the organization further established two other tourist attractions—the Chinese Garden and the Japanese Garden—in the western part of the island to provide more greenery to a largely industrial area then in the mid-1970s (Barnard & Heng, 2014, pp. 292–293). More recently, Gardens by the Bay opened in 2011 and is now a major tourist attraction. Gardens by the Bay is "the culmination of the greening projects […] It showcases high horticulture technology and is a model for the amalgamation of a multitude of disciplines ranging from architecture to bioengineering that the government has used for greening Singapore" (Barnard & Heng, 2014, pp. 303–305). The Garden City aims to both make Singapore more livable and pleasant, as well as to develop its tourism industry.

Another example of how tourism and nation building evolve simultaneously in Singapore is the presence of the STB behind the many cultural development plans of the city (Chang & Lee, 2003; Ministry of Information & the Arts, 2000; Ministry of Information Communication and the Arts (MICA), 2008; Ooi, 2018). One of the strongest rationale for the establishment of the Singapore Art Museum, the Asian Civilisations Museum, National Museum of Singapore and the National Gallery Singapore, together with the Esplanade and other cultural institutions is tourism. Since the 1990s, the tourism authorities are engaged in devising cultural policies to promote the arts and cultural life in the city, such as organizing festivals and arts events, commissioning public art installations, and bringing in concerts and clubs to enhance the entertainment life in the city (Ooi, 2005a, 2018). The arts are meant to entrench the cultural identity of the community, and with the help of the STB, also make Singapore into a global art city that attracts millions of visitors (Ooi, 2011). The founding Prime Minister Lee Kuan Yew lamented in 2006 that: "The Singapore that we had – very orderly, very wholesome, very clean – is not good enough […] Singapore must aim to inject more fun and buzz and create a lively night scene that can make it the 'Paris of South-east Asia'" (Peh, 2006).

Besides environmental and cultural development, tourism development in Singapore is also tied to its food safety standards and food culture. Food safety standards are embedded in the wider project of modernizing Singapore. The making of Singapore into a First World nation requires basic health standards that will attract investments, satisfy residents and draw visitors. Many of the food hygiene strategies make Singapore into a more comfortable and safer place, and strategies have direct implications for tourism and other aspects of society (Ooi & Tarulevicz, 2019). The STB considers "brand awareness among Singaporeans and residents important. This includes a comprehensive domestic tourism program, strategic outdoor advertising and communications platforms such as signs and display panels in immigration check-points, airports and districts with high tourist traffic" (STB, 2010). Brand awareness for Singaporeans works to reinforce ideas about the nation-state and nationalism, with food representing a critical medium through which these messages are illustrated (Tarulevicz, 2013). For the STB the intention is "to establish Singapore as one of the most compelling shopping and dining destinations in Asia, where every visitor's shopping and dining experience will be an enjoyable and unique one that exceeds expectations." (STB 2010) Food is positioned as central to the tourist experience. According to the STB: "Food has a sacred status in Singapore. Your trip here is not completed till you've tried the various cuisines and signature dishes" (STB 2010). That sacred status relies on locals understanding the brand and its importance.

How tourism relates to Singaporean food cultures and issues has evolved over the years, reflecting the changing tourism and national social development strategies. Like basic utilities such as water and electricity, food safety and food supply are perennial concerns there. So even with the high hygiene standards, in 2019, the Singaporean authorities established a new agency—the Singapore Food Agency (SFA) —to strengthen Singapore's food safety standards and food security (Seah, 2019). This new entity brings various food-related functions of the Agri-Food and Veterinary Authority, the National Environment Agency and the Health Sciences Authority together. The economy and society will not function if food supply is not secured and food standards are not maintained. Tourism will be adversely affected if visitors cannot trust what they eat, for instance.

Singapore has changed its tourism positioning several times. In the 1960s and 1970s, Singapore was "Instant Asia," where one could find an array of Asian cultures, peoples, festivals, and cuisine conveniently exhibited in a single destination. In the 1980s, "Surprising Singapore" positioned Singapore by placing contrasting images of modernity and Asian exoticism together. The co-existence of East and West, old and new were highlighted then. In these early decades of tourism development in Singapore, food and eating were already featured in the branding campaigns. Underlying the tourism food activities was that food in Singapore is safe to eat. And during that time, as mentioned, tourism development priority was on building more tourist attractions, such as the zoo, bird park, themed areas and gardens (Lee, 2004).

In the second half of the 1990s, Singapore was promoted as "New Asia – Singapore." This campaign marked a significant change to the country's tourism development strategy. While the brand message changed only in subtle ways; "Surprising Singapore" promised pockets of unexpected diverse and distinct ethnic cultures in a modern city, the replacement brand "New Asia – Singapore" offered ethnic cultures fused into modern development—the tourism development strategy for Singapore was markedly different (Ooi, 2004). It was also during that time that the "New Asia" kitchen was invented. The Singapore local newspapers *The Straits Times,* for instance, compared "New Asia cuisine" to fusion cuisine invented in California in the mid-1980s, and confessed that "New Asia cuisine" was "coined here about 18 months ago" (The

Figure 2. Uniquely Singapore Advertisement 2009. Source: Singapore Tourism Board.

Straits Times, 1997). The emphasis of the New Asia branding is the growing confidence of the region, and that Singapore has embraced the best of the East and West, old and new. Singapore should be seen as the hub and capital of a rising area of the world, and the city connects the world to the region (Lee, 2004; National Tourism Plan Committees, 1996; Ooi, 2004; Teo & Chang, 2000). Strategies are devised to build on the island's cultural, business, medical and education tourism sectors; visitors come to Singapore for leisure but also for purposes beyond holidaying. Working with the Economic and Development Board, the tourism authorities were and are actively canvasing for conventions and exhibitions to be held there, and encourage global companies, organizations and universities to set up regional offices/campuses. For such a strategy to work, food security and food safety are assumed to be non-issues; Singapore is a safe, stable and secure place, with culinary temptations.

While some advertisements target tourist and local alike some, such as this advertisement showing a European man enthusiastically tucking into local food on a table groaning with food, are international in focus (Figure 2). The central figure is clearly a tourist, evident in part from what he is eating. He reaches with his hand and chopsticks toward the food and the viewer; he appears to be about to attempt to eat shaved ice with chopsticks. There fourteen dishes on the table, with another about to be delivered. The dishes are an incoherent mix of snacks, main dishes, and desserts; a plate of chili crab sits next to a bowl of ice-kacang. For a tourist audience it is a portrayal of plenty and exoticism, rather than culinary chaos.

The food and environment mark this as a hawker center (Lau Pa Sat Festival Pavilion), architecturally exotic. Excess is acceptable in this Othered space. At a time when carbon-footprint-conscious globalism accompanies advocacy of low-fat, low-cholesterol, low-salt, low-sugar austerity, the change of location offered by travel can also offer an escape from the moral economy of food politics. The worries of calories and carbon footprints can be temporarily, and temporally, suspended. The text on the advertisement: 'It's easy to see why diet books seldom make it to the Singapore best-sellers list"—reinforces the distancing from the everyday and positions Singapore as a space in which excess is acceptable. Excess featured in other advertisements as well, such as the 2001 example which asks: "Why must a day have only three meals?"(Figure 3) Since New Asia in the 1990s, food tourism has evolved together with the wider national economic and tourism development strategy. There are at least three inter-related approaches used.

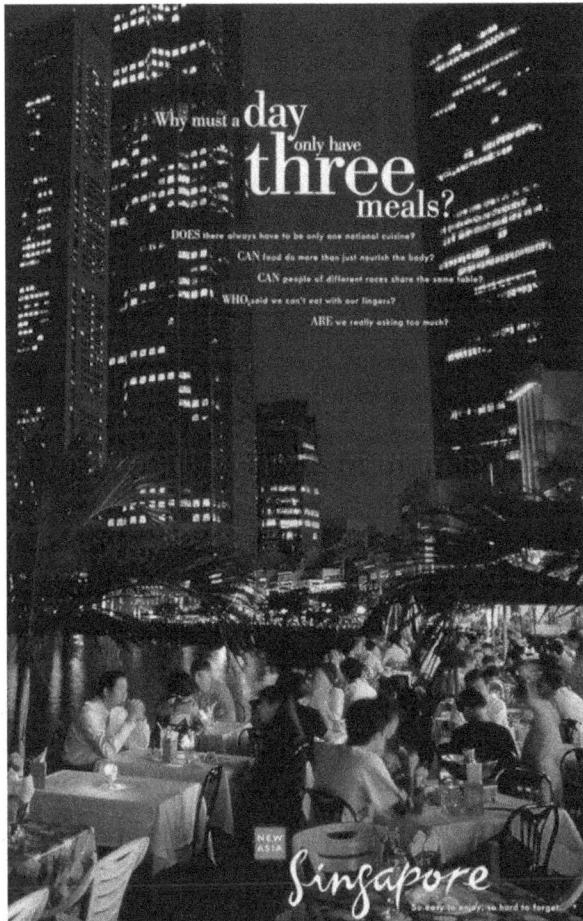

Figure 3. New Asia Advertisement 2011. Source: Singapore Tourism Board.

Food destination approach 1: popularizing a rich Singaporean food culture

The first approach focusses on creating awareness of Singaporean food and food culture, and suggests that the tropical island has cheap and delicious food. Attempts are made to introduce Singapore-style hawker culture to the world. For instance, STB worked with famous Danish gastronomic entrepreneur and chef Claus Meyer to open Nam Nam Restaurant and participated in the Copenhagen Cooking festival in the Danish capital, featuring Singapore hawker fare, including *roti prata* and chili crab (Singapore Tourism Board, 2014).

The popularization of Singaporean food culture around the world was taken to a higher level in 2018. In the Singapore Film Commission and STB-supported Hollywood movie, *Crazy Rich Asians*, various aspects of Singaporean culture were showcased. Food featured prominently (Loh, 2018; Yip, 2018). Nick Young—the English accented romantic hero of the film—explains the cuisine, acting as a guide and teacher for the audience. When the characters enter the Hawker Center Nick offers a reassuring explanation to the visual chaos: "Each of these hawker stalls sells pretty much one dish, and they've been perfecting it for generations." He goes on to tell his American girlfriend, and the audience, that "this is one of the only places in the world where street food vendors actually earn Michelin stars" (2018). Nick also helps the audience negotiate class and race. He is elite, yet comfortable with humble workers, evidenced by his use of vernacular language and the warm embrace he gives one the hawkers. Nick's comfort in this environment helps make the exotic seem accessible. The food choices are conventional—satay, laksa, beer, noodles, crabs—items that tourists are encouraged to sample as local fare. Other food moments in the film also use conventional or familiar foods, such as making dumplings. When Nick's mother, Eleanor Wong, inspects the catering at her house party, we are shown Peranakan desserts, while the kueh (cakes) may be unfamiliar, they are a very accessible food for a Western palate. The food choices in the film are thus constructing anticipatory desire for future tourists. In fact, the appeal of kueh is stressed elsewhere as well. On a list of best kueh on the popular food recommendation website *Hungry Go Where*, Kaya Kueh Ko Swee, the same type as featured in the *Crazy Rich Asians* kitchen scene, are especially recommended for visitors (Foo, 2018):

> These small green-coloured rice cake balls might not look like much, but take a bite and you'll experience a textbook example of "party in your mouth". Coated with grated coconut, the chewy rice cake balls implode into a heavenly combination of liquid palm sugar and pandan juice once you bite into them. If you've got a foreign friend in town and you want to impress them with our local kuehs, this should be your top pick.

Crazy Rich Asian has emerged as a powerful marketing tool. For instance, CNN dedicated a story on "The 'Crazy Rich Asians' guide to Singapore" (Kolesnikov-Jessop, 2018). Besides visiting the iconic Raffles Hotel, Gardens by the Bay and Marina Sands Bay integrated resort, two hawker food centers made it to the list:

> Hawker food is a big part of Singapore's culture, bringing together people from all walks of life – including the super-wealthy – to enjoy simple, tasty food … While the book refers to the historic Lau Pa Sat hawker center (*18 Raffles Quay*), where the main protagonists head for satay on their arrival in Singapore, the movie picked the touristy Newton Food Center for that scene.

The article then proceeded to describe both food centers, and recommended dishes to try out, including chili crab, oyster omelet and "Hokkien popiah (the equivalent of a stuffed savory crepe with chopped prawns, eggs, bean sprouts and lettuce)" (Kolesnikov-Jessop, 2018). The movie also showcased eating and drinking activities in fancy restaurants, such as having cocktails by the 57th floor infinity pool overlooking the city at Marina Bay Sands. Food culture in Singapore is diverse, rich, and safe.

At the premier of the movie in Hollywood in August 2018, STB hosted a three-day event, featuring Singapore food, drinks, chefs and artists. Not to give the wrong impression, Regional Director of STB Americas pointed out (Loh, 2018): "Not opulent luxury rich – but how rich we are culturally, and how rich we are in terms of our lifestyle offerings in Singapore." Few visitors get to fly to Singapore and sleep in a first-class bed, like Nick Young and Rachel Chu do in the film. It may be possible to get a two-dollar bowl of Michelin star noodles, but the Singapore of *Crazy Rich Asians* is not intended to attract backpackers.

Food is used to mediate the meaning of culture, and requires explanation of cultural navigation, provided by media, what Frost et al. (2016, 10) describe as "co-creation." In an article in the American food magazine *Bon Appetite* (Whitney, 2018) the hawker center scene discussed above is described in the following terms: "That scene serves as an introduction to Singapore for many moviegoers in America, and if there's one clear takeaway, it's that you should want to eat there." *Bon Appetite* continues to do this work, providing a behind-the-scenes video about hawker center foods in *Crazy Rich Asians*, a foodcast interview with stars about their Singaporean food experiences, and links to previous *Bon Appetite* articles about Singapore which includes a Singapore Restaurant Hot List. With luscious imagery and explanations it would be hard not want to eat in a Singapore that is at once safe and exotic, familiar and different, affordable and luxurious. In fact as Chang and Pang (2017, p. 220) have shown in their analysis of visitor expectations at Universal Studios in Singapore, visitors now have gustatory expectations of local food.

Food destination strategy 2: international recognition and inscription

Today, hawker centers are tourist attractions (Henderson, Ong, Poon, & Xu, 2012), promoted as safe places to eat, in which hawkers are licensed and food hygiene standards maintained. It is also compulsory for all cooked food stalls to display their food hygiene ranking by the authorities. In taking a step further in 2018, the Prime Minister announced that Singapore's hawker culture will be nominated for inscription into UNESCO's Representative List of the Intangible Cultural Heritage of Humanity (Zaccheus, 2018). The hawker culture in Singapore is considered an intrinsic part of Singaporean heritage and identity. A successful inscription into the UNESCO list will, according to the Prime Minister, "let the rest of the world know about our local food and multicultural heritage" (Zaccheus, 2018).

UNESCO offers international recognition and is a tactic in the accreditation approach to destination branding. The accreditation approach to destination branding uses "appropriate independent and external references to shape the image and perceptions of a place in a desired direction" (Ooi, 2014, pp. 232–233). The accreditation approach addresses two inherent challenges in destination branding. The first is the lack of credibility in the self-promotion of a destination; using an "independent" and external body (e.g. UNESCO) to vouch for the destination would be more credible and effective. The second challenge is that the uniqueness of the destination brand story may be lost to diverse audiences around the world as they have different backgrounds and knowledge; an internationally established and widely understood set of lenses may be used to help diverse audiences around the world interpret the destination and tourist products (e.g. a Guggenheim-branded museum). The Botanical Gardens in Singapore has recently been inscribed as a UNESCO heritage site, lifting the status of the tourist attraction. The recognition celebrates the park's universal values. The hawker center culture in Singapore is deemed to be at the same level of importance and interest.

Food destination approach 3: celebrity chefs and restaurants as destinations

The third approach is also an accreditation tactic in destination branding. High food hygiene standards have allowed Singapore to engage with the international gastronomy industry and create a class of restaurant-destinations, that is restaurants and food places that are the main attractions for certain visitors coming to Singapore (Henderson, 2011; Henderson et al., 2012; Ooi & Pedersen 2017; Tresidder 2015). With the support of the STB, Singapore was the first Southeast Asian nation to be rated by the Michelin Guide, and in 2016 (the first year of rating) several restaurants and a hawker stall were awarded Michelin stars. Without saying, the food must also be safe. The Michelin Guide accredits the quality of the food, and inevitably the food safety standard of the country. Chan Hon Meng, owner of the hawker stall, Hong Kong Soya Sauce Chicken Rice and Noodle, won and has come to signify the image of a hawker for international audiences (Henderson, 2016; 2017). Overnight, he became a celebrity chef. He is the star of a short film produced by Michelin and has been widely interviewed. In the film, Chan toasts with a flute of champagne "Cheers to all us hawkers! I really think that it an honour that I can represent every hawker in Singapore internationally and bring our cuisine to the world." Chan has big dreams, for the world to discover Singaporean cuisine and for "more of the undiscovered local hawker talent" to be recognized (Michelin Guide Singapore 2016). Marketed as the world's cheapest Michelin-starred meals, Chan's soya sauce chicken noodle or rice costs only two Singapore dollars, although as he regularly sells out, the cost in time is far greater—queues for lunch begin early in the morning—and include many international visitors. Chan is also bringing his noodles to the world and Melbournians were told to "brace yourself for the arrival of Hawker Chan – the world's cheapest Michelin-starred meal, period." (Cody, 2017). This image of the hawker in Singapore has come a long way from the 1960s, yet that historic image remains important to the marketing of street food. The Michelin Guide affirms the idea that the city-state has quality food.

Michelin awarded stars to more traditional fine dining establishments as well. In fact, Singapore is increasingly known for its award-winning fine dining, thanks, in part, to the efforts of the STB. Such restaurants are international in approach and cuisine, with alumni of well-established cuisine-defining restaurants such as Heston

Figure 4. Cassis plum exterior. Source: Author.

Blumenthal's *The Fat Duck* and Spanish restaurants such as *El Celler de Can Roca* and *Zuberoa*. Beyond the international iconography of Tetsuya branded restaurants, there are also local super-stars. Pastry chef Janice Wong, epitomizes the Singaporean celebrity chef role, and she is regularly featured in international travel infotainment media, such as the television shows *Destination Flavour* (2017) and *Poh's Kitchen* (2011). Wong and her dessert the "Cassis Plum" was, for example, featured as a recreation challenge in the Australian edition of *MasterChef* (2015) (Figure 4). The dessert appears to be a simple sphere, but is a complex cassis sorbet shell, with elderflower yoghurt foam, plum liqueur and a yuzu heart Wong is an artist using edible materials, and her craft has become a short-hand for high-end Asian dining, making food acceptably exotic but culturally safe (Figure 5).

There is, in Krishnendu Ray's (2016, p. 183) terms a "global hierarchy of taste," in which the food of certain cultures are considered cuisines, and command a higher cultural capital. Writing in the *Atlantic* Joe Pinsker (2016) described this hierarchy as "the more capital or military power a nation wields and the richer its emigrants are, the more likely its cuisine will command high menu prices." For Singapore this holds partly true. It is no coincidence that the restaurant (*Odette*) that is most frequently identified as the best fine dining option in Singapore, specializes in "Modern French Cuisine." As Amy Trubek (2000, 3) shows the "discourse and practice of the French, or what is said and done, has always provided the foundation for action far beyond France: the French invented the cuisine of culinary professionals." In the words of the *Odette* itself, that cuisine "is guided by Chef Julien's lifelong respect for seasonality, terroir and artisanal produce" (2019). Terroir is a complicated concept in a city-state reliant on imported foods. Claims to seasonality in an equatorial nation, are also different. But the trifecta of seasonality, terroir and artisanal, are recognizable in the global hierarchy, and the principles are liberally adopted or localized in Singapore.

With a majority Chinese population, Singapore has multiple examples of Chinese fine-dining. Chinese food, with its many regional variances, is a rich category of dining allowing for the familiar and the exotic within. Local food is also increasingly expressed as fine dining, although in more contested ways. Two fine dining establishments, Michael Lee's *Candlenut* and Violet Oon's *National Kitchen by Violet Oon* highlight the fracture lines. Both of them are rigorously promoted by the tourism authorities because of their Peranakan cuisine. That is the cuisine of descendants of

Figure 5. Cassis plum interior. Source: Author.

early Chinese immigrants who came to the Malay archipelago and Singapore, inter-married with Malays and Indonesians, and maintained their Chinese names and some cultural practices while speaking Malay and often dressing in Malay styles (Rudolph, 1998). The women were known as "Nonyas" and thus Peranakan cuisine is often known as Nonya cuisine (Ong, 2016), a fusion of Peranakan Chinese and Malay cuisines, generally emphasizing the application of Chinese culinary techniques to Malay flavors.

Michael Lee is recipient of a Michelin star, and is the only holder of one for Peranakan cuisine (see Tarulevicz & Hudd, 2018 on Peranakan cuisine). He interprets traditional Peranakan dishes as is evident in his use of buah keluak. Buah Keluak is the seed of the kepayang tree and must be cured before eating. Traditionally this was done by boiling, immersing in ash and burial in the ground. Lee deploys the ingredient traditionally and in new ways, as one review noted:

> While other Peranakan restaurants doggedly adhere to time-honoured recipes, Lee isn't afraid to take the occasional departure from tradition. Buah keluak appears thrice on his a la carte menu: first in its classic role as a rich curry base for ayam buah keluak; then as an earthy hit of umami in rice that is fried until it's gloriously imbued with wok hei; and most unexpectedly, a buah keluak ice cream with Valrhona chocolate and a hint of chilli. It's a combination that works and successfully teases out the nutty cocoa notes of the versatile nut. (Kok, 2018)

While some chefs have been criticized for using traditional Peranakan ingredients in unconventional ways, such as serving buah keluak as a sauce for spaghetti—which purists argue is a foreign, introduced ingredient out of place in Peranakan cuisine—others counter that Peranakan cooking, as a mix of Malay and Chinese, has always borrowed from other cultures (Lam, 2017). For Lee tradition and reinterpretation are not in conflict and his website notes that he "is constantly looking for ways to both preserve and innovate the flavours of his youth" (Lee, 2019). But also that this is not always well received: "As an innovative Peranakan chef, Malcolm is well aware of resistance to his dishes. 'Doing Peranakan food in Singapore is tough,' he laughs. 'It can make you feel like a target board'" (Lee 2019). Regardless of the local criticisms, foreign visitors and foreign Michelin evaluators provide credence to Lee by celebrating his cooking. Cultural safety cohabits with literal safety.

The *National Kitchen by Violet Oon* at the National Gallery Singapore is not explicitly connected to exhibition content beyond its status within the national gallery, rather it is designed to represent "Singapore's history as the crossroads of the world" (Oon, 2017). National Gallery Singapore is a major tourist draw and the restaurant is meant to complement and enhance the art experience for visitors. Violet Oon, restaurateur and former food critic, is of Peranakan ethnic heritage—the restaurant reflects this Peranakan heritage, in cuisine and esthetics. It is decorated with Peranakan tiles and old photographs of family members, evoking the colonial era. Historically, Peranakans were an elite group in Singapore, often working as clerks or overseers in British firms, and this is reflected in the inclusion of several British colonial dishes. The classic Singapore Sling, a drink concocted for European women in the famous colonial Raffles Hotel, and Coronation Chicken epitomize this. Most dishes, however, are classic Peranakan, such as Buah Keluak Ayam (spicy stew of chicken and buah keluak nut), Nonya Achar (mixed vegetable pickle), and Kueh Beng Kah (tapioca cake topped with gula melaka sirup and coconut milk). In Nonya cuisine, a dish like Bak Chang, while similar to Chinese glutinous rice packages, can be made with peaflower and Southeast Asian flavorings such as pandan leaves (Tarulevicz & Hudd, 2018).

National kitchen suggests that the visitor will experience a national cuisine, one that includes dishes from *all* ethnicities in Singapore. That is a problematic claim. Culturally, Peranakan food, despite its mix of Chinese and Malay, is read as Chinese. As Chua and Rajah (2003, p. 95) note, for non-Peranakans, the cuisine "resonates on the register of Chinese cuisine, as a marker of Chinese ethnicity, in spite of the hybridization." The key reason they suggest is that although there are Malay flavors, there are Chinese ingredients, most notably pork, a product not permissible for the majority of Malays who are Muslim. The cuisine thus "remains a 'Chinese' cuisine because of the presence of pork" (Chua & Rajah, 2003:95). The power of this cuisine lies in its metaphorical value, reflecting both the cultural mix of Singapore's peoples as well as the "cultural legacies of the Straits Chinese communities" in the region (Lee, 2001, p.2). Nonetheless, while Peranakan cuisine exists as a marker of Chinese culture for Singaporeans, it also occupies a place as a marker of creolization—an early fusion culture, a kind of acceptable mingling that codifies ethnic categories in a palatable manner (Tarulevicz & Hudd, 2018). Marketing the nation, domestically and internationally, requires culture to be displayed in ways that are recognizable, but not necessarily identical to, Western notions of restaurant high culture. The form of fine dining, however, remains familiar.

In spite of wanting to be local, fine dining restaurants in Singapore use the international gastronomy language of curating and presenting the dining experience. As Rebecca L. Sprang (2000, p. 236) noted in her history of the invention of restaurants, one of the key purposes of restaurants "was to eclipse the kitchen, to pull a curtain of illusion across the real conditions of production, to aestheticize and tidy." Her description of Paris restaurants where "...customer's never witnessed the cooking of food, the chopping of ingredients, the plucking of feathers, or the draining of blood; instead they waited," and drank wine "until the waiter, a Charon-figure passing between worlds, appeared with a paradisiacal bounty of flavors, smells, textures, and sights" still resonates with many iterations of contemporary Singaporean fine dining. In order to be fine dining, the grotesquery of food production must be invisible, the safety assumed in invisible cleanliness. Home-grown restaurants become destinations in their own right, as eating in them is one of the primary reasons for visiting Singapore.

Conclusion

This paper outlines the history of food safety and of tourism in Singapore, paying particular attention to its textual construction, to show the ways they have become interconnected. Singapore's growth as a port city was accompanied by public health challenges and authorities and citizens alike focused attention of the regulation of public eating spaces, especially those relating to hawking. Officials, citizens and visitors lamented food safety issues, but many also celebrated the food and its important social functions. Ultimately hawker food has been transformed from a source of concern, to a key tourism marketing platform, where it has symbolic values around cultural diversity and acceptable excess. Tourism has spearheaded infrastructure provision with the development and "touristification" strategies of Singapore being in locked-step. The use of external measures, such as UNESCO status and Michelin rankings work to make food the reason for visiting Singapore. Media, from infotainment to films such as *Crazy Rich Asians* are used to showcase and explain this destination-worthy cuisine, and set up expectations for visitors. For providers of tourist services, and marketers alike, making sure that experiences are safe but also conform to imagined exoticisms can be challenging to navigate.

The presentation of food in popular culture draw attention to the tension between high and low food culture, between safe and unsafe, between street food and fine dining, and they are being negotiated in Singapore. Taste, its arbiters, makers and guardians, are raced, and part in Ray's (2016, p. 183) terms of a "global hierarchy of taste." Certain European traditions, especially French, continue to dominate the international framing of fine-dining, and while Singapore is not immune to this, it is also producing its own superstars, its Janice Wong's and Michael Lee's. Fine dining might assume cleanliness and make concerns about food safety seem peripheral, but it beholds us to remember that restaurants had that invisibility at their core from their inception. Off stage, does not mean unimportant. If anything, the off stage-ness, the assumption, is evidence of just how important safety is—in fine dining, and beyond. The titular microbial Russian roulette does not have to be played discursively or literally in contemporary Singapore.

Disclosure statement

No potential conflict of interest was reported by the authors.

ORCID

Nicole Tarulevicz (iD) http://orcid.org/0000-0002-9884-5057
Can Seng Ooi (iD) http://orcid.org/0000-0002-0824-3766

References

Acerbi, P. (2017). *Street occupations: Urban vending in Rio de Janeiro, 1850-1925*. Austin: University of Texas Press.

Barnard, T. P. (Ed.). (2014). *Nature contained: Environmental histories of Singapore*. Singapore: NUS Press.

Barnard, T. P., & Heng, C. (2014). A city in a Garden. In T. P. Barnard (Ed.), *Nature contained* (pp. 281–306). Singapore: NUS Press.

Calaresu, M., & van den Heuvel, D. (2016). Introduction: Food hawkers from representation to reality. In M. Calaresu & D. van den Heuvel (Eds.), *Food hawkers: Selling in the streets from antiquity to the present* (pp. 1–18). New York, NY: Routledge.

Chang, T. C., & Lee, W. K. (2003). Renaissance City Singapore: A study of arts spaces. *Area*, *35*(2), 128–141. doi:10.1111/1475-4762.00155

Chang, T. C., & Pang, J. (2017). Between universal spaces and unique places: Heritage in Universal Studios Singapore. *Tourism Geographies*, *19*(2), 208–226. doi:10.1080/14616688.2016. 1183141

Chua, B.-H., & Rajah, A. (2003). Food, ethnicity and nation. In B. H. Chua (Ed.), *Life is not complete without shopping: Consumption culture in Singapore* (pp. 93–120). Singapore: National University of Singapore Press.

Cody, J. (2017), *Good food guide website*. Retrieved from: https://www.goodfood.com.au/eat-out/ news/singapores-michelinstarred-street-vendor-hawker-chan-to-open-in-melbourne-20170623-gwx7wq

Destination Flavour. (2017). *SBS website*. Retrieved from: https://www.sbs.com.au/food/programs/ destination-flavour

Fitch, G. H. (2017). A strange spectacle (1913). In I. Manley (Ed.), *Tales of old Singapore: The glorious past of Asia's Greatest Emporium* (p. 155). Hong Kong: Earnshaw Books.

Foucault, M. (1976). *The will of knowledge: The history of sexuality volume 1*. (Robert Hurley, Trans.). London: Routledge.

Frost, W., Laing, J., Best, G., Williams, K., Strickland, P., & Lade, C. (2016). *Gastronomy, tourisn and the media*. Bristol: Channel View Publications.

Gold, J. (2007). Singapore street food. Retrieved from http://www.saveur.com/article/Travels/ Singapore-Street-Food

Gulrud, N. M., & Ooi, C.-S. (2015). Manufacturing green consensus: Urban greenspace governance in Singapore. In L. A. Sandberg, A. Bardekjian, & S. Butt (Eds.), *Urban forests, trees, and greenspace: A political ecology perspective* (pp. 77–92). New York, NY: Routledge.

Hahn, E. (2017). The Rat City. In I. Manley (Ed.), *Tales of old Singapore: The glorious past of Asia's Greatest Emporium* (p. 20). Hong Kong: Earnshaw Books.

Hawkers in Singapore. (1924, August 23). *The Straits Times*, p. 9.

Health Officer Defines His Proposal. (1924, August 21). *The Straits Times*, p. 9.

Henderson, J. (2016). Local and traditional or global and modern? Food and tourism in Singapore. *Journal of Gastronomy and Tourism, 2*(1), 55–68. doi:10.3727/216929716X14546365943494

Henderson, J. (2017). Street food, hawkers and the Michelin Guide in Singapore. *British Food Journal, 119*(4), 790–802. doi:10.1108/BFJ-10-2016-0477

Henderson, J., Ong, S.-Y., Poon, P., & Xu, B. (2012). Hawker centres as tourist attractions: The case of Singapore. *International Journal of Hospitality Management, 31*(3), 849–855. doi:10.1016/j.ijhm.2011.10.002

Henderson, J. C. (2011). Celebrity chefs: Expanding empires. *British Food Journal, 113*(5), 613–624. doi:10.1108/00070701111131728

Hon, J. (1990). *Tidal fortunes: A story of change, The Singapore River and Kallang Basin*. Singapore: Landmark Books.

Kiat, G. H. (1924, August 26). The hawker question. *The Singapore Free Press and Mercantile Advertiser*, p. 12.

Kok, D. (2018). Candlenut. Retrieved from https://thepeakmagazine.com.sg/gra-2018-awards-of-excellence/candlenut/

Kolesnikov-Jessop, S. (2018). The "Crazy Rich Asians" guide to Singapore. Retrieved from https://edition.cnn.com/travel/article/crazy-rich-asians-travel-guide-singapore/index.html

Kong, L. (2007). *Singapore hawker centres: People, places, food*. Singapore: National Environment Agency.

Lam, S. (2017). The slow death of Peranakan cuisine? Retrieved from http://www.channelnewsasia.com/news/singapore/the-slow-death-of-peranakan-cuisine/3421788.html

Lee, G.-B. (2001). *Nonya favourites*. Singapore: Periplus.

Lee, M. (2019). *Candlenut restaurant website*. Retrieved from: https://comodempsey.sg/restaurant/candlenut

Lee, P. (2004). *Singapore, tourism and me*. Singapore: Pamelia Lee.

Lertputtarak, S. (2012). the relationship between destination image, food image, and revisiting Pattaya, Thailand. *International Journal of Business and Management, 7*(5), 111–122. doi:10.5539/ijbm.v7n5p111

Lin, Y.-C., Pearson, T. E., & Cai, L. A. (2011). Food as a form of destination identity: A tourism destination brand perspective. *Tourism and Hospitality Research, 11*(1), 30–48. doi:10.1057/thr.2010.22

Lockhart, R. H. B. (2017). Satai (1936). In I. Manley (Ed.), *Tales of old Singapore: The glorious past of Asia's Greatest Emporium* (p. 153). Hong Kong: Earnshaw Books.

Loh, G. S. (2018). Singapore takes over Hollywood as Crazy Rich Asians premieres. Retrieved from https://www.channelnewsasia.com/news/lifestyle/crazy-rich-asians-premiere-singapore-actors-talent-party-10607358

Maclaurin, T. (2004). The importance of food safety in travel planning and destination selection. *Journal of Travel & Tourism Marketing, 15*(4), 233–257. doi:10.1300/J073v15n04_02

Making Night Hideous. (1910, December 14). *The Straits Times*, p. 10.

Manley, I. (Ed.) (2017). *Tales of old Singapore: The glorious past of Asia's Greatest Emporium*. Hong Kong: Earnshaw Books.

Master Chef. (2015). *Channel 10 website*. Retrieved from: https://10play.com.au/masterchef/recipes/cassis-plum/r190614xbmak

Mendiola García, S. (2017). *Street democracy: Vendors, violence, and public space in Late Twentieth-Century Mexico*. Lincoln: University of Nebraska Press.

Michelin Guide Singapore. (2016). *Guide website*. Retrieved from: https://guide.michelin.com/sg/en

Ministry of Information & the Arts. (2000). *Renaissance city report: Culture and the arts in Renaissance Singapore*. Singapore: MITA.

Ministry of Information Communication and the Arts (MICA). (2008). *Renaissance city plan III*. Singapore: Ministry of Information, Communication and the Arts.

Municipal Matters. (1924, June 4). *The Straits Times*, p. 8.

National Tourism Plan Committees. (1996). *Tourism 21: Vision of a tourism capital*. Singapore: Singapore Tourism Promotion Board.

Odette. (2019). Resturant website. Retrieved from http://www.odetterestaurant.com/

Ong, J-T. (2016). *Nonya heritage kitchen: Origins, utensils and recipes*. Singapore: Landmark Books.

Ooi, C.-S. (2002). *Cultural tourism and tourism cultures: The business of mediating experiences in Copenhagen and Singapore*. Copenhagen: Copenhagen Business School Press.

Ooi, C.-S. (2004). Brand Singapore: The hub of New Asia. In N. Morgan, A. Pritchard, & R. Pride (Eds.), *Destination branding* (pp. 242–262). London: Elsevier Butterworth Heinemann.

Ooi, C.-S. (2005a). State-civil society relations and tourism: Singaporeanizing tourists, touristifying Singapore. *Sojourn - Journal of Social Issues in Southeast Asia, 20*(2), 249–272.

Ooi, C.-S. (2005b). The orient responds: Tourism, orientalism and the national museums of Singapore. *Tourism, 53*(4), 285–299.

Ooi, C.-S. (2011). Subjugated in the creative industries: The fine arts in Singapore. *Culture Unbound: Journal of Current Cultural Research, 3*(2), 119–137. doi:10.3384/cu.2000.1525.113119

Ooi, C.-S. (2014). The making of the copy-cat city: Accreditation tactics in place branding. In P. O. Berg & E. Björner (Eds.), *Branding Chinese mega-cities: Policies, practices and positioning* (pp. 232–248). Cheltenham: Edward Elgar.

Ooi, C.-S. (2018). Global city for the arts: Weaving tourism into cultural policy. In T. Chong (Ed.), *The state and the arts in Singapore* (pp. 165–179). Singapore: World Scientific.

Ooi, C. S. & Pedersen, J. S. (2017). In search of Nordicity: How new Nordic cuisine shaped destination branding in Copenhagen. *Journal of Gastronomy and Tourism, 2*(4), 217–231. doi:10. 3727/216929717X15046207899375

Ooi, C.-S., & Tarulevicz, N. (2019). From third world to first world: Tourism, food safety and the making of modern Singapore. In E. Park, S. Kim, & I. Yeoman (Eds.). *Food tourism in Asia* (pp. 73–88). Singapore: Springer Singapore.

Oon, V. (2017). National kitchen by Violet Oon. Retrieved from https://violetoon.com/national-kitchen-by-violet-oon-national-gallery-singapore/

Peh, S. H. (2006, August 19). Lively city means more tourists and jobs: MM - Wholesome and orderly place is no longer good enough or a new Singapore. *The Straits Times*, p. 1. Retrieved from http://eresources.nlb.gov.sg/newspapers/Digitised/Issue/straitstimes20060819-1

Pinsker, J. (2016). The future is expensive Chinese food. *Atlantic*. Retrieved from https://www.the-atlantic.com/business/archive/2016/07/the-future-is-expensive-chinese-food/491015/

Ray, K. (2016). *The ethnic restraurateur*. New York, NY: Bloomsbury.

Report of the committee appointed to investigate the hawker question in Singapore. (1924). Singapore: Government Printing.

Rickly-Boyd, J. (2013). Existential Authenticities: Place Matters. *Tourism Geographies, 15*(4), 680–686.

Rubbish Strewn Roads. (1911, July 29). *The Straits Times*, p. 10.

Rudolph, J. (1998). *Reconstructing identities: A social history of the Babas in Singapore*. Aldershot: Ashgate.

Seah, S. (2019, January 17). Explainer: New laws to pave the way for creation of Singapore Food Agency. *Today*. Singapore. Retrieved from https://www.todayonline.com/singapore/explainer-new-laws-pave-way-creation-singapore-food-agency

Singapore Tourism Board. (2008). Visitor arrival statistics. Retrieved from https://www.stb.gov.sg/statistics-and-market-insights/marketstatistics/2005vas.pdf

Singapore Tourism Board. (2010). *Brand overview*. Retrieved from https://app.stb.gov.sg/asp/des/des05.asp

Singapore Tourism Board. (2014). Authentic and innovative Singapore flavours featured at Copenhagen Cooking. Retrieved from https://www.stb.gov.sg/news-and-publications/lists/newsroom/dispform.aspx?ID=526

Singapore Tourism Board. (2018). International visitor arrivals statistics. Retrieved from https://www.stb.gov.sg/statistics-and-market-insights/marketstatistics/visitorarrivals2018.pdf

Singapore Tourism Board. (2019). History. Retrieved from https://www.stb.gov.sg/about-stb

Sprang, R. (2000). *The invention of the restaurant: Paris and modern gastronomic culture.* Cambridge, MA. Harvard University Press.

The Straits Times. (1997, August 3). Fusion cuisine, or confusion? *The Straits Times*, p. 9.

Tarulevicz, N. (2013). *Eating her curries and Kway: A cultural history of food in Singapore.* Champaign, IL: University of Illinois Press.

Tarulevicz, N. (2015). I had no time to pick out the worms: Food adulteration in Singapore. *Journal of Colonialism and Colonial History, 16*(3), 1–24. doi:10.1353/cch.2015.0037

Tarulevicz, N. (2018). Hawkerpreneurs: Hawkers, entrepreneurship, and reinventing street food in Singapore. *Revista de Administração de Empresas, 58*(3), 291–302. doi:10.1590/s0034-759020180309

Tarulevicz, N., & Hudd, S. (2018). From natural history to national kitchen: Food in Singaporean museums. *Digest, 6*, 18–44.

Teo, P., & Chang, T. C. (2000). Singapore: Tourism development in a planned context. In C. M. Hall & S. Page (Eds.). *Tourism in South and Southeast Asia: Issues and cases* (pp. 117–128). Oxford: Butterworth-Heinemann.

The comforts of Singapore hotel life. (1876, May 6). *The Straits Times*, p. 2.

The hawkers case: Hawkers and patrons condemn abolition. (1924, June 21). The Singapore and Mercantile Advertiser, p. 16.

Tresidder, R. (2015). Experiences marketing: A cultural philosophy for contemporary hospitality marketing studies. *Journal of Hospitality Marketing and Management, 24*(7), 708–726. doi:10.1080/19368623.2014.945224

Trubek, A. (2000). *Haute cusine: How the French invented the culinary profession.* Philadelphia, PA: University of Pennsylvania Press.

Verne, J. (2017). One large warehouse (1881). In I. Manley (Ed.), *Tales of old Singapore: The glorious past of Asia's Greatest Emporium* (p. 18). Hong Kong: Earnshaw Books.

Warren, J. (1986). *Rickshaw Coolie: A people's history of Singapore, 1880-1940.* Singapore: National University of Singapore Press.

Whitney, A. (2018). Dumplings, kaya toast, and chili crab: Inside the food of *Crazy Rich Asians*. Retrieved from https://www.bonappetit.com/story/crazy-rich-asians-food-singapore

Wise, M. (Ed.). (1985). *Travellers' tales of old Singapore.* Singapore: Marshall Cavendish International (Asia) Private Limited.

Wong, L. K. (1978). Singapore: Its growth as an entrepot port, 1819-1914. *Journal of Southeast Asian Studies, 19*(1), 50–84. doi:10.1017/S002246340000953X

Yip, W. Y. (2018, August 22). Made-with-Singapore Crazy Rich Asians a big hit. *The Straits Times*. Retrieved from http://str.sg/oW5z

Your food will cost you more. (1924, May 28). *The Singapore Free Press and Mercantile Advertiser*, p. 9.

Zaccheus, M. (2018, August 20). Singapore hawker culture to be nominated for UNESCO listing. *The Straits Times*. Retrieved from http://str.sg/oWkh

Visitor diversification in pilgrimage destinations: comparing national and international visitors through means-end

Ricardo Nicolas Progano, Kumi Kato (ID) and Joseph M. Cheer (ID)

ABSTRACT

In contemporary society, spirituality has dissociated from the tenets of organized 'official' religion, resulting in a rise of 'private' spirituality, defined by each individual's beliefs. In this context, visitors from a great variety of national backgrounds are increasingly visiting pilgrimage sites across the globe, even if they have little to no cultural connections to them, bringing with them a diverse range of values to the pilgrimage site. Despite the growing presence of international visitors from across the globe, nationality has not been a studied factor when researching tourism in pilgrimage-related destinations. In order to bridge this research gap, the present study's objective is to examine visitor diversification in pilgrimage tourism through a study of similarities and differences of values among domestic and international visitors. Utilizing means-end as a qualitative research methodology, the two most numerous nationalities were sampled: Japanese and Australians. Fieldwork was conducted in the Nakahechi trail of Kumano Kodo, an ancient pilgrimage site located in Tanabe city (Japan) developed for international tourism. The Nakahechi route is a popular route for both domestic and international visitors due to its cultural significance, easy access and moderate challenge. Results showed a variety of similarities and differences between the sampled nationalities, demonstrating a growing diversification in sacred sites which incorporates a complex range of elements related to leisure, sports, intercultural exchange, nostalgia, escapism and relaxation, beyond a continuum of contemporary spirituality and traditional religion. In conclusion, it was observed that nationality is a fundamental factor for studying pilgrimage tourism in contemporary society. As pilgrimage sites continue to develop into international destinations, nationality is an important factor that requires further attention from academics. Results also have practical implications for local administrations aiming to develop their pilgrimage resources to international visitors.

摘要

在当代社会, 灵性已经脱离了有组织的"官方"宗教的信条, 导致了由每个人的信仰定义的"私人"灵性的兴起。在这种背景下, 来自不同国家背景的游客越来越多地前往世界各地的朝圣地点, 尽管他们与这些地方几乎没有文化联系, 但却为朝圣地带来了各种各样的价值观。尽管来自世界各地的国际游客越来越多, 但在研究朝

圣目的地的旅游时, 民族并不是一个研究因素。为了弥补这一研
究空白, 本研究的目的是通过研究国内外游客价值观的异同来考
察游客在朝圣旅游中的多样化。本研究以方法-目的为定性研究方
法, 抽样调查了日本和澳大利亚两个拥有极多民族的国家。野外
工作是在熊野科多兽 (Kumano Kodo) 的中边 (Nakahechi) 古道
进行的, 这是一个位于田边市(日本)为国际旅游而开发的古代朝圣
地。由于它的文化意义, 容易到达和适度的挑战, 中边路线是国内
外游客的热门路线。结果显示抽样民族之间有各种各样的相似之
处和不同之处, 表明圣地日益多样化, 除当代精神和传统宗教连续
体之外, 还包括与休闲、体育、文化交流、怀旧、逃避和放松有
关的一系列复杂因素。综上所述, 民族是当代社会研究朝圣旅游
的一个基本因素。随着朝圣地点继续发展成为国际目的地, 民族
是一个重要的因素, 需要学术界的进一步关注。研究结果对地方
管理机构开发面向国际游客的朝圣资源也有实际意义。

Introduction

Extant scholarly literature regarding visitation to pilgrimage sites suggests increasing diversification of visitors underlined by processes whereby traditional religious beliefs have tended to lose significance in contemporary societies (Heelas, 2006; Houtman & Mascini, 2002). As a consequence, visitors to pilgrimage sites exhibit greater tendencies to not be motivated by traditional religious frameworks, but instead are driven by a variety of motivations, beyond religiosity. Indeed, different motivations display a greater tendency to coexist with religious motivations. Pilgrimage sites have increasingly become tourism commodities, as visitors are drawn to them in increasing numbers. Consequently, greater access to pilgrimage sites has emerged (Mori, 2005; Stausberg, 2011), with destinations keen to exploit the desire for tourists to undergo memorable and Instagram-worthy experiences, thus contributing to their growing popularity. Embarking on a pilgrimage has emerged as a key tourism theme in contemporary society (Collins-Kreiner & Wall, 2015), especially with shifts toward the desire for more immersive experiences that give the traveler a sense of meaning and provides opportunity for personal development and reflection.

While there are many significant pilgrimage sites across the globe, the Asia-Pacific region has by far the greatest number of pilgrims and travelers, both international and domestic, attending pilgrimage sites (United Nations World Tourism Organization, 2011). Also, according to the United Nations World Tourism Organization & Global Tourism Economy Research Center (2016), the Asia-Pacific has become the second most visited region after Europe, as well as being the fastest growing. However, research on pilgrimage tourism in the Asia-Pacific has tended to be surprisingly sparse given the significance of pilgrimage sites for some of the countries in the region (United Nations World Tourism Organization, 2011). In Japan, pilgrimage is both a centuries-old tradition and an important aspect of contemporary tourism development, principally in rural areas. In particular, different pilgrimage-related destinations have been aiming to boost both domestic and international tourism in order to achieve regional revitalization (Mori, 2005; Okamoto, 2015). This complements the Japanese government's nascent desire to optimize tourism, particularly inbound tourism as an economic diversification measure that can help address the burden of a rapidly ageing population, and decline in some traditional economic sectors (Prime Minister of Japan and Cabinet, 2016).

Although pilgrimage-related destinations are visited nowadays by an increasing number of international visitors, less attention has been paid to visitor nationality. However, research points out that nationality is an important factor influencing perception of tourists (Pizam & Sussmann, 1995), tourist behavior (Pizam & Reichel, 1996), travel information acquisition (Gursoy & Umbreit, 2004), travel motivation (Kim & Lee, 2000) and host-guest interactions (Reisinger & Turner, 1998), among other variables. The reason for these behavior differences is derived from the different cultural background of each country (Pizam & Sussmann, 1995). Regarding methodology, research is mainly carried out through statistical techniques. However, researchers question the validity and reliability of survey-based methodological tools for motivation and values research in cross-cultural context (Watkins, 2010; Watkins & Gnoth, 2011). This approach has been noted to only provide a list of superficial motivations, not touching deeper themes, as well as the researcher risking to list motivations that may not be the most relevant to the respondents (Jewell & Crotts, 2002). These issues are especially important when undertaking a national segmentation study. How can the researcher be sure that the listed motivations will be relevant to all the studied groups? At the same time, different nationalities may undertake travel for the same motives, but how can it been confirmed that all the visitors share the same underlying, deeper motivation for it? To solve these issues, qualitative methodological approaches, such as means-end (Watkins, 2010; Watkins & Gnoth, 2011) have been suggested as they address the mentioned questions by deeply exploring themes suggested by the participants themselves. Researchers who study visitors through nationality as a classification factor (Watkins & Gnoth, 2011) successfully utilized means-end. At the same time, means-end approach was applied in pilgrimage tourism as well (Kim et al., 2016; Kim & Kim, 2019). Therefore, it was considered that means-end would prove to be an adequate methodology for the present research.

This research aims to study visitor diversification in pilgrimage tourism through an analysis of similarities and differences among domestic and international visitors, in a context of contemporary spirituality. Fieldwork was conducted in the Nakahechi trail (中辺路) of Kumano Kodo (熊野古道), in Tanabe city, Japan. The location was selected due to following rationale. Firstly, Tanabe has been actively aiming to attract international visitors, particularly Western. Secondly, pilgrimage routes and sites constitute its main tourism resource, as prefectural documents show (Wakayama Tourism Agency, 2018). Thirdly, the selected section of Nakahechi route is a popular route because of its cultural significance, easy access and moderate challenge. The two most numerous nationalities were studied, Japanese and Australian, utilizing means-end as research methodology.

Research background

Pilgrimage in contemporary society

Modern narratives of secularization often predict the disappearance or at least the marginalization of religion, defined as an established religious system of doctrines and rituals (Zinnbauer et al., 1997). These predictions are based on the development of modern scientific knowledge and rationalization (Heelas, 2006; Houtman & Mascini, 2002). Still, religion remains as a fundamental aspect of modern life. As a corollary to the evolutionary trends in religiosity, the way spirituality is understood and practiced has also experienced radical changes in contemporary times (Heelas, 2006; Houtman & Mascini, 2002; Okamoto, 2015). Okamoto (2015) argues that religious dogma extended a significant influence over pre-modern societies in spheres such as politics, ethics and education. Because of this powerful influence over society, people shared a common spirituality and made sense of the world around them in a similar way to their fellow citizens. However, due to secularization processes in modern societies, religion lost its dominant influence in societies. In consequence, not under the beliefs of the 'officially' sacred of religious institutions (Di Giovine & Choe, 2019), the individual is free to accept or reject certain aspects of its own religion, as well as combine eclectic elements in its own private, individual spirituality. In contrast to religion, spirituality refers to phenomenon associated with personal feelings of transcendence, extra sensorial perception or meaning of life (Zinnbauer et al., 1997). This phenomenon is called the 'privatization of religion' (*shuukyou no shijika* – 宗教の私事化) (Okamoto, 2015, p. 16). This 'privatization' has led to the expansion of alternative spiritualties, such as the New Age movement (Houtman & Mascini, 2002). In Japan, this movement is known as Spirit World (*Seishin Sekai* – 精神世界), and began to appear in the early 1980s. Media plays a central role on the expansion of the Spirit World movement, presenting its themes in a non-affirmative entertainment format or for life improvement. In this context, the word spirituality (*supirichuaritei* – スピリチュアリティ) gained popularity because it has a foreign fashionable aspect to it (Horie, 2009). The 'privatization of religion' has implications for pilgrimage sites as well. Traditionally, pilgrimage sites were under religious administration, which controlled the site itself, as well as the narratives and meanings surrounding it. However, in contemporary society, individuals have given new narratives and meanings to pilgrimage sites, which coexist with the traditional one. As a result, a diversification in the purpose for visiting pilgrimage sites routes occurred, with secular or spiritual motives and behavior replacing or coexisting with authorized religion (Blom et al., 2016; Di Giovine & Choe, 2019; Okamoto, 2015). Following the above trend, the definition of 'pilgrimage' has expanded to include secular pilgrimage sites (Collins-Kreiner, 2010; Reader, 1993) related to popular culture, politics or historical events (Di Giovine & Choe, 2019; King, 1993; Okamoto, 2015; Stausberg, 2011). This process also expanded the scope of academic research on pilgrimage, which was previously centered on religious pilgrimage. This approach is based on functionalist and group dynamics theories, which originate from anthropology (Reader, 1993). For example, Hyde and Harman (2011) describe the journey of Australian and New Zealanders to the Gallipoli battlefields in Turkey, and how the participants' travel motivations are intermingled with spiritual and secular aspects.

Pilgrimage and tourism in Japan

Japanese local authorities and religious organizations realize the importance of their pilgrimage resources and actively promote them for tourism development (Reader,

Figure 1. View from the Takahara community.
Source: Authors.

2014). The UNESCO World Heritage Sites have also a relationship with tourism development in pilgrimage sites. In the 2000s, Japan experienced a nationwide 'World Heritage boom', as local communities realized the potential of economy revitalization through a UNESCO designation (Matsui, 2014). The first Japanese pilgrimage route designated was the 'Sacred Sites and Pilgrimage Routes in the Kii Mountain Range' in 2004, encompassing a large system of routes and temples located in the Kii Peninsula. Contemporary spirituality also reinterpreted traditional sites and given new meaning to them. In particular, the concept of *iyashi* (癒し) has experienced high notoriety in Japan since 1988 (Matsui, 2013), being associated with different products and services, including traveling. *Iyashi* can be understood as physical and mental restoration, or a holistic healing. This '*iyashi* boom' has also influenced tourism promotion and activities in pilgrimage sites (Mori, 2005). Kumano Kodo has also been promoted as a place for *iyashi* by its local tourism organizations by creating services such as wellness-related walking tours based on the cultural and geographic features of Kumano Kodo (Figure 1).

Case study

Kumano Kodo is a multi-site pilgrimage route structured around three Grand Shrines (*taisha* - 大社): Kumano Hongu Taisha (Tanabe city), Kumano Hayatama Taisha (Shingu city) and Kumano Nachi Taisha (Nachikatsura town). Each of these grand shrines is dedicated to one of the three deities of Kumano (*Kumano Gongen* – 熊野権現), which are also associated with three particular Buddhas. These main pilgrimage sites are linked through three main routes. The Kiiji route, which connects to Kyoto, is comprised three sub-routes: Nakahechi route, Kohechi route and Ohechi route. The Iseji

Figure 2. Map of Kumano Kodo routes, highlighting Nakahechi route.
Source: Kumano Tourism Bureau (reproduced with permission).

route connects the Kumano region with the Ise Grand Shrine, located in Mie prefecture. Finally, the Omine Okugake route connects the Kumano Hongu Taisha with the Yoshino region, in Nara prefecture. Kumano Kodo is associated with Shugendo (修験道), a Japanese mountain faith that appeared as a syncretic religious system at the end of the Heian period (794–1185). Its main premise is the acquirement of supernatural powers through ascetic practices in sacred mountains and their usage for the benefit of society (Miyake, 1966) (Figure 2).

Tanabe city, in Wakayama prefecture, is situated on the west-central part of the Kii peninsula, and includes important tourist resources such as 60 kilometers of pilgrimage trails comprised of the Nakahechi and the Kohechi routes, the Kumano Hongu Shrine and Oyunohara, the former location of Kumano Hongu Shrine. The area is also filled with important Onsen, with the most famous being Yu no Mine. Its segment of the Nakahechi route, which links the Takijiri Ouji shrine to Kumano Hongu Taisha, is the most popular route for tourists. In 2004, properties related to Kumano Kodo were inscribed as the 'Sacred Sites and Pilgrimage Routes in the Kii Mountain Range' World Heritage Site under the category of cultural landscape. To coordinate tourism policies, a local Destination Management Organization, named Tanabe Kumano Tourism Bureau (hereafter, the Bureau), was established in April 2006. As part of its tourism policies, Tanabe city decided to focus on Western tourists and small groups,

Table 1. Domestic and internationals visitors who lodged in Tanabe city between 2015 and 2018.

Year	Total visitors	Domestic visitors	International visitors	Unidentified visitors
2015	443,532	369,085	21,536	52,915
2016	407,427	311,043	30,958	65,426
2017	441,686	307,065	36,821	97,800
2018	444,211	297,386	43,939	102,886

Source: Wakayama Tourism Agency (2015, 2016, 2017, 2018).

estimating that they would have long stays and spend more money. This proved to be right: during fiscal year 2016, the average international visitor spent 2.50 nights in Tanabe, spending an average of 24,008 yens, compared with the Japanese visitor spending 1.45 nights and 10,854 yen on average (Kumano Tourism Bureau, personal interview, Tanabe City, 2016). The Bureau made it easier to travel around the area by setting signs in English, restaurant menus and bus timetables. It also conducted workshops with the local community in order to improve English conversation skills. In 2010, the Bureau established its travel agency, Kumano Travel, which includes an online reservation system. These policies were successful in attracting Western visitors to Kumano. According to data from the Bureau's official online booking site, from April, 2016 to March, 2017, Australia was at the top, with 1,039 visitors (17.65%), followed by United States with 972 (16.51%), and France with 418 (7.10%) (Tanabe Kumano Tourism Bureau, 2018). However, domestic tourists, while in decline, still make up for the majority of overnight visitors. Yet, due to official statistics not counting day-trippers and a large number of unidentified visitors, their number might be bigger than reported (Table 1).

Methods

About means-end

This qualitative research method theorizes that consumers make choices because the specific attributes of their chosen option can help them achieve desired values through the consequences of those choices. The researcher aims to determine the links between attributes, consequences and values (A-C-V) to build means-end chains (MEC) (Reynolds & Gutman, 1988), suggesting that consumers think about product attributes in terms of personal consequences. Attributes are relatively concrete meanings that represent the physical or observable characteristics of a given product. Attributes can be abstract as well when they refer to intangible characteristics such as style, brand or perceived value. Next, consequences reflect the perceived costs or benefits related to a specific attribute. Finally, values represent beliefs or ends that the consumers seek to attain through consumption behavior (Miles & Rowe, 2004). To carry out means-end approaches, the laddering interview is mostly utilized. It involves asking the participant why each attribute is important to them through the simple probe of asking 'why' (Miles & Rowe, 2004). The process continues by moving towards more abstract concepts (consequences and values) until the participant cannot respond any further. The same procedure is carried out for each elicited attribute. After the interviews are conducted, there are transcribed and then content analyzed to identify the different A-C-V of each participant and breaking down all responses into individual summary codes. From the data obtained, an aggregate implication matrix is constructed, which represents the linkages between concepts identified in the laddering interviews. Both direct and indirect relations are represented. Based on this data, next is the construction of the Hierarchical Value Map (HVM), a visual representation of the results found in the aggregate implication matrix. To determine which connections will be represented, a cut-off value is established (Gengler & Reynolds, 1995; Miles & Rowe, 2004). The HVM traditionally is constructed in a tree-like format, but centralized formats also exist (Gengler et al., 1995).

Sampling and data collection

Australian and Japanese visitors were sampled because it would cover both domestic and international visitors, and thus be in line with the research objective. Secondly, because they are most numerous nationalities, they will be easier to locate, facilitating fieldwork. Thirdly, a study on them would be an important practical contribution for the local DMO. Fourthly, the researchers are proficient in both English and Japanese, the native languages of the participants. Because the research aims to obtain in-depth data, soft interviews were carried out. Participants also were asked to fill a simple questionnaire on their socio-demographic data. In order to standardize the education level classifications, it was decided to follow the International Standard Classification of Education 2011 (ISCED) utilized by the UNESCO Institute for Statistics (2012). For this research, level 0 was not considered relevant to be included, and Master and Doctoral levels were combined in a 'post-graduate' level. It was decided that interviewing tourists in lodging facilities would provide an appropriate space to conduct interviews, where visitors had spare time. In order to overcome access and financial obstacles, it was decided that one researcher stayed in a lodging facility as a temporary worker. The selected accommodation is located in the Takahara community, Tanabe city. This small community is a popular spot for visitors who walk the Nakahechi trail. They tend to spend the night here after hiking for around 4 hours from Takijiri Oji, where the journey towards the Kumano Hongu Taisha begins.

Fieldwork was carried out during August 2017, November 2017 and March 2018, covering the main tourism seasons. Visitors were approached at different times of the day, and those who accepted to be interviewed were read an interview protocol, in order to obtain their informed consent. After this, the participants completed the self-administered survey. Finally, laddering interviews were conducted in the participants' native language. The interviews were fully recorded and notes were taken. Each participant was asked to elicit up to three attributes about Kumano Kodo that were important for them. The selection of three attributes was based taking into consideration interview fatigue, travel fatigue and participants' time constrains. 25 interviews were conducted for each sampled country, leading to a total of 50 interviews. Recorded interviews were transcribed into digital files and analyzed through the six-step thematic analysis suggested by Braun and Clarke (2006), which has been noted to be useful for tourism studies (Walters, 2016). In order to facilitate data management and analysis, the assistance of Computer Assisted/Aided Qualitative Data Analysis

Table 2. Occupation of Australian participants.

Occupation	n=
Teacher	2
Public employee	5
Professional	5
Student	3
Company employee	4
Self-employed	3
Other	2

Source: author.

Software was utilized. Finally, the centralized HVM format was utilized as it is visually easier to interpret (Gengler et al., 1995)

Participants' profiles

Majority of Australians were between 55 and 64 years old (n = 14), with the rest of the participants being evenly distributed among the remaining age ranges. The majority of the participants were women (n = 14). In relation to their marital status, most participants were married (n = 13), while the rest reported to be single (n = 5) or other (n = 7). The participants' level of education was high, with most them holding a bachelor or equivalent (n = 11), and a post-graduate level (n = 9). The rest of the participants reported to hold a post-secondary non-tertiary education (n = 3), upper secondary (n = 1) or lower secondary (n = 1). Participants reported a variety of occupations (Table 2).

Most of the participants travelled with one companion (n = 15), while the rest travelled with 3 (n = 2) or 4 (n = 8) companions. There were no solo travelers. Regarding the type of relationship with their companions, they traveled with their partner (n = 15), friends (n = 12) or family (n = 1). Most of the Australian participants were traveling to Kumano Kodo for the first time (n = 21). Three visited once before, while only one visited 3 or more times. Regarding their length of stay during the current trip, participants stayed an average of 5 days and 4 nights in Kumano Kodo. Regarding the Japanese, participants, most of them were an age range of 55–64 (n = 9), with the rest of them were ranges of 45–54 (n = 8), 18–24 (n = 4), 25–34 (n = 2) and 35–44 (n = 2). There were more female (n = 18). Regarding marriage status, participants were almost evenly divided between single (n = 11) and married (n = 14). All of them stated to have Japan as their country of residence. The participants also reported a high level of education, with the selected education levels being bachelor or equivalent (n = 19), shorty cycle tertiary education (n = 4) and post-secondary non-tertiary education (n = 2). The Japanese participants also showed a range of occupations (Table 3).

In relation to their travel party composition (including themselves), most of them informed to travel with in groups of 2 (n = 13), with other participants indicating travel parties of 1 (n = 5), 4 (n = 4), and 3 (n = 3). Participants reported to have been travelling with their family (n = 13), friends (n = 6) and partner (n = 1). Most of the Japanese reported to have not travelled to Kumano Kodo before their current trip (n = 20), while the ones who did (n = 5) visited it once (n = 4) or twice (n = 1) before. Finally, the average length of stay for their current travel in Kumano Kodo was 3 days and 2 nights.

Table 3. Occupation of Japanese participants.

Occupation	n=
Public employee	4
Student	2
Company employee	9
Self-employed	2
Part-time	1
Unemployed	1
Other	6

Source: author.

Data analysis

Thematic analysis was conducted following Braun and Clarke (2006). After recording, interviews were transcribed into individual digital files. Thematic analysis used the assistance of CAQDAS. Australians showed 50 elements (12 attributes, 24 consequences and 14 values) and Japanese, 41 elements (11 attributes, 18 consequences and 12 values). From the data obtained, an implication matrix was built for each group. The numbers are written in fractional form with direct relations to the left of the decimal and indirect relations to the right of the decimal. Based on previous research (Reynolds & Gutman, 1988), the cut-off value was set at 2, considering that 25 sample interviews were made for each group. Both direct and indirect relations were counted. As mentioned before, the centralized format (Gengler et al., 1995) was adopted for easier reading. Numbers were added to each link between elements, to show have many times each element lead to the next one (Figures 3 and 4).

Limitations

Due to time constrains and the challenging nature of conducting qualitative research in mountainous areas, only two nationalities were analyzed. Said factors also limited the number of sampled participants, although previous studies suggest at minimum 20 interviews per subgroup (Blake et al., 2004). Thus, it is considered that the sampled number of participants is still satisfactory. Also, the present research was only conducted in the Takahara community, but previous research has showed that pilgrims may change their attitudes towards their travel as they progress through the pilgrimage (Norman, 2009). Finally, the sampled visitors did not include one-day tourists.

Findings

Japanese HVM

Japanese participants mentioned several times the importance of the attribute of nature and its strong correlation with relaxation, the most prominent value of their HVM. In particular, the participants mentioned several times that being in a natural place such as the tranquil mountainous ranges, small shrines and rural villages of Kumano gave them a different experience from their everyday life, which was often described as busy and stressful, situated in large metropolitan areas such as Osaka or Tokyo. Coming to a natural landscape such as Kumano Kodo provided them an opportunity to get away from their stressful urban lifestyles and relax in nature.

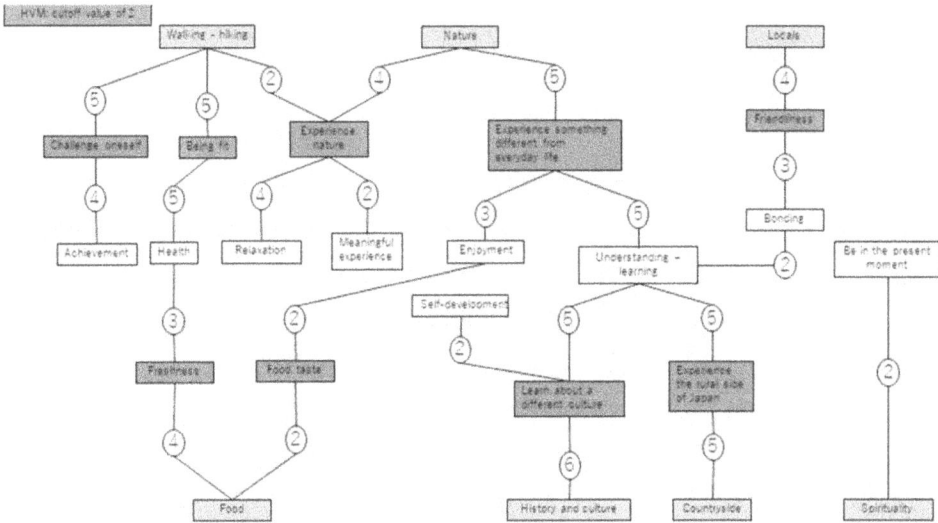

Figure 3. Hierarchal value map of Australian participants.
Source: Authors.

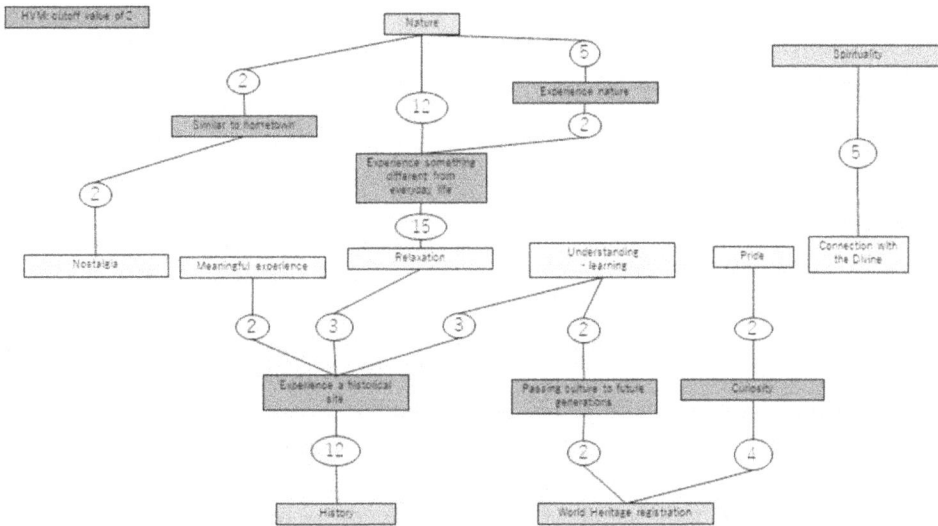

Figure 4. Hierarchal value map of Japanese participants.
Source: Authors.

Well, (nature is important) because I live in the city. Naturally, when I take some distance from my workplace, I can hear the sound of the wind and the voice of the mountain. I can be away from everyday life (Japanese N14).

By being in nature, I want to relax and be healed (*iyasaretai* – 癒されたい) (Japanese N20).

Relaxation, however, was not connected to spirituality *per se*: the natural character-istics of the Kumano landscape were cited as the attribute that led to relaxation, but not its associated spiritual aspects,. The importance of relaxation in pilgrimage sites reflects previous studies that demonstrates how trends such as *iyashi* influenced Japanese pilgrimage in recent years (Matsui, 2013; Mori, 2005).

The attribute of history was the second most salient one, and also led to the greatest number of values: relaxation, understanding - learning and meaningful experience. Through experiencing a historical site such as Kumano Kodo, the participants mentioned that they could learn about the lifestyle and faith of the Japanese from the past. Interestingly, they also mentioned that the historical and cultural characteristics of Kumano led them to a sense of relaxation, as a contrast to modern architecture and lifestyle:

> It's hard (to explain). I simply feel calm, you know? Modern buildings and lifestyles make me tired. Perhaps there were good things in the past (Japanese N18).

Next, the WH registration itself proved to be an attractive characteristic. The Japanese mostly mentioned a sense of pride, after being drawn by curiosity towards Kumano when they learned of its registration. In this sense, the participants, previously unaware of the WH registration of Kumano Kodo, felt national pride towards the place because of its newly-acquired world prestige. Literature shows similar findings, as nationalism and tourism have been noted to have strong connections (Pretes, 2003). Japanese participants also mentioned that, after the registration, they understood the importance of protecting Kumano Kodo so future generations can learn about it. The general importance of the WH registration for the Japanese is in line with previous studies that analyzed the 'World Heritage boom' phenomenon (Matsui, 2014). Regarding the attribute of spirituality, the Japanese mentioned religious-specific words such as 'gods' (*kamisama* – 神様) or 'Buddhas'. However, there were no mentions of Kumano-specific deities. Because of this, it was directly linked to the value of "connection with the Divine":

> It is very important (being together with the gods and Buddhas). For me, the gods and Buddhas are important (Japanese N22).

Interestingly enough, spirituality was not understood as a way to relaxation, but as a way to commune and connect with the Divine. This is more in line with traditional conceptions of religion, which offers an interesting contrast to contemporary discourses on spirituality (Heelas, 2006; Houtman & Mascini, 2002; Okamoto, 2015). Despite this, in comparison to the value of relaxation, it was not a prevalent element in the HVM; thus, it cannot be stated that the Japanese participants showed a particular interest in traditional religious values, although the element is indeed present. Finally, the Japanese showed a small connection within nature and nostalgia. Participants narrated that the natural landscapes of Kumano, with small villages hidden between misty valleys, brought them nostalgic memories of their childhoods' hometowns. As the participants now lived in bustling urban centers, there was a stark contrast between both landscapes. Robertson (1995) notes that these warm and nostalgic feelings towards the countryside, which evoke idealized images of traditional farmhouses among rivers and mountains, connote a desirable lifestyle characterized by rustic simplicity, as the following quote from one of the participants show:

> I am in my fifties now, but this is perhaps like the scenery I saw during my teens (…) I feel I can go back to the environment where I was raised (…) I feel I can go back to

myself and take back my years. It is a nostalgic (*natsukashii* – 懐かしい) feeling (Japanese N8).

Australian HVM

In comparison, the Australian HVM showed a greater variety of elements, with the value of understanding being the most salient one by being linked to four different attributes. The two most mentioned attributes were nature and walking. Walking in particular was a central attribute for the Australians, with a number of them being avid outdoor enthusiasts before their travel to Kumano. Its importance is shown in the fact that it led to four values: achievement, health, relaxation and meaningful experience. Nature was also a salient attribute for the Australian participants, although it did not show the same predominance as in the Japanese HVM. While the Australians also showed an interest in nature for relaxation, they also connected it to the value of understanding - learning, as they learnt about a different ecosystem while they walked through the Kumano trail. Nature was connected also to the values of relaxation, in a similar to the Japanese HVM, and to meaningful experiences:

> I feel very calm in nature. I'm attracted to the beauty of nature. It is my go-to place for relaxation (Australian N19).

> Yes, I feel connected to the wholeness of life in all its forms, including rocks, trees, spirits of earth and sky (Australian N11).

Participants were eager to deepen their knowledge of Japanese culture and rural areas through their travels. The historical characteristics of Kumano played an important role in facilitating this. Also, the fact that Kumano is located in rural areas of Japan provided an opportunity to observe and experience the 'traditional' side of the country, different from the bustling metropolis of Tokyo and Osaka and experience the 'whole' country. This connection between learning and travel, an important aspect for the Australian participants, is in line with previous research that highlights that learning in tourism contexts transcends traditional school-based passive knowledge transfer, as the visitors learn by engaging in different experiences (Falk et al., 2012). In the present study, participants mentioned similar situations, as they learned not only by reading information on the area, but by engaging with its environment and community.

> It makes it more whole, like you are seeing Japan in its glory, the business in the cities and the tourists flocking to the big sights. And now, we didn't see anyone yesterday while we were walking. So, you are getting a different sense of Japan (Australian N4).

> It's the idea of trying to broaden my horizons and understand other cultures. And same with the religion fits into that cultural thing, Shinto, Buddhism and stuff like that. It is also very interesting (Australian N13).

A smaller number of Australians mentioned that learning about a different culture was important for their self-development. In a similar note, Morgan (2010) notes that travel provides opportunities to encounter Otherness, whether in terms of different cultures (such as in the present research) or nature. These encounters carry important transformative potential, in particular if the travels are conducted by foot, like in pil-grimage, because it promotes a contemplative mood.

The Australian participants showed an important appreciation towards the locals they met along during their journey in Kumano, including the general community and the accommodation staff. The most significant chain led to their friendliness and then to the value of bonding, which was also connected to the value of learning about a different culture. In other words, the Australian visitors mentioned how friendly and open the locals were towards them

> To understand something, you need to be able to relate to people, and if they are friendly, that helps (Australian N25).

The Australians also mentioned the importance of Japanese food, which was per-ceived as fresh and tasty, leading to health and enjoyment respectively. These findings are in line with previous research on the perception of Japanese cuisine by Westerns (Jang et al., 2009). Finally, spirituality was comparatively less mentioned and led to the value of being in the moment. No mentions of divine figures were made, reflecting trends of secularization and individualized quest of spirituality in contemporary pil-grimage (Mori, 2005; Okamoto, 2015).

Discussion

This study aimed to understand similarities and differences among domestic and inter-national visitors in a pilgrimage destination. The HVMs of both nationalities show interesting differences and similarities between them. Variances are of two kinds: dif-ferent connections from a same element, or a complete absence of certain elements. Regarding the first kind of differences, nature led to different values to each national-ity, as the Japanese mainly associated it with relaxation. The Australians, on the other hand, showed a greater number of consequences and values related to nature. Naturally, the Australians did not linked nature to nostalgia, since they spent their childhood away from Japan. The historical attributes of Kumano Kodo were partially perceived differently by each nationality. Both nationalities linked the historical and cultural attributes to the value of understanding - learning, although their chains dif-fer: Japanese mentioned the consequence of 'experience a historical site' while the Australians showed a strong importance towards 'learn about a different culture'. The Japanese also linked the cultural attributes to relaxation, a chain completely absent from the Australian HVM. Finally, the Japanese linked historical attributes to having a meaningful experience, while the Australians did it with nature. These differences may be attributed to how each nationality values landscapes: Australians emphasizing pris-tine nature, while the Japanese appear to appreciate man-made environments as well. Fox and Xu (2017) reached similar results in their study, showing that nationality was an important factor for attitudes towards nature. Regardless, both nationalities found relaxation in the cultural and natural features of the Kumano landscape.

The absence of certain elements between the HVMs is also of interest, and can be explained mostly by travel patterns and cultural differences. The Japanese did not mention the locals as a relevant attribute, while the Australians did so, despite of language and cultural differences. This indicates that the way of viewing and interacting with the local communities may differ according to the nationality. This observation is partially supported by previous research by Reisinger and Turner (1998), who found that cultural background is a central factor in host-guest interactions. However, they noted that dissimilar cultural backgrounds produced difficulties in establishing relationships and expressing feelings. The present research differs in this point, as Australians expressed bonding experiences with the friendly local community, despite clear cultural differences and language barriers. On the other hand, the Japanese participants seldom mentioned the locals as an important attribute of Kumano Kodo.

The Japanese participants also did no mention of walking, while this was a central attribute to the Australian participants. In consequence, many of its related consequences and values are absent as well in the Japanese HVM. The reason for these differences may be found in dissimilar travel patterns and mobility that both nationalities displayed. Japanese reported to mostly travel by car inside the Kumano area, and had comparatively shorter trips, which is related to the general difficulty of the Japanese population to take long holidays (Prime Minister of Japan and Cabinet, 2016). Cultural attitudes towards walking may also be of relevance. The leisure practice of bushwalking, the equivalent of hiking in other cultures, is a central aspect in modern Australian identity (Hamilton-Smith, 1993). The importance of Aboriginal perceptions regarding spirituality and natural landscapes, which also influence non-Aboriginal Australians, is also of relevance (Trigger & Mulcock, 2005). The Japanese also have an important connection to hiking, although not walking due to the mountainous terrain of the country. Mountains play a central role in Japanese religions are places where gods and the dead dwell, but the practice of hiking for leisure or scientific research is a Western concept introduced in the late 19th century (Koizumi, 2001), and thus may not be as culturally central as bushwalking for Australians. With the nonappearance of walking, derived values such as health and achievement were also absent from the Japanese HVM. This also explains the low importance given to locals by the Japanese participants, as their short visits would not allow them to engage with the community. These findings are in line with previous studies, which support that slower forms of travel tend to favor closer engagement with the destination's natural features and local communities (Dickinson et al., 2011). This result shows that each nationality may engage the pilgrimage through different types of mobilities, according to their cultural background and travel patterns, which in turn changes how their travel experience. Finally, Australians naturally showed no pride towards a Japanese WH registration; although it is important to note that they did not mention the WH registration at all, even when tourism marketing of Kumano Kodo repeatedly mentions the UNESCO designation. So, while it is highly improbable that Australians were unaware of it, both nationalities may place a different degree of importance on the WH registration, which in turn is reflected in the research findings. As explained before, the Japanese society went recently through a 'World heritage boom' which influenced also pilgrimage-related destinations (Matsui, 2014), which may explain the presence of this attribute in the Japanese HVM.

Regarding similarities, both HVMs show a predominance of secular values based on physical and psychological benefits over both traditional religion and contemporary spirituality. The Japanese did a few mentions of divine figures, with no specific deity mentioned, while the Australians did mentions of contemporary spirituality, such as being in the moment. Nevertheless, it was a minor attribute in comparison to values derived from the natural and cultural features of Kumano Kodo, which were the most appreciated attributes for both nationalities. Although spirituality was not a predominant attribute, findings are in line with previous research regarding spirituality in contemporary society, where individuals conform their own individualized spirituality and not strictly follow traditional religious authorities (Okamoto, 2015). Still, the relatively minor presence of religious or spiritual-related value differs from previous studies in pilgrimage tourism. For example, in the study of Kim et al. (2016), which also employed means-end, contemplation and self-reflection were the most salient values mentioned by pilgrims in Santiago de Compostela, Spain. Blom et al. (2016) also noted individual spirituality as a central motivation in Santiago de Compostela. The reason for this difference is not readily clear, although it may be related to the selected case study, as Santiago de Compostela and Kumano Kodo differ in various geographic and cultural aspects, as well as in their tourism development and length. Also, previous research has showed that pilgrims may change their attitudes towards their travel as they progress through the pilgrimage (Norman, 2009). During informal conversations, some of the participants mentioned this possibility as well. Therefore, conducting research in different locations in the same pilgrimage trail, such as the end of the journey were the Kumano Hongu Taisha is located, may produce different findings regarding spirituality.

Another similarity between both nationalities is the importance given to relaxation and its connections with the cultural and natural landscape characteristics of Kumano. This is in line with previous studies that mention the positive effects of both nature (Velarde et al., 2007) and outdoor hiking (Rodrigues et al., 2010) have on human mental wellbeing. Other studies on pilgrimage tourism also noted the importance of relaxation and wellness elements. For example, pilgrimage to certain locations for miracle healing is an ancient tradition found in different cultures even today. Some of the practices at the sacred sites involved holy springs or rivers, showing an early connection between pilgrimage and wellness. Centuries later, the practice of travel for health and wellness developed into spa tourism during the 18th and 19th centuries in Europe. Additionally, in contemporary society religious traditions have been commodified as wellness practices, such as yoga and Chinese medicine. Pilgrimage is no exception, as some travelers move to certain locations deemed to be more effective for achieving wellness-related benefits. Also, contemporary spirituality stresses the alleged holistic healing 'power' of certain landscapes related to religions. These different factors lead to an important intersection between wellness, spirituality and pilgrimage travel in our society. (Connell, 2011; Matsui, 2013; Stausberg, 2011). Progano and Kato (2018) noted the presence of wellness-related spiritual tourism elements is strong in Kumano Kodo, and can be principally seen in the concept of *iyashi*. This finding is supported by previous Japanese researchers on their studies about pilgrimage tourism in contemporary Japan (Matsui, 2013; Mori, 2005; Okamoto, 2015). Finally, both nationalities linked the value of understanding - learning to the historic aspects of Kumano Kodo in different aspects: for the Australians, it opened a window to Japanese culture; to the Japanese, they could learn about their own past. Similar results were reached by previous research, accentuating the linkage between knowledge and historical heritage present in sacred sites (Kim & Kim, 2019), and tourism in general (Morgan, 2010).

Conclusions

Means-end was utilized to study differences and similarities between domestic and international visitors traveling the same pilgrimage route, in order to understand visitor diversification in contemporary pilgrimage tourism. Research findings demonstrated that, in the current context of individualized spirituality, pilgrimage sites have become places with multi-layered, diversified meanings and behaviors according to the different nationalities that visit them. International tourists bring new meanings, behaviors and values, even though they may have few or no cultural links to the sacred sites they visit. In this research, Australian visitors attached numerous values to Kumano Kodo, describing it, for example, as a place for multi-cultural understanding, while the domestic tourists evoked feelings of relaxation among a demanding urban lifestyle. None of these interpretations strictly follow the traditional narratives of the Kumano region, but shows that pilgrimage sites still play a significant role in society. Interestingly enough, spirituality, even understood as in its contemporary individualized form, is not among the most mentioned element by any of the two sampled nationalities. While it is important to consider that spirituality may be expressed in a diversity of performances and experiences beyond mere verbal communication, the research findings suggest that the diversification process in sacred sites goes beyond a continuum of contemporary spirituality and traditional religion, and incorporates a diverse and complex range of elements related to leisure, sports, intercultural exchange, nostalgia, escapism and relaxation, among others. Some previous studies back up this point, exploring the intersections of pilgrimage with some of these themes (Di Giovine & Choe, 2019; Lois-Gonzalez & Santos, 2015; Nakai, 2011) Thus, while contemporary spiritualty does play a role, research on pilgrimage sites can be expanded from it, which opens the field for numerous opportunities in both tourism development and academic investigation. As Eade and Sallnow (1991) note, major sacred sites such as Kumano Kodo possess a universalistic character that can absorb the diversity pilgrims' hopes, motivations and spirituality, as an empty vessel would. In particular, when international tourists are analyzed, a wide range of themes emerge when they bring their cultural and personal background such as needs, interests and expectations to a pilgrimage site, increasing diversification. The study also demonstrated the importance and applicability of nationality as a fundamental variable factor for studying pilgrimage tourism. In a context of individualized spirituality, with pilgrimage sites increasingly becoming international destinations, nationality may become an important factor for academics to study. However, these differences should be contextualized to avoid essentialist statements on nationalities.

The present study has implications for future research. Firstly, research on international and domestic tourists may be carried out in other Asian pilgrimage sites. As stated before, the Asia-Pacific region has become the second most visited region in the world (United Nations World Tourism Organization & Global Tourism Economy Research Center, 2016), and has the greatest number of pilgrims and travelers for religious events for both international and domestic tourism (United Nations World Tourism Organization, 2011). Therefore, research based on national segmentation carried out in the many sacred sites in Asia would provide a vast and interesting research field that would not only be useful for policy-makers, but also help to contribute to the number of academic studies done on non-Western sites. There is tremendous potential for academic fieldwork in this study area that is still untapped. Secondly, as inbound tourism continues to expand in pilgrimage-related destinations, host-guest interactions also yield an interesting subject. The present study's results already showed that the local community was perceived differently by both nationalities. However, it is important to understand that the present study's results only showed what the participants mentioned but not necessarily *do*. Further research employing participant observation or ethnography might shed light on these subjects. As pilgrimage tourism continues to expand, studies about the community's views on tourists (or tourism in general) will be an important academic to study. This specific field remains largely understudied, so there is a great potential for future research. The locals may possibly perceive the tourists based on other factors, such as gender, age and length of stay, which can be researched in the future. For example, previous research (Terzidou et al., 2008) showed that locals' religiosity is a factor of importance in this subject. Thirdly, the variances in mobility between tourists of different nationalities visiting a pilgrimage site may be as well of interest for future research. Dejbakhsh et al. (2011) mentioned differences according to nationality regarding travel spatial behavior in areas such as accommodation location, mode of transport, length, direction, type and pattern of movements. The present study also suggested that Japanese and Australian have different spatial behaviors: Australians tended to travel by foot in the Nakahechi trail and had longer stays, while the Japanese showed less inclination to walking and mentioned using automobiles, as well as having shorter stays overall, which could be attributed to holiday patterns. These results may motivate future studies in mobility and nationality in pilgrimage tourism.

These conclusions are relevant for policy-makers related to pilgrimage destinations. Firstly, the different views on the local community by each nationality are of relevance. Because pilgrimage-related destinations are often in regional areas that are not historically used to receive international guests, monitoring the interactions between locals and guests is of particular importance to ensure the community supports further tourism development. It is also important to understand what type of benefits the locals expect from tourism. Progano (2018) found that the residents prefer a small development that favors close interactions with visitors. Thus, a balance in development may be of utmost significance for sustainable tourism policies, contrasting to cases of overtourism. Secondly, the different emphasis put on walking is also of interest, which is related to the travel patterns and cultural background of each nationality. Perhaps some of these subjects are beyond the control of regional tourism bodies, but nonetheless they require careful consideration when promoting local tourism destinations. For example, while the proximity of a large urban center such as Osaka may provide a large number of potential visitors, the shorter holidays and easy access to the destination can potentially lead to a growth of one-day visitors, who have a smaller economic impact. Thirdly, the attribute of countryside was important for the Australian participants. This information would be of special relevance for public bodies that aim to revitalize regional economies through tourism development. Showcasing the countryside areas of Japan as a complement to urban destinations may help direct a portion of visitors to the regional areas. This way, urban and rural destinations would avoid having to compete between them and instead form collaborative projects. Finally, the sector of wellness and health tourism in pilgrimage sites holds, in the researchers' opinion, a great potential. As noted before, Kumano Kodo has a strong emphasis on wellness-related spirituality (Progano & Kato, 2018). Results showed that, while Japanese are certainly attracted to pilgrimage sites for wellness purposes, foreigners are also a potential market.

Disclosure statement

No potential conflict of interest was reported by the author(s).

Funding

Funded by Ministry of Education, Culture, Sports, Science and Technology.

ORCID

Kumi Kato (iD) http://orcid.org/0000-0002-1783-8155
Joseph M. Cheer (iD) http://orcid.org/0000-0001-5927-2615

References

Blake, B. F., Saaka, A., & Sidon, C. (2004). *Laddering: A "how to do it"* manual—with a note of caution. Cleveland State University.
Blom, T., Nilsson, M., & Santos, X. (2016). The way to Santiago beyond Santiago. Fisterra and the pilgrimage's post-secular meaning. *European Journal of Tourism Research*, *12*, 133–146.
Braun, V., & Clarke, V. (2006). Using thematic analysis in psychology. *Qualitative Research in Psychology*, *3*(2), 77–101. https://doi.org/10.1191/1478088706qp063oa
Connell, J. (2011). *Medical tourism*. CABI.

Collins-Kreiner, N. (2010). The geography of pilgrimage and tourism: Transformations and implications for applied geography. *Applied Geography*, 30(1), 153–164. https://doi.org/10.1016/j.apgeog.2009.02.001

Collins-Kreiner, N., & Wall, G. (2015). Tourism and religion: Spiritual journeys and their consequences. In S. D. Stanley (Ed.), *The changing world religion map: Sacred places, identities, practices and politics* (pp. 689–707). Springer.

Dejbakhsh, S., Arrowsmith, C., & Jackson, M. (2011). Cultural influence on spatial behaviour. *Tourism Geographies*, 13(1), 91–111. https://doi.org/10.1080/14616688.2010.516396

Dickinson, J. E., Lumsdon, L. M., & Robbins, D. (2011). Slow travel: Issues for tourism and climate change. *Journal of Sustainable Tourism*, 19(3), 281–300. https://doi.org/10.1080/09669582.2010.524704

Di Giovine, M. A., & Choe, J. (2019). Geographies of religion and spirituality: Pilgrimage beyond the 'officially' sacred. *Tourism Geographies*, 21(3), 361–383. https://doi.org/10.1080/14616688.2019.1625072

Eade, J., & Sallnow, M. J. (1991). Introduction. In J. Eade & M. J. Sallnow (Eds.), *Contesting the Sacred: The anthropology of Christian pilgrimage* (pp. 1–19). Wipf and Stock Publishers.

Falk, J. H., Ballantyne, R., Packer, J., & Benckendorff, P. (2012). Travel and learning: A neglected tourism research area. *Annals of Tourism Research*, 39(2), 908–927. https://doi.org/10.1016/j.annals.2011.11.016

Fox, D., & Xu, F. (2017). Evolutionary and socio-cultural influences on feelings and attitudes towards nature: a cross-cultural study. *Asia Pacific Journal of Tourism Research*, 22(2), 187–199. https://doi.org/10.1080/10941665.2016.1217894

Gengler, C. E., Klenosky, D. V., & Mulvey, M. S. (1995). Improving graphical representation of means-end results. *International Journal of Research in Marketing*, 12(3), 245–256. https://doi.org/10.1016/0167-8116(95)00024-V

Gengler, C. E., & Reynolds, T. J. (1995). Consumer understanding and advertising strategy: analysis and strategic translation of laddering data. *Journal of Advertising Research*, 35(4), 19–33.

Gursoy, D., & Umbreit, W. T. (2004). Tourist information search behavior: cross-cultural comparison of European Union state members. *International Journal of Hospitality Management*, 23(1), 55–70. https://doi.org/10.1016/j.ijhm.2003.07.004

Hamilton-Smith, E. (1993). In the Australian Bush: Some reflections on serious leisure. *World Leisure & Recreation*, 35(1), 10–13. https://doi.org/10.1080/10261133.1993.10559134

Heelas, P. (2006). Challenging secularization theory: The growth of "new age" spiritualties of life. *The Hedgehog Review*, 8(1–2), 46–58.

Houtman, D., & Mascini, P. (2002). Why do churches become empty, while new age grows? Secularization and religious change in the Netherlands. *Journal for the Scientific Study of Religion*, 41(3), 455–473. https://doi.org/10.1111/1468-5906.00130

Horie, N. (2009). Spirituality and the spiritual in Japan: Translation and transformation. *Journal of Alternative Spiritualties and New Age Studies*, 5(11), 1–15.

Hyde, K. F., & Harman, S. (2011). Motives for a secular pilgrimage to the Gallipoli battlefields. *Tourism Management*, 32(6), 1343–1351. https://doi.org/10.1016/j.tourman.2011.01.008

Jang, S. C., Ha, A., & Silkes, A. C. (2009). Perceived attributes of Asian foods: From the perspective of the American customers. *International Journal of Hospitality Management*, 28(1), 63–70. https://doi.org/10.1016/j.ijhm.2008.03.007

Jewell, B., & Crotts, J. C. (2002). Adding psychological value to heritage tourism experiences. *Journal of Travel & Tourism Marketing*, 11(4), 13–28. https://doi.org/10.1300/J073v11n04_02

Kim, B., Kim, S. S., & King, B. (2016). The sacred and the profane: Identifying pilgrim traveler value orientations using means-end theory. *Tourism Management*, 56, 142–155. https://doi.org/10.1016/j.tourman.2016.04.003

Kim, B., & Kim, S. S. (2019). Hierarchical value map of religious tourists visiting Vatican City/Rome. *Tourism Geographies*, 21(3), 529–550. https://doi.org/10.1080/14616688.2018.1449237

Kim, C., & Lee, S. (2000). Understanding the cultural differences in tourist motivation between Anglo-American and Japanese tourists. *Journal of Travel & Tourism Marketing*, 9(1–2), 153–170. https://doi.org/10.1300/J073v09n01_09

King, C. (1993). His truth goes marching on: Elvis Presley and the pilgrimage to Graceland. In I. Reader & T. Walter (Eds.), *Pilgrimage in popular culture* (pp. 92–104). The Macmillan Publishing.

Koizumi, T. (2001). *The birth of hiking—Why did people start climbing mountains?* Nakamatsu Shinsho (in Japanese).

Lois-Gonzalez, R. C., & Santos, X. M. (2015). Tourists and pilgrims on their way to Santiago: Motives, Caminos and final destinations. *Journal of Tourism and Cultural Change*, *13*(2), 149–164. https://doi.org/10.1080/14766825.2014.918985

Matsui, K. (2014). *Sacred sites for tourism strategy: The Nagasaki churches and the commodification of place*. Tsukuba Daigaku Shuppan Kai (in Japanese).

Matsui, T. (2013). *Language and marketing: A social history of the 'healing boom' in Japan*. Sekigakusha (in Japanese).

Miles, S., & Rowe, G. (2004). The laddering technique. In G. Breakwell (Ed.), *Doing social psychology research* (pp. 305–343). Blackwell Publishing.

Miyake, H. (1966). Genjutsu in Shugendo: Its mechanism and world view. *Tetsugaku*, *48*, 47–70 (in Japanese).

Morgan, A. D. (2010). Journey into transformation: Travel to an "other" place as a vehicle for transformative learning. *Journal of Transformative Education*, *8*(4), 246–268. https://doi.org/10.1177/1541344611421491

Mori, M. (2005). *The modernization of Shikoku Henro: From a 'modern pilgrimage' to a 'healing journey*. Sogensha (in Japanese).

Nakai, J. (2011). The nostalgic value of travelling Kumano Kodo. *Bulletin of the Faculty of Sociology, Ryukoku University*, *39*, 43–53 (in Japanese).

Norman, A. (2009). The unexpected real: Negotiating fantasy and reality on the road to. *Santiago. Literature & Aesthetics*, *19*(2), 50–71.

Okamoto, R. (2015). *Pilgrimage to sacred sites: From world heritage to anime setting*. Chuokoron Shinsha (in Japanese).

Pizam, A., & Reichel, A. (1996). The effect of nationality on tourist behavior: Israeli tour-guides' perceptions. *Journal of Hospitality & Leisure Marketing*, *4*(1), 23–49. https://doi.org/10.1300/J150v04n01_03

Pizam, A., & Sussmann, S. (1995). Does nationality affect tourist behavior? *Annals of Tourism Research*, *22*(4), 901–917. https://doi.org/10.1016/0160-7383(95)00023-5

Pretes, M. (2003). Tourism and nationalism. *Annals of Tourism Research*, *30*(1), 125–142. https://doi.org/10.1016/S0160-7383(02)00035-X

Prime Minister of Japan and Cabinet. (2016). *Tourism vision for supporting the Japan of tomorrow*. Retrieved October 29, 2018, from http://www.mlit.go.jp/common/001126598.pdf (in Japanese).

Progano, R. N., & Kato, K. (2018). Spirituality and tourism in Japanese pilgrimage sites: Exploring the intersection through the case of Kumano Kodo. *Fieldwork in Religion*, *13*(1), 22–43. https://doi.org/10.1558/firn.36137

Progano, R. N. (2018). Residents' perceptions of socio-economic impacts on pilgrimage trails: How does the community perceive pilgrimage tourism? *Asian Journal of Tourism Research*, *3*(2), 148–178. https://doi.org/10.12982/AJTR.2018.0014

Reader, I. (1993). Introduction. In I. Reader & T. Walter (Eds.), *Pilgrimage in popular culture* (pp. 1–25). The Macmillan Publishing.

Reader, I. (2014). *Pilgrimage in the marketplace*. Routledge.

Reisinger, Y., & Turner, L. (1998). Cross-cultural differences in tourism: A strategy for tourism marketers. *Journal of Travel & Tourism Marketing*, *7*(4), 79–106. https://doi.org/10.1300/J073v07n04_05

Reynolds, T. J., & Gutman, J. (1988). Laddering theory, method, analysis and interpretation. *Journal of Advertising Research*, *28*(1), 11–31.

Robertson, J. (1995). Hegemonic nostalgia, tourism and nation-making in Japan. *Senri Ethnological Studies*, *38*, 89–103.

Rodrigues, A., Kastenholz, E., & Rodrigues, A. (2010). Hiking as a wellness activity: An exploratory study of hiking tourists in Portugal. *Journal of Vacation Marketing*, *16*(4), 331–343. https://doi.org/10.1177/1356766710380886

Stausberg, M. (2011). *Religion and tourism: Crossroads, destinations and encounters*. Routledge.

Tanabe City. (2016). *Tanabe city statistics book Heisei 28 November edition*. Tanabe (in Japanese).

Tanabe Kumano Tourism Bureau. (2018). *Management situation of travel business*. Kumano Tourism Bureau (in Japanese).

Terzidou, M., Stylidis, D., & Szivas, E. M. (2008). Residents' perceptions of religious tourism and its socio-economic impacts on the island of Tinos. *Tourism and Hospitality Planning & Development*, *5*(2), 113–129. https://doi.org/10.1080/14790530802252784

Trigger, D., & Mulcock, J. (2005). Forests as spiritually significant places: Nature, culture and 'belonging' in Australia. *The Australian Journal of Anthropology*, *16*(3), 306–320. https://doi.org/10.1111/j.1835-9310.2005.tb00313.x

UNESCO Institute for Statistics. (2012). *International Standard Classification of Education ISCED 2011*. UNESCO Institute for Statistics.

United Nations World Tourism Organization. (2011). *Religious tourism in Asia and the Pacific*. UNWTO.

United Nations World Tourism Organization & Global Tourism Economy Research Center. (2016). *Asia tourism trends 2017 edition*. UNWTO.

Velarde, M. D., Fry, G., & Tveit, M. (2007). Health effects of viewing landscapes: Landscape types in environmental psychology. *Urban Forestry & Urban Greening*, *6*(4), 199–212. https://doi.org/10.1016/j.ufug.2007.07.001

Wakayama Tourism Agency. (2015). *Heisei 26 tourist motivation research report*. Wakayama Tourism Agency (in Japanese).

Wakayama Tourism Agency. (2016). *Heisei 27 tourist motivation research report*. Wakayama Tourism Agency (in Japanese).

Wakayama Tourism Agency. (2017). *Heisei 28 tourist motivation research report*. Wakayama Tourism Agency (in Japanese).

Wakayama Tourism Agency. (2018). *Heisei 29 tourist motivation research report*. Wakayama Tourism Agency (in Japanese).

Walters, T. (2016). Using thematic analysis in tourism research. *Tourism Analysis*, *21*(1), 107–116. https://doi.org/10.3727/108354216X14537459509017

Watkins, L. (2010). The cross-cultural appropriateness of survey-based value(s) research. *International Marketing Review*, *27*(6), 694–716. https://doi.org/10.1108/02651331011088290

Watkins, L., & Gnoth, J. (2011). The value orientation approach to understanding culture. *Annals of Tourism Research*, *38*(4), 1274–1299. https://doi.org/10.1016/j.annals.2011.03.003

Zinnbauer, B. J., Pargament, K. I., Cole, B., Rye, M. S., Butter, E. M., Belavich, T. G., Hipp, K. M., Scott, A. B., & Kadar, J. L. (1997). Religion and spirituality: Unfuzzing the fuzzy. *Journal of Scientific Study of Religion*, *36*(4), 549–564. https://doi.org/10.2307/1387689

The materiality of air pollution: Urban political ecologies of tourism in Thailand

Mary Mostafanezhad

ABSTRACT

Between February and April of each year, air pollution blankets much of northern Thailand and severely impacts the livelihoods of tourism practitioners and farmers in the region. Environmental narratives among lowland, urban residents attribute haze to biomass burning among highland farmers. Highland farmers, however, contend that environmental governance regimes have lengthened and exacerbated what is now regarded as the annually recurring haze crisis. Drawing on ethnographic fieldwork among urban tourism practitioners, rural farmers, and natural scientists, I demonstrate how, as a physical and symbolic entity, air pollution circulates between urban and rural spaces in ways that reshape tourism and urban-rural social relations. In doing so, I bring emerging work at the intersection of urban political ecology and new materialism to bear on tourism to reveal the more-than-human sociality of air pollution in northern Thailand.

摘要

在每年的2月到4月间, 空气污染覆盖了泰国北部的大部分地区, 影响着旅游业和农业约60%人口的生计。低地和城市居民的环境叙事将雾霾归因于高原地区农民的生物质燃烧。然而, 高原地区的农民认为, 环境治理制度已经拖延甚至加剧了现在被视为每年反复出现的雾霾危机。利用城市旅游从业者、乡村农民和自然科学家的民族志田野调查, 我论证了空气污染作为一个物理和象征实体, 如何在城市和农村空间之间循环, 从而重塑旅游业和城乡社会关系。在此过程中, 我将城市政治生态学和新唯物主义交叉领域的新兴研究引入到旅游领域, 以揭示泰国北部空气污染的超越人类的社会性。

Introduction

Along the crowded Walking Street Market in Chiang Mai, Thailand, vendors sell t-shirts that proclaim, "I Survived the Air in Chiang Mai" (Save Chiang Mai From Burning, 2020) and glass jars with notes that read: "Sorry for the air pollution, you must deserve better air than this" to tourists. These ironic souvenirs are meant to garner tourists' as well as public support for urban dwellers' struggle to combat intense air pollution during the widely dubbed "smoky season" which occurs annually between February and

April. During each smoky season haze smothers much of northern Thailand, contributes to the hospitalization of more than 81,000 residents, costs the region an estimated 50 billion baht (USD 1.5 billion) in tourism revenue (The Nation, 2010; National News Bureau of Thailand, 2012), and impacts the livelihoods of approximately 60 percent of the population working in the tourism and agriculture industries. Lowland, urban dwellers attribute haze to highland, rural farmers' expanded use of fire which threatens the livelihoods of urban based tourism practitioners by deterring tourist arrivals. Farmers' livelihoods are also, albeit differently, imperiled by the haze. Intensified market integration and increased competition from global markets have compelled many farmers to transition from small-scale subsistence to large-scale and plantation agriculture and increase their dependence on lucrative non-timber forest products (NTFP) such as *hed thob*, a popular mushroom in the region (Hayward, 2018). Recent policy responses include the 2015 burning ban which could severely impact their livelihoods. Yet, despite definite reactions to the haze, pervasive uncertainty over the exact causes and potential solutions to what is now described as the "haze crisis" persists throughout the region. This uncertainty has led to a range of responses including calls to carry female cats from house to house to be doused with water. The revival of this traditional ritual to entice the rain was proposed by the former governor of Lampang (a northern Thai province) who contended at a crisis meeting in Chiang Mai, "We have to do everything, even superstitious means, to bring the rain". In a similar vein, the former Mayor of Chiang Mai called for BBQ style restaurants to cut back on their smoke in order to help ameliorate the haze crisis (Sukplang, 2007). Residents claim that government officials have sought to minimize publicity of the haze crisis to protect the lucrative tourism industry. Reflecting this concern, Bunnaroth Buaklee, a member of the newly formed Chiang Mai Breathe Council explained that the council's "first mission is to work on the Air Quality Index standard, emphasising citizens' awareness of health effects — without having tourism distracting us" (City News, 2019). This relationship between tourism and agricultural livelihoods reflects the ongoing negotiation of urban anxieties over, on one hand, urbanization and economic development and, on the other, the significant role of agricultural production in the region.

I bring urban political ecology (UPE) and new materialialism to bear on tourism and the haze crisis in northern Thailand to argue that the materiality of haze as a physical (particulate matter (PM)) and narrative entity engenders livelihood contestations that are re-shaping social relations in the region. Drawing on ethnographic fieldwork among urban tourism practitioners as well as natural scientists and rural farmers, I demonstrate how haze circulates between urban and rural spaces in ways that challenge methodological cityism by triggering resource conflicts over the exchange value of land (for rural farmers) and landscapes (for urban tourism practitioners) (Adey, 2014; Connolly, 2019; Newell & Cousins, 2015). Environmental narratives stabilize the logic through which governance mechanisms are justified and enacted (Roe, 1991; Fairhead & Leach, 1995). They also comingle with rumor and the boundaries between them are often blurred. Rumor includes but is more than narrative and discourse (Samuels, 2015). It accounts for the circulation of unconstrained information that is deemed important as well as how this information conjures collective imaginaries that reflect social inequalities (Butt, 2005). Rumors of haze are particularly indefinite" in so far as significant uncertainty exists around its causes and effects. As Moulin argues, rumors "come to life as people try to understand events, facts, or perceptions that are unclear, unsaid, uncertain in the face of weak information" as they become a way to "fill in the gaps and to suggest particular understandings of events within a community" (Moulin, 2010: 351). Significantly, rumors do not require validation by scientific "experts", community members, or others, but instead, index value of specific types of ecological knowledge for individual and collective actors. Yet, while rumors are also possible truths (Samuels, 2015), they should not be conflated with evidence. Political ecologies of rumor are deeply enmeshed with environmental narratives as well as perception and affect. In the context of environmental change, rumors and narratives collaboratively draw on socio-historical events that serve the interests of the urban elite (Kull & Laris, 2009). Thus, while environmental narratives are often told as "stories", factual or otherwise (Roe, 1991), "perhaps more than any other narrative category, rumor illuminates how the assignment of truth to narratives is invested with power" (Samuels, 2015: 238). By accounting for the collaborative role of environmental narratives and rumor in UPE, I demonstrate how the haze crisis and its more-than-human actants including PM itself mobilizes social action in ways that perpetuate historically salient distinctions between urban-rural residents with critical implications for forest policy and tourism development in the region.

Urban political ecologies and the materiality of air pollution

Over the past several decades, UPE has grown in visibility as a framework from which scholars examine the interconnected social, political, economic and ecological processes that shape uneven urban landscapes as well as mediate socio-ecological relations of power (Angelo & Wachsmuth, 2015; Gandy, 2003; Graham, 2015; Heynen, 2014, 2016; Newell & Cousins, 2015; Rice & Tyner, 2017; Robbins, 2012; Swyngedouw & Heynen, 2003). Yet, surprisingly little UPE scholarship examines the networked matrix of urban-nature relations. Moreover, UPE scholars have increasingly called for research that extends beyond methodological cityism (Angelo & Wachsmuth, 2015, Connolly, 2019 and Newell & Cousins, 2015) and argued that to study urbanization processes is also to follow loose threads that often lead one well beyond the urban boundaries of the city proper (Angelo & Wachsmuth, 2015: 23). I investigate how blame that urban residents attribute to rural farmers for the haze crisis mediates environmental governance mechanisms and regulatory regimes that reveal urban-rural livelihood entanglements. Despite growing awareness of the noxious effects of air pollution globally and the significance of how air links urban and rural spaces, few political ecologists have made air their primary object of analysis relative to other vital resources such as water and food (Bennett, 2010; Graham, 2015; Harper, 2004; Véron, 2006). Rather, air is often regarded by social scientists as a "neutral container" or "immaterial ether" that is "only present when made visible by abnormal circumstances" (Comaroff, 2017). Despite this gap, there are several notable exceptions among scholars focused on the materiality of air as well as the cultural politics of air in social theory (Adey, 2014; Comaroff, 2017; Gissen, 2015; Graham, 2015; Harper, 2004; Lieto, 2017; Miller, 2005). Integrated with post-humanist and materialists turns, this work challenges "conventional conceptions of the urban, and of the Society/ Nature binary" and "posits urban processes as complex multi-scaled metabolisms" (Graham, 2015: 194).

In northern Thailand, the collective narrativization of haze reflects its more-than-human materiality as a complex entity that is physically and socially produced. Haze is physically produced through the incorporation of rapid urban development, periodic temperature inversions, and rural practices of burning biomass, among other biophysical and industrial processes, the specific permutations of which have yet to be fully understood. The social production of haze encompasses a range of socio-cultural and political-economic processes. Together, these processes mediate the construction of environmental knowledge and narratives, some of which are enrolled in policy. For instance, high levels of PM in 2015 triggered the institutionalization of a blanket burning ban that threatens farmers with hefty fines for using fire to prepare their fields or to eliminate agricultural waste, thus impinging upon and criminalizing their livelihoods. In this way, the imaginary space of air and its symbolic meaning can be "valued or devalued, salutary or threatening" (Comaroff, 2017: 607). Air pollution in the form of haze and narratives of its causes and effects has become a site of contestation in ways that demonstrate how "materials do not just reveal cultural categories— they transmit them" (Carse, 2014: 391). Air is "quite literally coextensive with our material lives and physical bodies: it extends our representations into us. As such, air blurs the boundaries between metaphor and flesh" (Comaroff, 2017: 607). Narratives of the cause and effect of PM become part of the sociality and well as materiality of air as they fill in the gaps left between these boundaries and are enrolled in resistance and accommodation strategies during times of crisis. Thus, PM can create new and flare up past socio-ecological relations (Moulin, 2010). As Carse (2014: 391) argues, "moving beyond the notion that meanings are attached to people and materials", the relationship between shared meanings, categories and their attachments to people and things is recursive and often shaped by natural and built landscapes that become reified as "environmental". In this way, materiality refers to both the physical and symbolic qualities of a thing (Bennett, 2010). While anthropologists have long argued that the circulation of material objects such as bracelets, pigs and canoes create coherent cultural spaces (Carse, 2014: 391), we know much less about how the flow and form of environmental materialities (e.g. air, dust, rain, and haze) reshapes socio-political space and its corollary governance regimes that impact on lives and livelihoods.

Northern Thailand's haze crisis

In northern Thailand, livelihood struggles are at the center of debates over the haze crisis. Today, 80 percent of the poor (7.3 million people) live in rural areas, making poverty as well as agricultural livelihoods a predominately rural experience (World Bank, 2016). Thailand has the largest wealth gap in the world and this disparity is exceptionally unequal between the north as well as northeast and central regions (Chaitrong, 2012; Lindsay, 2019; Yuthamanop, 2011). These inequalities are felt in everyday urban-rural dynamics that have deep historical roots in the region. As early as 1200 AD, Peter Grave notes that for "non-Buddhist swidden agricultural and foraging groups, the uplands of continental Southeast Asia provide a marked geopolitical contrast to the socio-political transformations of the Buddhist lowland" (Grave, 1995: 243). Throughout the ensuing years, upland-lowland relations continued to be mediated by social distinctions in culture and livelihoods (Winichakul, 1997). By the 1950s, the state had begun governing swidden farmers in the highlands of northern Thailand through a range of surveillance and policy mechanisms that have stifled their access to markets (Scott, 2009). Additionally, during the Vietnam War conflict between upland and lowland dwellers escalated when the state identified highlanders as communists who used the thriving opium trade to buy weapons (Forsyth, 1995). By the 1960s and 70 s, narratives of the so-called "hill tribe problem" were widely reflected in policy documents and functioned as a technology of governmentality (McKinnon, 2008). Since this period, lowland Thais have used the ethnic category of "hill tribe" to refer to ethnic minority groups that live in the mountainous areas and as a core sociological category of difference in the region (Stott, 1991). Three primary factors drove these policies including "hill tribe" opium production, shifting cultivation and environmental degradation and the threat that the mobile ethnic minority populations were perceived to pose to national security (McKinnon, 2008). In the 1970s, environmental degradation of the highlands was blamed on upland shifting cultivation which was seen as the cause of declining soil fertility as well as sedimentation and water shortage in the lowlands (Forsyth, 1995). There is a long and rich literature that addresses how Thailand's upland minorities have been blamed for environmental degradation in the region for decades as well as the prevalence of urban bias in contestations over agriculture and the environment (Tapp, 1988; Rigg, 2014; Walker, 2001; Walker & Farrelly, 2008). Since the 1980s and the end of northern Thailand's insurgencies, the region has shifted from a notorious site of armed rebellion and opium production to "a tourism promoter's dream with 'hill-tribe treks' on the tourism 'to do' lists of those seeking a manageable combination of ecological immersion and cultural authenticity" (Walker & Farrelly, 2008: 373). This shift is reflected in the concurrent growth of tourism (17.7% of the GDP) and broader processes of "deagrarianization" (between 1975 and 2014, employment fell from 73 to 32 percent) (Hall et al., 2011).

Methods

Between 2017 and 2019, I conducted ethnographic research including participant observation and 70 semi-structured interviews in the Chiang Mai and Mae Hong Son provinces in northern Thailand. These provinces represent the two most established tourism sites in the region, with urban centers that are severely affected by seasonal air pollution. My research sought to understand how people perceive the causes, effects, and solutions to the recurrent haze crisis. Semi-structured interviews include 30 urban tourism practitioners including individuals who depend primarily on tourism for their livelihoods (e.g. trekking and tour guides, guest house and hotel owners, and souvenir vendors) and live and work in urban centers as well as 20 interviews with natural scientists and 20 interviews with rural farmers. Research participants were identified using convenience and snowball sampling. Urban tourism practitioners and rural farmers are particularly well suited to comment on their experience of air pollution because it directly affects their lives and livelihoods.

Qualitative interviews were conducted in Thai by the author as well as several local research assistants. All interviews were audio recorded, transcribed and input into NVivo, a qualitative data analysis software package that facilitates efficient data storage and analysis (e.g. coding) as well as enables the user to visually identify connections in the data to develop theoretical and methodological models. Data from interviews and fieldnotes from participant observations were recursively coded for recurrent themes using grounded theory approaches (Charmaz, 2006; Glaser & Strauss, 2017). Grounded theory coding includes an initial and a focused coding stage (Charmaz, 2006). During the initial coding stage, I highlighted words and phrases with analytical import and in the second stage I selected the most useful initial codes and tested their explanatory power with extensive data. For instance, in the first stage I highlighted *in vivo* descriptions of how people described the causes of seasonal air pollution; forest scavengers, hunters, swidden farmers, poverty, and agribusiness were all recurrent themes in my interviews. In the second stage of focused coding, I synthesized these descriptions as "attributing blame". When I returned to the data, I used theoretical sampling meaning that I selected new cases for study (e.g. additional rural residents) to further investigate emergent themes (e.g. differing attributions of blame among rural and urban residents) (Bernard, 2011: 430). I then continued to search for previously unidentified categories and further refined my analytical framing to account for the properties of my coding categories such as the relationship between urban and rural narratives of the causes of seasonal air pollution. I continued this iterative process until I reached the point of data saturation—or when no additional theoretical insights or properties could be identified (Charmaz, 2014: 113). Throughout this process, I continuously moved between data, codes, and analytical interpretations such as the circulation of blame narratives. These recurrent themes are reflected in the subsections below including uncertainty and blame, resource socialities and urban and rural livelihoods.

Positionality in this project took many forms. Positionality refers to the relationship between the researcher and research participants and it has been extensively addressed in social science literature where scholars have outlined key themes that influence this relationship such as race, gender, class, power, privilege, representations, interpretive conflict, and ethical dilemmas (Sanghera & Thapar-Björkert, 2008). As a non-Thai, female academic from a western country, I conducted research among a range of participants such as rural residents and natural science professors. In some cases, such as in rural villages, my "outsider" position as a female *farang* (westerner) and an educated academic was likely perceived to be of one of prestige and status. In other cases, such as interviews with academics, these power differences were less prominent and in some cases, academia provided a common "insider" status. Further, while Thai academics also worked on this project as research assistants, they too were not necessarily seen as "insiders" in all cases (e.g. in rural villages) given the power differences typically associated gender as well as disparate levels of education and class. Thus, as Sanghera and Thapar-Björkert (2008) point out, the relationship between the key informants and researchers is an ever-evolving relationship. While it is not possible to eliminate power differences in research, one way we sought to address uneven power relationships was to develop rapport with research participants to identify commonalities and build trust (Schensul et al., 2012: 34).

Circulations: Uncertainty and blame

The haze crisis in northern Thailand is riddled with uncertainty as the actual causes of the increased burning are not completely understood by residents and natural scientists alike (Sirimongkonlertkun, 2014). This uncertainty has left a gap that is often filled by wide-ranging narratives that seek to place culpability: forest scavengers, hunters, swidden farmers, agribusiness (especially Charoen Pokphand, Thailand's largest private company), and swidden farmers in Laos and Myanmar have all been identified as potential culprits (Bach & Sirimongkalertkal, 2011). While some narratives are validated by scientists' data such as the transboundary effect and burning of agricultural waste in maize production areas, fire burning practices in the region are notably diverse and data does not fully account for the range of fire burning practices in the region which are a challenge to monitor with precision. For instance, fire has also long been a non-anthropogenic feature in the highlands and the long dry season and high levels of material contribute to frequent forest fires in the region. Additionally, both lowland and upland rural farmers have numerous uses for fire in forested areas; fire is known to be used to rid land of agricultural waste, to clear land for cultivation, to hunt, and to scavenge for non-timber forest products (NTFP) such as the lucrative *hed thob* mushroom. This diversity of fire usage is intensified by the mixed reliability in air quality monitoring instrumentation and a lack of reporting. For example, Aroon, a middle-aged natural scientist and leading air pollution researcher from a Chiang Mai based university explained:

> Chiang Mai air pollution is an annually recurring event. We have been living with haze pollution for a long time. But over the last 20 years, the particulate sensor system, sensor calibration and other equipment have not been updated, and people could not access official data. They accessed data on TV and radio which was screened and then reported by news reporters. But nowadays, in the data world, we are in a period of information freedom and everyone can easily access data on the internet.

Significantly, natural scientists also suspect that in addition to Chiang Mai's location in a low basin, urban based pollution from waste burning, traffic and construction also exacerbate the problem by making the air more toxic than previous haze episodes (Jeensorn et al., 2018).

Over the last two decades, many farmers throughout the region transitioned to commercial maize cultivation for animal feed, making it one of the most profitable boom crops in the region (Hall et al., 2011; Sirithian et al., 2018; Yap et al., 2017). Driven by broader shifts in the global political economy of agriculture, this transition contributes to an increase in the use of fire because more land must be cleared for planting and farmers typically burn the maize stalks to clear their land of accumulated agricultural waste (Arunrat et al., 2018; Bruun et al., 2016). While alternative land and agricultural waste clearing techniques have been proposed by a range of governmental and non-governmental organizations (e.g. biochar and intensive mulching) (Shafer,

2016), most are costly as well as time and labor intensive (Joseph et al., 2015). For instance, when asked what he would do if he didn't burn agricultural waste, an ethnic Lahu village leader in Mae Chaem explained, "I don't know where to keep huge piles of brushwood if I don't burn them". While some alternatives to maize cultivation are technically viable, most are difficult to incorporate into rural agricultural communities, as fire continues to be the most cost effective and fastest way to clear land and remove agricultural waste (Murdiyarso et al., 2004: 52).

Beyond maize, narratives of highlanders scavenging for *hed thob*, a lucrative and highly prized black mushroom is rekindled each year as soon as the air begins to thicken. *Hed thob*, is described by residents as akin to "receiving merit in the mouth" (Chiang Mai City News, 2015). Urban tourism practitioners describe how highlanders burn the forest to collect *hed thob* which is rumored to grow better in the ashes as well as to be more easily identified after the forest has been cleared by fire. As Chongrak, a 50-something male tourism practitioner who runs a small fruit smoothie stands in the center of the Old City of Chiang Mai explained in no uncertain terms: burning the forest "provides mushrooms for the people. Once, the rain comes, then the mushrooms can come up and they pick those mushrooms and then sell them at the market". A gregarious British restaurant owner in his mid-40s from the UK, who, while having never been to the highlands, participated in the circulation of this narrative: "sometimes, they will go from province to province just to go and find some of these forests to light up and gather these mushrooms." The circulation of environmental narratives regarding the mushroom collectors is widespread throughout Thailand and by 2019, there were widespread campaigns to deter consumers from buying *hed thob* to reduce demand.

Significantly, many highlanders contend that while they do collect *hed thob*, this is mostly for personal use. For example, Sopa, a female head of an ethnic Lahu village explained how she believe only about 20 percent of the *hed thob* rumors are true. She explained,

> Your point about burning for mushrooms is true but it is not a significant point of burning. Gathering wild mushrooms accounts for just 20 percent of the fires and haze. They [Highlanders] burn thick leaves to clear the ground surface, hed thob grows underground. It is easier to see mushrooms after burning a large number of leaves. Mostly, local people [i.e. Thais] come to this mountain rather than the Highland people. We are busy with our fruit trees. Villagers here have gathered mushrooms for household consumption, not for commercial products. I found that only local people come here for mushroom scavenging. What I am trying to say it they have practiced and have more skills than Highland people. They are very good at this (laugh)… If I find them, I gather them for my family. For our own consumption, I gathered them only for food, probably we eat them only one or two dishes. That's it.

Sopa's observation that it is lowland Thais which are primarily responsible for *hed thob* scavenging and its associated fires in her area is significant for how it demonstrates the multidirectional circulation of rumors.

Haze production narratives often circulate beyond national borders. Government actors, natural scientists, farmers, and other residents have all argued that farmers in Myanmar and Laos are the primary culprits of haze production and urban tourism practitioners often reiterate this possibility. For example, a Lahu farmer in a village in

Mae Chaem noted how, despite the fact that people in his village burn the forest and agricultural waste regularly, that the haze "might come from neighboring countries". He further explained "I have seen the hotspot map at the District Office showing many locations of active fires that are detected in the neighboring countries rather than in Chiang Dao. There is a Thai –Myanmar border point at Aruno-tai village which is very close to Chiang Dao and winds mostly blow haze towards Thailand and in the Highlands, haze might spread over the hill. Aroon, the Chiang Mai based natural scientist corroborated these points, noting that the solution to seasonal haze needs to include a transboundary component: "I think the root cause of haze pollution comes from neighboring countries", he explained. While many natural scientists agree that transboundary haze is one of several potential causes of seasonal air pollution, it does not fully explain seasonal haze in the north and most natural scientists believe that the majority of haze is domestically produced (Sirimongkonlertkun, 2014). Residents also critique the notion that haze is mostly coming from outside Thailand, a city bike tour guide in his mid-30s jokingly described how people love to blame the Burmese, while, as he suggested, Thai farmers are really to blame: "Now imagine when you have that mountain on fire, that mountain on fire, that mountain on fire, and not just here, just in this valley. All the way to Burma. All the way into Burma. People love to blame the Burmese. They're doing much worse than us. Bullshit, we're [Thailand] doing just as bad." Thus, narratives that identify Myanmar as the culprit may be mediated by historical narratives of cross-border blame for economic and ecological transboundary problems (Chongkittavorn, 2001; Lamb, 2014). This is not to mistake evidence for rumor, but instead to highlight the sociality of air pollution and its narratives.

Significantly, circulations of uncertainty and blame are not unidirectional. Highlanders are also aware of the potential impact of urban pollution on the haze crisis. For example, Sopa, the Lahu village leader explained:

> In the city, carbon monoxide from vehicles and factories significantly contribute to haze pollution. Also, a researcher found that fertilizer factories are one of the big factors contributing to the haze rather than forest fires because those factories release haze every day, but villagers here just burn weeds two months a year. I understand that burning is not good, but it is necessary for our subsistence farming.

Urban based causes of the haze crisis are an important counterpoint to the circulation of blame among Highlanders. These uncertainties are significant for how they breed crisis judgments.

Crisis: Resource socialities

Urban tourism practitioners' judgment of seasonal haze as a crisis is mediated by resource socialities and conflicts over the exchange values of landscape and land. On one hand, cultural geographers have long demonstrated that landscapes, as physical, subjective and objective entities, are mediated by language and representation and are therefore not neutral backdrops of experience, but rather should be understood as an "ideologically charged 'way of seeing'" (Thompson et al., 2013: 2). On the other hand, land is similarly made meaningful through social relations and production, yet it is a resource with unique qualities insofar as its value is created through practices of territorialization, regulation and private property establishment (Hall, 2013).

In northern Thailand, rural farmers often perceive land primarily as a resource for the production of agricultural commodities to be sold in urban markets. Urban tourism practitioners see the productive potential of land embedded in its potential as a landscape (including air) that can be commodified for tourists' consumption. Thus, the exchange value of the landscape within the tourism industry is affected most acutely by the polluted air that canvasses the city and deters its touristic consumers. As a resource, the landscape requires "regimes of exclusion that distinguish legitimate from illegitimate uses and users, and the inscribing of boundaries through devices such as fences, title deeds, laws, zones, regulations, landmarks and story-lines" (Li, 2014: 217). Tourism facilitates the establishment of new exchange values for intangible "products" such as landscape and air that are in tension with rural farmers' commodity production. Additionally, both tourism and agricultural commodities "come into value by using—and obviating—non-capitalist social relations, human and non-human" (Tsing, 2013: 21). For Anna Tsing, mushrooms, the object of her analysis, are valuable not because of substantive changes to their character but rather as a result of new social relations of exchange and forms of inequalities through the accumulation process. Flowing in and out of capitalist commodity status, mushrooms are sorted, assessed and shift between gift and commodity forms. This insight is significant for understanding the "irreducibly social nature of resources" in tourism and agriculture as well as how social interactions "govern their availability and allocation" (Bridge, 2009: 1217). During the annual smoky season, haze critically erodes the allure of the region by limiting visibility of the landscape as well as the air quality. Scholarship on visibility and air pollution in tourism destinations corroborates these observations in their descriptions of how air pollution compromises the commodity value of the touristic experience (Boyle et al., 2016; Eusébio et al., 2020; Zajchowski et al., 2019). Travel guidebooks, social media and popular press widely report on the negative effects of the haze with firm warnings to tourists not to visit the region during this period.

Residents of northern Thailand often try to make sense of resource conflicts in cultural terms. Burning is widely described by urban tourism practitioner as a "way of life" and/or as "part of the culture of the people". As Narong, a 40-something tour guide from Chiang Mai explained to me: "First you have to talk about the fact that it's cultural. It's built into their society. They [Highlanders] just do it. Whether it's harmful or not, they've been doing it for years, almost centuries. It's part of their culture here". This sense of cultural essentialism or the perception of culture as biological extends well beyond burning practices. Stereotypes of highlanders as "naturally" irresponsible, uneducated and destructive exist alongside their re-presentation in the tourism industry as the bearers of local culture, authenticity and morality. For instance, Lek, a 20-something Thai trekking guide based just outside the Old City of Chiang Mai explained how, while burning is part of Highlanders' culture, it is also driven by corruption. This narrative exists alongside that of Zado, a 55-year-old Lahu headman whom, when he sat down with the village assistant to talk about local burning practices, he seemed anxious due to the highly political nature of the haze crisis. His tone was sincere as he explained: "During the smoky season the haze comes everywhere in the village and this happens every year. So, when we burn crop fields, it is certain that haze will increase little by little. But we cannot stop burning, it is our way of life- ...". Thus, in the context of seasonal haze in northern Thailand, a tourism practitioners' crisis can be farmers' economic opportunity.

Contestations: Urban-Rural livelihoods

Haze production narratives among tourism practitioners in northern Thailand are shaped by the extent to which they entangle urban and rural livelihoods. In tourism, the beautification of the city is part of the product being sold to tourists as an "experience". The expansion of tourism often includes beautification practices to reduce the presence of pollution and poverty. Véron describes how in Delhi, urban elites have driven efforts to beautify the world city by purifying urban air through the removal of disenfranchised residents, their vehicles and communities (Véron, 2006). In a similar way, Chiang Mai based tourism practitioners are affected by the impact of air pollution on tourism. Aat, a coffee shop owner in the Old City of Chiang Mai explained that his business depends on tourists and that the haze had a big impact on his business. He lamented, "Chiang Mai is a big tourist city. If less tourists come here, everything will be slow and stop." Aat is in his mid-40s and, like many Chiang Mai residents, had come to depend on tourists for his livelihood.

Tensions between the "traditional" and "modern" Thai culture are reflected in widespread ambivalence around urbanization. In Pai, Pim, a female masseuse who is employed at a shop along the Pai Night Market enthusiastically noted how life was getting difficult for people in Thailand, especially in the rural areas. In her mid-20s, she lamented the increased dependence on waged labor and growing consumerism throughout Pai. These social shifts, she suspected, were responsible for the haze. Additionally, she suspects that Highlanders burn the forest because they have no choice: "They cannot afford to buy fertilizer, so they burn the forest and it becomes fertilizer for them and they get the mushrooms from the burnt forest ground." This sentiment was echoed by highland farmers themselves who frequently explain that they cannot afford tractors or fertilizer and need to burn to support their way of life. For example, Sopa, the Lahu village headwoman explained: "we need to burn, it is necessary for our livelihood." She further explained how: "Our ancestors believed that burning increased soil fertilizers .... We don't need any chemical fertilizers because of rich soils and fertilizer stored below the ground." Sud, a 48-year-old farmer from Mae Chaem similarly noted that lowland farmers don't need to burn and instead, often plow the soil. Yet, for highland farmers, he explained, "we have to burn". Traditional burning practices, he noted, only produce smoke for a couple of days.

Forest and agricultural waste burning, Sopa further pointed out, was becoming an increasingly sensitive topic for residents who were aware of the controversy surrounding haze in the urban areas. Kiet, a male farmer in his mid-30s corroborated these sentiments. He explained that people in his village burn the forest because it is Highlanders' way of life:

> If they are Hill Tribe people, they will understand and agree with me. We always burn for subsistence agriculture. Some of us live in the Highlands and some live near the river. [If

they live by the river] they are very fortunate to have nice farmland so that crops can be easily grown. Unfortunately, Hill Tribes who live in the Highlands must wait for the rainwater. Some migrate to town and then change their livelihoods because they get into debt. This is the life cycle of Hill Tribes nowadays.

While urban tourism practitioners are often sympathetic to the livelihood struggles of rural farmers, they also explain that they believe that Highland farmers are motivated to burn the forest by greed. These considerations are often overlooked by urban residents who, like Chai, a female handicraft shop owner in her mid-30s commented: "I don't think they care, I think they care about themselves only. They care only how they can find the wild products and sell them to make money … If they want to burn something, they will do it." In this context, urban tourism practitioners frequently describe education as an important part of the solution. Rural farmers and especially ethnic minority highlanders are often perceived to be uneducated, and thus unaware of the effects of the haze on urban dwellers. This rumor was rearticulated by Pim, a souvenir shop owner in Chiang Mai who explained: "Give them education on how to manage and add value from their farm. They can make natural fertilizer or charcoal. The government sector needs to give them an education. Villagers can't think of anything complicated that differs from their lifestyle". The narrative of the "uneducated" highlander has historical resonance in northern Thailand where education has long served as a social category of distinction, the structural conditions of which continue to be struggled over.

Towards an urban political ecology of tourism

Environmental narratives of the causes and effects of the haze crisis are stitched together with historically salient narratives as well as crisis judgments that circulate throughout the region. Particulate matter triggers the exaggeration of social distinctions which are reenacted through narratives of the causes and effects of the haze crisis in ways that perpetuate limited visions of the rural idyll in the context of a rapidly changing countryside (Tubtim & Hirsch, 2017). The everyday observations of haze are described through the sensed affect and effect of particulate matter in urban tourism space. The visibility of Doi Suthep, Chiang Mai's most sacred and visited temple sits above the city as a guide. For instance, outside of Thapae Gate where a cancelled #rightobreathe rally was meant to take place, a 50-something woman wearing a mask explained why she decided to protest the haze: "The reason that I came today is because I felt that it's heavier than past years. I can't even see Doi Suthep. I woke up, 'Oh my, I can't see roads, I can't see Doi Suthep'. That's why I have to speak out. But they (i.e. the government) do not allow us to do the campaign. I think they are afraid of hurting Chiang Mai's image. Is this about tourism? The government is afraid of the effect on the tourism industry. Chiang Mai is a touristy province with a lot of tourists. But then what? Are we buffalo which have to breath in the smoke like this?" These observations echo what Davies describes as practices of "slow observation" through which residents witness gradual changes to their environments as a barometer for understanding chronic pollution (Davies, 2018). In this way, haze and the narratives that mediate people's experience of it are shaped by resource conflicts over the exchange value of land (for rural farmers) and landscapes (for urban tourism practitioners).

In the context of air pollution and other toxicities, these knowledges are culturally embedded in ways that sometimes "contradict public health concepts of environmental health and, in turn, differentially shape people's interactions with the environment" (Harper, 2004: 295). Narratives of agricultural and forest burning take shape through urban and rural livelihood practices that differently value land. In Madagascar, Kull illustrates how the "fire problem" is less about the frequency of fire than it is a representation of ongoing conflicts over resource use (Kull, 2004). In this way, the materiality of resources is deeply entangled in discursive formations of power that seek to make sense of uncertainty by identifying perpetrators and victims which take place within historically situated social relations (Smith & Dressler, 2020). By occupying the space between knowledge and belief, rumors can be viewed as a commentary on the urban as an undefined, and complex process.

Deeply uneven entanglements of power govern rural and urban livelihoods and drive environmental policy. Urban tourism practitioners increasingly impinge on rural residents' livelihoods through their support for governance, securitization, and regulatory mechanisms such as the burning ban. Yet, many highlanders contend that the burning ban has only extended the burning period and the intensity of the fires. For instance, a 48 year old farmer in the Mae Chaem district explained that he did not agree with the burning ban and contended that if the government allowed farmers to burn in March, the haze situation would improve: "In the past, I think the smoke was not as bad as now. After they instituted the burning ban, I think there is more smoke. In the past, they burnt for a certain time and there was a lot of smoke for two days, then the smoke was gone. Currently, some people burn before the burning ban period, some still burn the forest for mushrooms. Some burn after the burning ban. This extends the burning period".

In addition to the burning ban, satellite surveillance and livelihoods transition programs now monitor highland farmers. The effects of these surveillance strategies are reminiscent of programs that threatened to evict Highlanders from their land for opium production by creating an atmosphere of fear that allowed "local officials to take unfair advantage of villagers in the uplands", thus making it "an old story" with new characters (Tapp, 2010). Today, this story is driven by narratives of the causes, effects and solutions to the haze crisis and reflects how cities are entanglements of linked social networks that are simultaneously urban and rural as well as human and physical. In addition to government responses, regulatory responses been enacted by NGOs and non-profits to reorient rural farmers to new crops and/or technologies. Several NGOs in Chiang Mai work with university and/or government initiatives focused on air pollution reduction. Additionally, citizen science groups measure PM throughout the city. These groups seek to make sense of the haze crisis through independent air quality measurements. For instance, Craig Houston, a Scottish aeronautical engineer who has lived in Chiang Mai for three years with his Thai wife and young son began building a network of sensors across the city in 2017. Houston explained how he "has taken it upon himself to set up a detailed network of air pollution sensors that are both low cost and accurate" (Stuart, 2018). In a related effort, a group of

Thai engineers developed Dust Boy, an air quality measuring system and app. Dust Boy is the first civil society project to monitor air pollution in the country. Greenpeace explained how Dust Boy reflects a transition to a private air quality index (AQI) monitoring system (Kittikongnaphang, 2018). Houston and Dust Boy developers continue to surveil AQI levels in Chiang Mai, despite their limited influence on policy initiatives. These civil society interventions are in part a reaction to ineffective haze policy. For instance, Sirikit, an atmospheric scientist at the National Astronomical Research Institute of Thailand (NARTI) explained: "it is important to learn their [Highlanders'] perspectives because they play an important role in a policy formulation. As I used to work in the Pollution Control Department (PCD), PCD strictly follows the government policy, we are practitioners. But the policymakers never know all the details. Sometimes, a leader just commands 'seven days-fires out smog pollution' ... it is not a solution at all. We need long term protection, not only a campaign." Thus, current environmental governance regimes are often riddled with uncertainty and provoke contestations among urban and rural residents in ways that impinge upon their increasingly entangled livelihoods.

Conclusion

Streaming on smartphones throughout northern Thailand, Thai hip-hop artists express their frustration on Youtube through lyrical rhymes of PM2.5 and its noxious effects. As residents try to make sense of the widespread uncertainty of the causes and effects of the haze crisis, environmental narratives circulate throughout the region in ways that are mapped out by historically situated urban-rural relations. It is in this context that environmental narratives become entangled in the more-than-human materiality of air pollution as both a physical and discursive entity. The materiality of air pollution is symptomatic of an increasingly urbanized nature where new meanings of the environment and new values of land and landscapes take shape through entangled urban and rural spaces. Market incentives have expanded and diversified traditional uses of fire as well as become an obstacle to tourism development. The tourism industry is particularly vulnerable to environmental degradation, especially where the landscape is seen as a primary "sight" for domestic and international touristic consumption. Contestations between the urban-based tourism practitioners and rural farmer are highlighted through a UPE of the recurring haze crisis. Residents seek to reconcile these contestations, yet currently no alternatives to biomass burning for rural farmers have proven viable.

Environmental narratives also mediate governance mechanisms and regulatory regimes, as well as the proliferation of urban-based green grassroots haze amelioration initiatives. In this way, air pollution is not only a comment on the sociality of urbanization, but also highlights the processual nature of the urban. While scholars have addressed urban metabolism and exchange between nature and society, the state plays a critical role in the social and ecological forces of urban metabolic processes; regulatory regimes of livelihood practices in rural areas of northern Thailand have long been targeted by the state. These regulatory regimes are linked to ethnicity, class and locale and responses to the haze crisis have further exaggerated these distinctions.

This article contributes an ethnographic approach to a UPE of tourism. It also provides a nuanced understanding of how narratives of air pollution and its causes and effects enable and foreclose access to rights and resources in northern Thailand. In other words, environmental narratives influence material perceptions of environmental change that drive governance regimes. Ethnographically informed understandings of how differently positioned actors (e.g. tourism practitioners, farmers, natural scientists) make sense of environmental change in the context of uncertainty over its causes and effects can play a critical role in environmental policy, politics, and practice. Finally, as anthropogenic environmental crises intensify globally, there is a critical need for future UPE scholarship to account for the role of environmental materialities in the reshaping of social relations in tourism landscapes.

Disclosure statement

No potential conflict of interest was reported by the author(s).

References

Adey, P. (2014). *Air: Nature and culture*. London: Reaktion books.

Angelo, H., & Wachsmuth, D. (2015). Urbanizing urban political ecology: A critique of methodological cityism. *International Journal of Urban and Regional Research*, *39*(1), 16–27. https://doi.org/10.1111/1468-2427.12105

Arunrat, N., Pumijumnong, N., & Sereenonchai, S. (2018). Air-pollutant emissions from agricultural burning in Mae Chaem Basin, Chiang Mai Province, Thailand. *Atmosphere*, *9*(4), 145. https://doi.org/10.3390/atmos9040145

Bach, N. L., & Sirimongkalertkal, N. (2011). Satellite data for detecting trans-boundary crop and forest fire dynamics in Northern Thailand. *International Journal of Geoinformatics*, *7*(4), 47.

Bennett, J. (2010). *Vibrant matter: A political ecology of things*. Duke University Press.

Bernard, R. H. (2011). *Research methods in cultural anthropology: qualitative and quantative approaches* (5th ed.). AltaMira Press.

Boyle, K. J., Paterson, R., Carson, R., Leggett, C., Kanninen, B., Molenar, J., & Neumann, J. (2016). Valuing shifts in the distribution of visibility in national parks and wilderness areas in the United States. *Journal of Environmental Management*, *173*, 10–22. https://doi.org/10.1016/j.jenvman.2016.01.042

Bridge, G. (2009). Material worlds: Natural resources, resource geography and the material economy. *Geography Compass*, *3*(3), 1217–1244. https://doi.org/10.1111/j.1749-8198.2009.00233.x

Bruun, T. B., Neergaard, A., Burup, M. L., Hepp, C. M., Larsen, M. N., Abel, C., … Mertz, O. (2016). Intensification of upland agriculture in Thailand: Development or degradation? *Land Degradation & Development*. https://doi.org/10.1002/ldr.2596

Butt, L. (2005). Lipstick girls" and "fallen women": AIDS and conspiratorial thinking in Papua, Indonesia. *Cultural Anthropology*, *20*(3), 412–442. https://doi.org/10.1525/can.2005.20.3.412

Carse, A. (2014). The year 2013 in sociocultural anthropology: Cultures of circulation and anthropological facts. *American Anthropologist*, *116*(2), 390–403. https://doi.org/10.1111/aman.12108

Chaitrong, W. (2012). Thailand reflects Asia's growing income inequality. Business. Retrieved from   http://www.nationmultimedia.com/business/Thailand-reflects-Asias-growing-income-inequality-30179912.html

Charmaz, K. (2014). *Constructing grounded theory*. London: Sage.

Charmaz, K. (2006). *Constructing grounded theory: A practical guide through qualitative analysis*. Sage.

Chongkittavorn, K. (2001). Thai-Burma relations. In Zaw, A., Arnott, D., Chongkittavorn, K., Liddell, Z., Morshed, K., Myint, S., & Aung, T. T. (Eds.), *Challenges to democratization in Burma: Perspectives on multilateral and bilateral responses*. Stockholm, Sweden: International Institute for Democracy and Electoral Assistance. 117–130.

City News. (2015). Hed thob mushroom prices skyrocket. City Life Chiang Mai. Retrieved from Chiang Mai City Life website http://www.chiangmaicitylife.com/news/hed-thob-mushroom-prices-skyrocket/.

City News. (2019). The Chiang Mai Breathe Council has been set up to fight pollution. Chiang Mai City Life. August 21, 2019. https://www.chiangmaicitylife.com/citynews/local/the-chiang-mai-breathe-council-has-been-set-up-to-fight-pollution/.

Comaroff, J. (2017). On the materialities of air. *City*, *21*(5), 607–613. https://doi.org/10.1080/13604813.2017.1374776

Connolly, C. (2019). Urban political ecology beyond methodological cityism. *International Journal of Urban and Regional Research*, *43*(1), 63–75. https://doi.org/10.1111/1468-2427.12710

Davies, T. (2018). Toxic space and time: Slow violence, necropolitics, and petrochemical pollution. *Annals of the American Association of Geographers*, *108*(6), 1537–1553. https://doi.org/10.1080/24694452.2018.1470924

Eusébio, C., Carneiro, M. J., Madaleno, M., Robaina, M., Rodrigues, V., Russo, M., Relvas, H., Gama, C., Lopes, M., Seixas, V., Borrego, C., & Monteiro, A. (2020). The impact of air quality on tourism: a systematic literature review. *Journal of Tourism Futures*. Ahead of print. https://doi.org/10.1108/JTF-06-2019-0049

Fairhead, J., & Leach, M. (1995). False forest history, complicit social analysis: rethinking some West African environmental narratives. *World Development*, *23*(6), 1023–1035. https://doi.org/10.1016/0305-750X(95)00026-9

Forsyth, T. J. (1995). Tourism and agricultural development in Thailand. *Annals of Tourism Research*, *22*(4), 877–900. https://doi.org/10.1016/0160-7383(95)00019-3

Gandy, M. (2003). *Concrete and clay: reworking nature in New York City*. MIT Press.

Gissen, D. (2015). Air apparent. *Architecture and Culture*, *3*(2), 133–135. https://doi.org/10.1080/20507828.2015.1067000

Glaser, B. G., & Strauss, A. L. (2017). *Discovery of grounded theory: Strategies for qualitative research*. Routledge.

Graham, S. (2015). Life support: The political ecology of urban air. *City*, *19*(2–3), 192–215. https://doi.org/10.1080/13604813.2015.1014710

Grave, P. (1995). Beyond the mandala: Buddhist landscapes and upland-lowland interaction in north-west Thailand AD 1200–1650. *World Archaeology*, *27*(2), 243–265. https://doi.org/10.1080/00438243.1995.9980306

Hall, D. (2013). *Land*. Polity Press.

Hall, D., Hirsch, P., & Li, T. (2011). *Powers of exclusion: Land dilemmas in Southeast Asia*. National University of Singapore Press.

Harper, J. (2004). Breathless in Houston: a political ecology of health approach to understanding environmental health concerns. *Medical Anthropology*, *23*(4), 295–326. https://doi.org/10.1080/01459740490513521

Hayward, D. (2018). From Maize to Meat: Placing maize production in Thailand within a global poultry value chain. Southeast Asia, Thailand: Greenpeace.

Heynen, N. (2014). Urban political ecology I: The urban century. *Progress in Human Geography*, *38*(4), 598–604. https://doi.org/10.1177/0309132513500443

Heynen, N. (2016). Urban political ecology II: The abolitionist century. *Progress in Human Geography*, *40*(6), 839–845. https://doi.org/10.1177/0309132515617394

Jeensorn, T., Apichartwiwat, P., & Jinsart, W. (2018). PM10 and PM2. 5 from haze smog and visibility effect in Chiang Mai Thailand. *Applied Environmental Research*, *40*(3), 1–10.

Joseph, S., Anh, M. L., Clare, A., & Shackley, S. (2015). *Socioeconomic feasibility, implementation and evaluation of small-scale biochar projects* (pp. 853–879). Earthscan London.

Kittikongnaphang, R. (2018). Do not reconcile the state! Chiang Mai has People AQI to measure public air quality. *Greenpeace Thailand*. Bangkok. http://www.greenpeace.org/seasia/th/news/blog1/people-aqi/blog/61316/#.Wrz2J65t570.facebook

Kull, C. A., & Laris, P. (2009). *Fire ecology and fire politics in Mali and Madagascar tropical fire ecology* (pp. 171–226). Springer.

Kull, C. A. (2004). *Isle of fire: the political ecology of landscape burning in Madagascar* (Vol. 245). Chicago: University of Chicago press.

Lamb, V. (2014). Where is the border?" Villagers, environmental consultants and the 'work'of the Thai–Burma border. *Political Geography*, *40*, 1–12. https://doi.org/10.1016/j.polgeo.2014.02.001

Li, T. M. (2014). What is land? Assembling a resource for global investment. *Transactions of the Institute of British Geographers*, *39*(4), 589–602. https://doi.org/10.1111/tran.12065

Lindsay, S. (2019). Thailand's wealth inequality is the highest in the world: What will this mean for the upcoming elections? *ASEAN Today*. January 16, 2019. https://www.aseantoday.com/2019/01/thailands-wealth-inequality-is-the-highest-in-the-world-what-does-this-mean-for-upcoming-elections/.

McKinnon, K. (2008). Taking post-development theory to the field: Issues in development research, Northern Thailand. *Asia Pacific Viewpoint*, *49*(3), 281–293. https://doi.org/10.1111/j.1467-8373.2008.00377.x

Miller, D. (2005). *Materiality*. Duke University Press.

Moulin, C. (2010). Border languages: Rumors and (dis) placements of (inter) national politics. *Alternatives: Global, Local, Political*, *35*(4), 347–371. https://doi.org/10.1177/030437541003500402

Murdiyarso, D., Lebel, L., Gintings, A., Tampubolon, S., Heil, A., & Wasson, M. (2004). Policy responses to complex environmental problems: insights from a science–policy activity on transboundary haze from vegetation fires in Southeast Asia. *Agriculture, Ecosystems & Environment*, *104*(1), 47–56.

Newell, J. P., & Cousins, J. J. (2015). The boundaries of urban metabolism: Towards a political–industrial ecology. *Progress in Human Geography*, *39*(6), 702–728. https://doi.org/10.1177/0309132514558442

Rice, S., & Tyner, J. (2017). The rice cities of the Khmer Rouge: an urban political ecology of rural mass violence. *Transactions of the Institute of British Geographers*, *42*(4), 559–571. https://doi.org/10.1111/tran.12187

Rigg, J. (2014). *More than the soil: Rural change in SE Asia*. Routledge.

Robbins, P. (2012). *Lawn people: How grasses, weeds, and chemicals make us who we are*. Temple University Press.

Roe, E. M. (1991). Development narratives, or making the best of blueprint development. *World Development*, *19*(4), 287–300. https://doi.org/10.1016/0305-750X(91)90177-J

Samuels, A. (2015). Narratives of uncertainty: The affective force of child-trafficking rumors in postdisaster Aceh, Indonesia. *American Anthropologist*, *117*(2), 229–241. https://doi.org/10.1111/aman.12226

Sanghera, G. S., & Thapar-Björkert, S. (2008). Methodological dilemmas: Gatekeepers and positionality in Bradford. *Ethnic and Racial Studies*, *31*(3), 543–562. https://doi.org/10.1080/01419870701491952

Save Chiang Mai From Burning. (2020). In Facebook page. May 1, 2020, from https://www.facebook.com/2300061783651287/posts/blue-sky-for-chiang-maiplease-support-clean-air-and-clear-sky-for-chiang-mai-by-/2317728348551297/.

Schensul, S. L., Schensul, J. J., & LeCompte, M. D. (2012). *Initiating ethnographic research: A mixed methods approach* (Vol. 2). AltaMira Press.

Scott, J. C. (2009). *The art of not being governed: An anarchist history of upland Southeast Asia.* Yale University Press.

Shafer, M. (2016). Biochar: a profitable solution to Thailand's haze problem. The Nation. http://www.nationmultimedia.com/opinion/Biochar-a-profitable-solution-to-Thailands-haze-pr-30281261.html.

Sirimongkonlertkun, N. (2014). Smoke haze problem and open burning behavior of local people in Chiang Rai province. *Environment and Natural Resources Journal, 12*(2), 29–34.

Sirithian, D., Thepanondh, S., Sattler, M. L., & Laowagul, W. (2018). Emissions of volatile organic compounds from maize residue open burning in the northern region of Thailand. *Atmospheric Environment, 176*, 179–187. https://doi.org/10.1016/j.atmosenv.2017.12.032

Smith, W., & Dressler, W. H. (2020). Forged in flames: Indigeneity, forest fire and geographies of blame in the Philippines. *Postcolonial Studies*, 1–19. https://doi.org/10.1080/13688790.2020.1745620.

Stott, P. (1991). Mu'ang and pa: elite views of nature in a changing Thailand. *Thai Constructions of Knowledge, 1*, 142–154.

Stuart, A. (2018). Explaining Chiang mai's smoke pollution with real data. Chiang Mai City Life. http://www.chiangmaicitylife.com/citylife-articles/explaining-chiang-mais-smoke-pollution-real-data/.

Sukplang, S. (2007). Smoke thickens in choking Thai North. Reuters. https://www.reuters.com/article/dcbrights-thailand-haze-dc/smoke-thickens-in-choking-thai-north-idUSBKK29196620070314?fbclid=IwAR2Q15VBVCGsJfy5kExTvfuUj8R11BqOAImpuD9ftGpeMgfHSYb9d5fGhUQ

Swyngedouw, E., & Heynen, N. C. (2003). Urban political ecology, justice and the politics of scale. *Antipode, 35*(5), 898–918. https://doi.org/10.1111/j.1467-8330.2003.00364.x

Tapp, N. (1988). Geomancy and development: The case of the white Hmong of North Thailand. *Ethnos, 53*(3–4), 228–238. . https://doi.org/10.1080/00141844.1988.9981371

Tapp, N. (2010). Letters: Response to eviction of upland peoples in Northern Thailand. *Critical Asian Studies, 42*(2), 327–329.

The Nation. (2010). Smog Eases in Most of the North After Rains. March 10, 2010. *The Nation.*

Thompson, I., Howard, P., Waterton, E., & Thompson, I. (2013). Introduction. In Waterton, E., & Atha, M. (Eds.), *The Routledge companion to landscape studies* (pp. 1–11). Routledge.

Tsing, A. (2013). Sorting out commodities: How capitalist value is made through gifts. *HAU: Journal of Ethnographic Theory, 3*(1), 21–43. https://doi.org/10.14318/hau3.1.003

Tubtim, T., & Hirsch, P. (2017). Decropping the Southeast Asian countryside. In *Routledge handbook of Southeast Asian development*. Routledge.

Véron, R. (2006). Remaking urban environments: the political ecology of air pollution in Delhi. *Environment and Planning A: Economy and Space, 38*(11), 2093–2109. https://doi.org/10.1068/a37449

Walker, A. (2001). The 'Karen Consensus', ethnic politics and resource-use legitimacy in Northern Thailand. *Asian Ethnicity, 2*(2), 145–162. https://doi.org/10.1080/14631360120058839

Walker, A., & Farrelly, N. (2008). Northern Thailand's specter of eviction. *Critical Asian Studies, 40*(3), 373–397. https://doi.org/10.1080/14672710802274078

Winichakul, T. (1997). *Siam mapped: A history of the geo-body of a nation.* University of Hawaii Press. https://doi.org/10.1086/ahr/100.2.477

World Bank (2016). Thailand: Overview. Retrieved from http://www.worldbank.org/en/country/thailand/overview Chaitrong 2012

Yap, V., Neergaard, A., & Bruun, T. (2017). To adopt or not to adopt?'Legume adoption in maize-based systems of Northern Thailand: constraints and potentials. *Land Degradation & Development, 28*(2), 731–741. https://doi.org/10.1002/ldr.2546

Yuthamanop, P. (2011). Income Inequality in Thailand. Bangkok Post Learning. Retrieved from http://www.bangkokpost.com/learning/learning-from-news/270964/income-inequality-in-thailand

Zajchowski, C. A., Brownlee, M. T., & Rose, J. (2019). Air quality and the visitor experience in parks and protected areas. *Tourism Geographies, 21*(4), 613–634. https://doi.org/10.1080/14616688.2018.1522546

Ontological mingling and mapping: Chinese tourism researchers' experiences at international conferences

Jundan Jasmine Zhang and Carol Xiaoyue Zhang

ABSTRACT

Within tourism studies, the 'critical turn' has evoked growing reflective and critical perspectives on the role of the researcher in producing knowledge. This has also led to calls for building an inclusive research community, particularly through including non-Western and non-positivist methodologies. While it is noted that non-Western scholarship has gained more visibility in the international tourism research community through publications in prestigious academic journals, few studies discuss non-Western scholars' interactions with other scholars in a qualitative and individual manner. Based on in-depth interviews with nineteen Chinese tourism scholars, we explore how their experiences at international conferences have shaped their positionality as 'Chinese researchers' in the international scientific community and thus contributed to their knowledge making. A process of mingling and mapping is shown in the narratives, where Chinese scholars attempt to find meanings of being at an academic conference and to understand the relations embedded in the conference space. Dynamics and reflexivity are seen in terms of how one goes around certain constructed binaries, such as 'Western/non-Western', 'male/female' and 'junior/senior'. Finally, such a process of mingling and mapping affects the participants' views on who will make the non-Western knowledge and how. With these voices from Chinese tourism scholars, we therefore contribute to the discourse of non-Western knowledge-making.

摘要

在旅游研究中,"批判性转向"引发了对研究者在知识生产过程中的作用的反思和批判。这也导致营建一个包容性研究社区的呼声,特别是包容非西方和非实证主义的方法论的研究社区。虽然有人指出,非西方学者通过在著名学术期刊上发表论文,在国际旅游研究界获得了更大的知名度,但很少有研究以定性和个性化的方式讨论非西方学者与其他学者的互动。通过对19位中国旅游学者的深入采访,我们探讨了他们在国际会议上的经验如何塑造了他们作为"中国研究者"在国际科学界的立场,从而为他们的知识创造做出了贡献。在叙事中,中国学者试图找到学术会议的意义,理解会议空间中蕴含的关系,呈现出一种交融与映射的过程。动态性和

反身性显现在一个学者如何围绕诸如"西方学者/非西方学者"、"男
性学者/女性学者"和"资浅/资深学者"等某些二元结构进行行动。
最后, 这种混合和映射的过程会影响参与者对于谁生产以及如何
生产非西方知识的看法。有了这些来自中国旅游学者的声音, 我
们为非西方知识生产的话语做出了贡献。

Introduction

There is a clear recognition that an inclusive tourism knowledge production is increasingly important. Reflecting on the ten years since the 'critical turn' tourism studies took, Morgan et al. (2018) urge tourism scholars to march on with the spirit of 'hopeful tourism' while also confronting a number of challenges, among which they mention that 'we must dare tourism to develop conceptualizations that include multiple cultural differences and worldviews that reflect and recognize the plurality of human practices, positions, and insights' (p. 185). Indeed, the themes of inclusion and pluralism have been central in the establishment and development of 'critical tourism studies' scholarship. To some extent, these themes are instrumental in fostering a growing awareness within tourism studies to recognize the ever-persistent divisions and binaries between tourism studies/tourism management, the colonial/the colonized, qualitative/quantitative, and Western/non-Western.

At the same time, it must also be acknowledged that our inquiries into the inclusive conceptualizations in tourism studies are still at an early stage, despite increasing engagements with studies of feminism studies, Marxist, and postcolonialism. For instance, we are reminded that in several aspects, tourism knowledge is still largely shaped by Anglo-Saxon paradigms, traditions, and ideals (Morgan et al., 2018). First of all, it is argued that even within the paradigm shifts from positivism and post-positivism towards more critical and constructivist paradigms, non-Western scholars' knowledge-making is still deeply dependent on Western thoughts (Mura & Khoo-Lattimore, 2018b; Wijesinghe et al., 2019). Secondly, tourism knowledge and traditions in non-English systems are rarely presented or studied in English scholarly articles, confirming that English as the field's *lingua franca* remains a major limiting factor for academic inclusiveness (Mura et al., 2017). Finally, yet importantly, the benchmarks for measuring the progress of inclusiveness in tourism studies are unclear. How can we know if the community of tourism research has become more open-minded, inclusive, and equal over time? And on whose terms?

The above three aspects all deserve more attention in the future, if our aim for a hopeful future, academia, and society is serious. However, for this special issue of *Recentering critical tourism studies*, it is the last point of the 'how' that we wish to explore and contribute to. As two Chinese-born female researchers who have been trained and have worked largely within Western education and academic systems, we are interested in the unclear, messy and ambiguous situations deriving from the process of judging, selecting and evaluating the very status of 'criticality' (the ability to constructively criticize the existing knowledge), as well as 'creativity' (the ability to contribute new knowledge and new ways of thinking to the existing knowledge, i.e. non-Western knowledge). We observe that, despite the previous studies reflecting on the inquiry-action nexus of non-Western knowledge (e.g. in Zhang, 2018 and Tucker &

Zhang, 2016), very little attention has been paid to the implications and dilemmas faced by scholars who live, study, and work across the Western/non-Western boundaries. We do notice that several studies attempt to visualize and measure the progress of non-Western scholarship in the international tourism research community, yet adopt an exclusively quantitative approach by describing the shifts in publication quantity results (Bao et al., 2014; Huang & Chen, 2016). While these studies are valuable, we maintain that it is also important to investigate questions concerning the process of non-Western scholars' knowledge production in a more qualitative and interpretative manner.

Emerging from conversations on the authors' own experiences at international tourism conferences, this article regards international conferences as an important yet neglected social space and context for understanding how non-Western scholars experience and perceive their own interactions with others in the academic world, and how these interactions and experience have influenced their knowledge-making process. Conferences are traditionally viewed from an organizational perspective, and most previous studies tend to focus on the motivations of conference attendees and the functionalities of the conferences, such as networking, getting feedback, developing career paths, keeping up with research trends, and encountering research interests (Mair & Frew, 2016). Recently in tourism studies, some scholars situate issues relating to knowledge production, such as gender and social justice, in the context of conferences (Munar et al., 2015; Walters, 2018). To some extent, these studies bring in new perspectives that regard conferences as a social space and open up new discussions concerning inclusion and diversity in the tourism research community. Following this line of discussion, we aim to examine the relationships between the personal conference experiences of 19 Chinese tourism scholars and their own perceptions of the roles of their identities in their knowledge-making.

Theoretical background

In this section, we present theories and concept that lay the foundation for the research. It contains three subsections: knowledge production/distribution; conference as social space and positionality at international conferences

Knowledge production/distribution

Although thinkers from all cultures and civilizations have discussed ideas, wisdom and beliefs throughout history, the discussion of knowledge and epistemology in a reflective and systematic way did not start until the 20th century, largely led by philosophers and social scientists in the West. In such discussions, it is fundamental to note a division: knowledge as science and knowledge as culture (Delanty, 2001). Similarly, as Latour (1993) describes it, the knowledge of things is placed on one side, and power and human politics on the other. Moreover, Machlup's (1980) work *Knowledge: Its Creation, Distribution and Economic Significance* notes that knowledge can be viewed as a product and that there is a market and demand for certain types of knowledge rather than others. When discussing the flows of knowledge, Machlup argues that the flows from transmitter to recipient happen through both space and time. Through materializing and rationalizing knowledge as something tangible and countable, the process of knowledge-making and distributing also becomes objective and thus manageable.

According to Foucault (1972), this rationalization is part of the archaeology of knowledge and historically plays a role in the intertwined relationships of knowledge and power. Through specifying each stage in the knowledge-production process, the flows of ideas, ideals and shared feelings are interrupted and isolated, and thus knowledge comes to be governed and controlled. Sometimes this is done by assigning different tasks to different individuals, and sometimes by constructing different tasks for the same individual but at different times. This mentality has come to be fundamental in the organization and management of time, space and agency in modern time. From this perspective, we can really start to see that the binary of the distribution and production of knowledge is institutionalized and part of the governmentality of knowledge.

This governmentality of knowledge may explain the lack of research on conference attendance's influence on academics' reflection on their knowledge production. Through the arranging and cataloging of knowledge-making into steps periodically and spatially, productivity and applicable values come to be the priority, while irrational factors such as feelings, ethics of research, and politics and power are detached from the process of knowledge-making. As a result, it is normalized that knowledge production happens at a particular place or site, such as a university, laboratory, or field site (Delanty, 2001; Latour, 1988). The conference is one of the places where scholars share and present the knowledge they have produced. Indeed, in a way, the conference itself is a by-product or invention in order to serve the rational organizations of knowledge. Therefore, studies on conference attendance have mostly taken an organizational perspective (e.g. in Rittichainuwat et al., 2001; Oppermann & Chon, 1995), missing the human attendees and the dynamics and complexities of intellectual interactions that are essential in the knowledge-making process.

Previously knowledge production and sharing have been analysed in management studies using socio-psychological frames. For instance, social exchange theory suggests that self-profit is the motivation for knowledge sharing behaviours (Liao, 2008). Yet again, in such frameworks, knowledge is viewed as a product. In this paper, we follow Foucault's (1972) view that knowledge is discursively formed through discourse. A discourse is produced through language (in the broadest sense) and is constantly reproducing itself through discursive practices. As a result, all forms of ideas are produced, circulated and disseminated through discourses. Hence, it is the competing discourses and their enactment of power that construct what we know as common knowledge and common sense. In this way, knowledge-making is not a conduct carried out in a certain period, at a certain space and by a particular individual, but rather the discursive practices and formations that interrelate subjects, place and time all at once.

Hence, many questions remained unanswered: what could be counted as non-Western knowledge? Who can be said to be the producer of non-Western knowledge and where and when does the knowledge become non-Western knowledge? And indeed, why is there even a thing called 'non-Western knowledge'? With these questions in mind we also start to see that the conference, as a space for the communication of different discourses, is spontaneously the space of knowledge production. Looking into how the conference as a social space enables and dis-enables individuals to express and access discourses, then, is important for understanding how individuals' knowledge production can be shaped through their conference attendance.

Conference as social space

Ford and Harding (2008) use Lefebvre's (1991) theory of spaces/places, and suggest that arrival at the physical territory of a conference endows one with an 'identity' at the conference. When conference attendees identify themselves at the actual place socially and collectively, the material and discursive spaces and places of where the conference activities happen become 'the conference'. On the other hand, 'it is place that also helps bring individuals into being, for place solidified a sense of embodiment' (Ford & Harding, 2008, p. 7). Therefore, it is the dialectic relations between the conference attendees and the conference as place/space that bring each of them into being, perhaps even before the conference 'actually' happens. For in practice, an academic conference often has particular theme(s) and targets. The making of an academic conference is thus built on a set of presumptions. Here there is a paradox in relation to the inclusiveness of a particular conference. On the one hand, the functionality sides of a conference require it to reach a certain number of attendees; on the other hand, its specific themes as well as its entrance fee do create boundaries between who can be 'in' or 'out'(Lloyd, 2015; Sweeting & Holh, 2015).

It is thus pertinent to say that conference space is constructed socio-culturally, economically, and politically. Why do we care? Again, the dialectical relationship between the individual and the conference results in the social construction of the conference space interrelating with how we as conference attendees identify ourselves with the broader scientific community. Ford and Harding (2010) observe that:

> We arrive at a conference knowing how to occupy the subject position of conference participant, monitoring ourselves to ensure we occupy it correctly, and in doing so render ourselves somewhat passive and receptive to whatever it is that we experience. (Ford & Harding, 2010, p. 509).

And, in their recent paper, Edelheim et al. (2018) ask what conferences do to the academic attendees, reflecting:

> By attending a conference in person, attendants subject themselves to the judgement of the collective, as well as to individual judgements of others, be it based on gender, age, appearance, or metrics like number of publications in highly ranked journals (Edelheim et al., 2018, p. 98).

It is the meeting between the individual and the collective that evokes the process of evaluating, reflecting, and positioning. Walters (2018) suggests that academic conferences can be regarded as texts and represent a community, conveying a particular social order. Bell and King (2010) also suggest that conferences 'constitute a key site of academic socialization that enables norms and values to be passed on from experts to newcomers. They thus provide a material means of inscribing regimes of cultural power onto the embodied subject' (Bell & King, 2010, p. 432). Notably, the masculine atmosphere and climate have been described and analyzed in recent years. For instance, in their report, Munar et al. (2015) note that women are subject to unequal treatment and social status at tourism conferences, especially concerning who can be the keynote speakers and hold honorary chair positions. Studies from other fields have also discussed gender inequality at academic conferences, noting that female participants are not as included or listened to compared to their male peers (King et al., 2018). The phenomena as such pose a question of what conference space can enable and dis-enable for long-term knowledge-making. Do academic conferences ignore or even reinforce the very problems they claim to criticize (Mair & Frew, 2016; Walters, 2018)? If tourism academia strives to be more open and inclusive, what kind of conference space is desirable? And who has a say in this?

King et al. (2018)believe academic conferences can potentially alleviate barriers and reproduce diversity in academic field, while Edelheim et al. (2018) also point out that a good conference containing 'thought-providing presentations' can create 'new thought paths' and therefore lead to future change within both the participants and the field (p. 101). However, as Mair (2015) observes, previous studies researching motivations for and factors influencing delegates' decision-making process are largely situated in a Western context (Mair & Thompson, 2009; Oppermann & Chon, 1995; Rittichainuwat et al., 2001). With increasing internationalization at academic conferences, attendees from other cultural backgrounds may bring new dynamics and perspectives regarding not only how conferences are organized, but also how they may function as a socio-cultural and geopolitical space for knowledge production. For example, Rowen (2019) argued there is a clear 'geopolitical effects of tourism scholarship' (p.3). Hence, interpersonal conference link to the broader knowledge production landscape and geopolitical dynamics.

Positionality at international conferences

Maher and Tetreault (1993) contend that the concept of positionality refers to 'gender, race, class, and other aspects of our identities [as] markers of relational positions rather than essential qualities' (p. 188). Positionality takes a more relational approach in examining individuals' roles in a certain social setting. It starts from and requires constant reflexivity, with researchers considering their subjectivities as negotiable in different situations in relation to others, be it the field site or other academic settings. While positionality in tourism research has received more attention in recent years, most studies have emphasized the researcher's positionality in relation to the 'researched' (Ateljevic et al., 2005; Bakas, 2017; Hall, 2010). Only a few have discussed the researcher's positionality in knowledge-making in a general sense (Khoo-Lattimore, 2018; Xiao, 2016; Zhang, 2017; Zhang, 2018). This is surprising, considering that positionality is an important concept in all three trends in the current epistemological and ontological inquires, namely the gender equality (Khoo-Lattimore, 2018), critiques on Western centrism within tourism studies from both decolonial and postcolonial perspectives (Chambers and Buzinde, 2015; Tucker & Zhang, 2016), and inter/trans/post-disciplinary debates (Hollinshead, 2010).

Drawing upon our theoretical understandings of the conference as the space manifesting the discursive practices of presumptions and interrelations within knowledge production, we may also consider conference attendance not only as shaped by the conference space but also as a form of contributing back to the broader discourse of knowledge-making. This is done through individual attendees' active self-positioning within the conference environment. Conferences are self-evident elements of being academics, as they are linked with many essential elements within our academic careers: networking, presentations, meetings and publications. On some occasions, special issue and publication opportunities are announced as part of the conference to popularize it among academics. Conferences organized by top journals are often recognized as high quality due to their exclusivity (Bell & King, 2010). Increasingly, popular workshops like publishing in top journals are organized by conferences to set out what is recognized as acceptable knowledge and approaches. Those discursive activities and objects of conferences produce highly complex social spaces oriented to the production of knowledge and the production of successful scholars (Edelheim et al., 2018).

Despite the style and objective of different conferences, they have their own social norms. By attending a conference, attendants subject themselves to the judgement by others. Unlike writing a paper, academics' identities at conferences are easily spotted as conference attending requires face-to-face interactions. Attendees therefore look for physical clues such as name, affiliation, age, race and gender etc. to find the right people they want to be around or away from based on their different motivations (Edelheim et al., 2018; Halford & Leonard, 2006). Those embodiments become symbols, which helps people to find the right people, who might contribute to their development as academics. This is because conferences enable academics and attendees to be counted and legitimated as scholars (Henderson, 2015). The outputs of conferences are thus layered to signify one's seniority in the field from conference papers to invited keynote speeches. Recognition through a keynote speech or fellowship at an exclusive conference like the International Academy for the Study of Tourism implies the success of one's academic career. That recognition often assumes that individuals have made a significant contribution to knowledge-making.

To sustain one's ideal academic self, individuals often perform a conference identity to fit the conference style and expected behaviour (Bell & King, 2010). While mingling and interacting with other delegates who each hold their own ontologies and epistemologies, it is inevitable that one will find a place in the net of relations and opinions. Who am I, and where do my thoughts and I belong? Or simply, who should I talk to? While 'positioning' is 'the key practice grounding knowledge organized around the imagery of vision … positioning implies responsibility for our enabling practices' Haraway (1991, p193), a question is worth raising: can we position ourselves? Our conference bodies perform as academics, but our values and judgments are not decided by individuals but broadly the academic field and general accepted roles. Social identities such as gender and race have stereotypes beyond academic field, which still significantly influence those from minorities groups (Edelheim et al., 2018). These questions thus are not simply related to one's self positionality as academics in a particular field, but what is to be a *recognized* academics in the field, here tourism studies. In the context of calling for more inclusive tourism studies and a re-centering of critical tourism studies towards non-Western epistemologies, we should therefore ask how non-Western scholars are positioning themselves in the intellectual community of tourism studies.

Methodologies and research design

Methodology should be more than just the procedures of selecting the most appropriate methods for collecting 'data' in the most efficient way or picking the 'camp' of which paradigm we subscribe to. It is rather about the whole process we as researchers and authors go through, reading, thinking, questioning, experimenting, challenging, interpreting, and negotiating with both internal and external factors. Hence, in this section we explain how our ontological and epistemological positions are influenced by both Western and Chinese philosophical schools.

Our ontological and epistemological positions are reliant on social constructionism theories and the 'School of Mind' in Chinese Neo-Confucian philosophy. While social constructionism shares similar views with constructivism, that reality is constructed by society and individuals, the former emphasizes the significance of social interchange between individuals in the making of reality and knowledge (Parker, 1998). Particularly, the linguistic practices (e.g. language, pictures, and other texts) embedded in our everyday discourse actively shape and constitute our realities (Parker, 1991). For instance, what we 'are' (as women, Chinese, non-Westerners, etc.) is in fact the result of how these existences are spoken of and discussed in different social contexts (Yu & Kwan, 2008). Along a similar line, the Yangming School of Mind (阳明心学) has challenged the rationalistic views of reality in the School of Principle (程朱理学), arguing that it is the mind that gives meaning to the world rather than the other way around (Chang, 1962). Furthermore, the social constructionism and the Yangming School of Mind share the view on knowledge that there is no pure knowledge 'out there' for us to simply observe and describe, but that knowledge is produced and reproduced through our actions and interactions. These thoughts also resonate with our theoretical framework, again highlighting the fact that methodological pondering occurs throughout the entire process of the design and carrying out of research.

Starting from these philosophical points, we view our examination of Chinese tourism scholars' conference experiences as exploring the fragments of reality that emerge from social interchange. Importantly, research design, methods, and analysis result from discussions and negotiation between us two authors. To some extent, our experiences of co-authorship also reflect a social constructionist approach that each of us contributes to the construction of this study, not only from our own perspectives but also through our interactions with one another. We also extend this process of co-creation of knowledge to our participants: that they as knowledge co-producers and collaborators are essential in how we think and write (Ren et al., 2017). Moreover, we adopt an interpretivist approach to read and analyze their narratives, aiming both to highlight their voices and perspectives as well as to situate them in the broader discussion of including non-Western tourism scholars in the tourism scientific community.

A qualitative approach was employed to collect materials through semi-structured interviews. Following Denzin (2001), we see interviews as 'a way of bringing the world into play' and 'a site where meaning is created and performed' (p. 25). It is the performative sensibility that enables the researchers and the participants to explore the different ways of presenting an interview text (ibid). A total of 19 participants were recruited through a combination of purposive sampling and maximum variation sampling techniques. As shown in Table 1, all participants were ethically Chinese and self-

Table 1. Profile of research participants.

Participant No.	Age	Gender	Current Location	Position	Conference experience
1	30s	Female	Sweden	Research fellow	>10
2	30s	Female	UK	Lecturer	Around 5
3	30s	Female	UK	Lecturer	Around 5
4	20s	Female	China	Lecturer	3
5	20s	Male	New Zealand	Lecturer	4
6	30s	Male	China	Assistant Professor	6-10
7	30s	Male	China	Assistant Professor	5-6
8	30s	Female	UK	Lecturer	3
9	30s	Female	China	Associate Professor	>10
10	30s	Female	China	Associate Professor	1
11	30s	Female	UK	Senior Lecturer	Around 10
12	40s	Male	Macao	Assistant Professor	>10
13	30s	Male	China	Assistant Professor	>10
14	40s	Male	China	Assistant Professor	>10
15	50s	Male	Hongkong	Assistant Professor	Many
16	30s	Female	Macao	Assistant Professor	7-10
17	40s	Female	Australia	Professor	>10
18	50s	Female	China	Professor	Many
19	50s	Female	China	Professor	Many

identified as Chinese tourism academics despite their geographical locations and affiliations. All participants had attended international conferences and their conference experiences are included in the table. It should be noted that those who had limited experience at international conferences indeed had well attended academic conferences within China. Here, international tourism conferences are discussed due to the fact that international conferences in English have been prioritized as a recognized and valuable conference. And by international conferences we mean those are outside of China with English as the official communication language. Table 1 also demonstrates our sample diversity in terms of their age, gender, geographical location, and career stage.

The interview questions consist of four parts: 1) descriptions of previous international tourism conference experiences; 2) understandings of⁄the conference as a social space; 3) reflections on positionality at the conference; and 4) reflections on knowledge production in relation to conference attendance. We prepared the interview questions in both Mandarin and English, and then cross-checked them to ensure accuracy. One of us conducted eight interviews and the other eleven, both independently. The majority of the interviews were conducted in Mandarin, and the others in English. The average interview duration was 45 minutes. The interviews were either audio-recorded when the interviewee consented, or documented through extensive notes taken during the interviews and double-checked with the interviewee afterwards. The interview material was then transcribed into texts.

According to Wolcott (1994), there are several different ways to transform data and thus different 'levels of data analysis', including description, analysis, and interpretation. We also agree with Riessman (2012) that participants' interview performances are sometimes done in a way that they hinder researchers' efforts to fragment the material into codable categories. We thus take narrative analysis to mean a set of methods and a combination of describing, analyzing, and interpreting the narratives (in both oral and graphic forms) we collected from participants. Mura and Sharif (2017) identify different types of narrative analysis in the social sciences, and categorize them under structural and post-structural analysis approaches. While trying to navigate the stories and representations, we based the analysis on various strategies from both structural and post-structural analysis. Hence, the following steps are taken:

a.   Examining the 'story' in the narration; the flows and plots
b.   Evaluating the parts of the story that reveal the narrator's attitude towards research questions and 'emphasizing the relative importance of some narrative units as compared to others' (Labov & Waletzky, 1997, p. 37)
c.   Attending to disruptions and contradictions; places where a text fails to make sense or does not continue (Czarniawska, 2004, p. 92)
d.   Identifying a dichotomy in the narratives and exposing it as false distinction (e.g. local/global, Chinese/Western, insider/outsider, etc.) (Czarniawska, 2004, p. 92)
e.   Making interpretations and resolutions for the results of Step b, while emphasizing the relevance of the results of Step d
f.   Reflecting on and examining the narrator's interaction with the interviewer, and the role of the interviewer, while re-examining the analysis a second time (Polkinghorne, 1995)

Taking these steps, the narratives are shattered, scrambled, and then re-organized and re-constructed, as a process of creating montage and collage. As Denzin (2001) describes: '…narrative collage allows the writer, interviewer and performer to create a special world, a world made meaningful through the methods of collage and montage' (p. 30). Furthermore, it should be noted that we also interviewed each other and include our own narratives on our experiences of international conferences in the material. Finally, the role of reflexivity is central in the whole process of interviewing, transcribing and analyzing, enabling us to be aware of our own participation in social dynamics and, in general, a worldmaking process (Tucker, 2009; J. Zhang, 2017).

Findings and discussion

Below are the four areas where the narratives cluster after the process of breaking down and reforming the interview materials. Here, it is extremely important to bring the interesting discussions that emerged from the narratives into the context of non-Western knowledge production in tourism studies, rather than simply describing and pinning down 'what has been said'. Through situating the discussions in a broader context and consequently starting a conversation with the existing discussion, we create a channel for the voices of Chinese tourism scholars and their knowledge.

What's the meaning of going to an international conference?

When talking about the meanings of conferences, what first comes to mind is often the purposes, reasons and motivations of attending them. While our participants' reasons for attending international conferences are very similar to what is written in the literature – namely networking, getting feedback, developing career paths, keeping up with research trends, and encountering research interests – there are also many valuable narratives that reflect issues under the surface of 'functionalities of conferences'. The 'meaning of going to an international conference' therefore becomes much more contextualized in the ethics of academic life, often tightly related to how knowledge is produced. For instance, one participant says:

> Intellectual inputs at the conferences are really important ... A lot of times our conferences don't necessarily provide those experiences ... But collectively we damage the conferences; we try to support only the people we know and we all only go to the sessions we think are relevant. It's like some papers are highly cited, while others are not. (Participant 17)

Though has not been able to experience intellectual stimulation or satisfaction at most of the conferences, this participant does not simply complain about their organization. Instead, she turns inward and points out that the lack of intellectual exchange is a result of less engaged attendees. By comparing how people attend conference presentations to how people cite articles, this participant highlights the pragmatic climate that still dominate not only tourism studies but the academic world in general; that producing knowledge has little to do with intellectual exchanges or contributions, but is rather driven by a rationalization and normalization of knowledge as a product. And we 'collectively ... damage the conferences' because we submit to the normalization of the knowledge-making system. However, another participant raises a different point regarding knowledge as product and the conference as a platform:

> I think knowledge is like a product, and people's goal is to explain or present their value through this product. For a long time we've thought everything from the West is good, so everyone uses Western theories. Now we put more emphasis on personal ontology. So, I can find my own theories and ways of producing knowledge. The conference then becomes a place where I can present myself also broaden my views. (Participant 2)

Different from Participant 17, this participant finds possibilities enabled by the conference space in a new era in which knowledge-making is increasingly personalized and individualized. Knowledge production in a neoliberal era may overly emphasize short-term values (e.g. rebound effects), but it also provides more equal and democratic ground for different ideas, and benefits individuals who are willing to learn. But intellectual exchange is not a matter of one-way learning or presenting; rather, it requires dialogue and communication. To have these, you have to be relevant, as stated by another participant:

> I believe any presentation should be relevant to other people, if you want to have a dialogue. We all need to contribute to form an academic community. So, what we say should try to generate relations to others as much as possible ... If you only talk about China, people will feel irrelevant. But if you link the Chinese issues with others' issues, then you will attract the audience (Participant 18)

For Participant 18, it is everyone's responsibility to make their own knowledge relevant to the other scholars, the academic community, and perhaps society. Although one can argue that this emphasis on relevance again reflects the rationalization of knowledge (more valuable when more applicable), these narratives are insightful in that they connect a number of relations together: those between presenter and audience, individual academics and academic community, and China and the rest of the world. The overlapping and intertwining connections between these relations are critical in transferring knowledge from one scale to another both within and beyond the conference space, and as Participant 18 indicates, can be either made or unmade depending on how much one wants to have a dialogue, or how much one doesn't want to be alone and isolated. Indeed, for one participant, 'generating relations' is a necessary step before attending an academic conference:

> I always want to gather as much information as possible on as many people as I can before I go to a conference, so that I will have more to discuss with them. It's a way for me to make the most out of a conference. If I'm unprepared, and the chance comes and I miss it ... I'll feel really embarrassed and unsettled. (Participant 16)

Still, a question extending from Participant 18's narrative is: Knowledge relevant to whom? Isn't this always a relative and subjective question? One participant makes yet another point about academic conferences, that nothing is relevant enough:

> I think, in comparison to the Internet, academic conferences offer little space for real exchange. I'm more industry-focused so I pay more attention to industrial news and have many exchanges with netizens on relevant forums. Some of them have become good friends and collaborators. I think that's the real value. (Participant 13)

A constant criticism regarding academic conferences is that they are almost always limited within their own bubble, without much contact with or relevance to the 'real world'. Participant 13 unfolds another layer of 'relevance' that perhaps adds to the discussion of what counts as knowledge, and what researchers' roles in our society are.

Yet, one can also wonder when negative emotions do arise, such as feelings of being left out and neglected, is it enough to say that an irrelevance of one's research or a lack of preparation is the reason? Pondering over whether she has established good contact with Western tourism scholars at international conferences, one participant recalls:

> Even when someone is interested in communicating, it's not really like the interaction they would have with one of their own (Western). There might also be some other purposes behind, that they may actually want to have contact with the Western senior researcher who supervises me, but not me. Also, I feel like most of the Western scholars don't really know about China. They tend to make connections between your topic and other political issues. So, the questions you get aren't really based on what you present but are people's own presumptions. (Participant 11)

Here Participant 11 experiences prejudice, ignorance, and perhaps even discrimination from other Western scholars, while she is obviously willing to have some real communication at the conference. What can be sensed here is her disrupted and contradictory feelings of wanting to be understood and seen but not believing it can/will happen, largely because of the imagined/invisible line between 'Chinese' and 'Western'.

How to cross 'the line'?

When attempting to position oneself in a conference's social space, a few participants brought up the issue of the existing division between Chinese and Western scholars, especially those with an Anglo-Saxon background. However, the dichotomy of 'Chinese' and 'Western' is also challenged and disrupted by other participants' narratives on other 'lines' such as hierarchical and gender divisions. While these binary positions are certainly interrelated, we suggest that it is the continuous discussion on 'how to cross the line' that makes the tensions between each of the binary positions more meaningful than the oppositions themselves.

First let us stay with Participant 11 and listen to her story about the line between 'we' and 'they':

> There is this phenomenon at conferences that is somewhat similar to zhadui (sticking together), that Chinese and non-Chinese people are in their own groups. I'm curious to see 'the other side' so I do want to cross the line. But it's quite difficult... Sometimes I feel I have come to their side, but for them I'm still on my side... Meanwhile, if I get too close to the line, I become distant from my own group and they think I've become too Western. (Participant 11)

> Zhadui (扎堆), literally meaning 'sticking into one pile', is a Chinese expression describing the phenomenon of many people forming a crowd not really for any particular meaning or purpose, but just for the feeling of security or a lively, bustling sense (renao, 热闹). Several other participants make the same observation and commented that "You will see Italians stayed all together as well, it is not a Chinese thing." (Participant 12)

Although it is clarified that sticking together in a group does not make one more 'Chinese' or 'Western', what Participant 11 describes as an invisible line is perceived to be one of the effects of the grouping phenomenon. Moreover, she portrays a rather graphic dynamic between her curiosity and attempts to 'cross the line' on the one hand and the internal press and potential risk of alienating herself from her 'own group' on the other. This dynamic directly affects how one socializes and establishes contacts with scholars 'on the other side'. However, some participants believe this is due to a lack of confidence:

> I think... whether one feels uncomfortable doesn't really have much to do with being Chinese, but is more likely due to whether you're familiar with the people and knowledge in your field. The most important thing is about sharing the same language (epistemologically rather than linguistically). (Participant 18)

Other researchers think the 'line' is in fact set by oneself:

> If we believe we're the same as the other scholars then we don't need to feel troubled. I'm not really concerned about my Chinese identity... I use Chinese concepts in my research not because of my Chinese identity but because of my ontology. (Participant 2)

The above narratives on the 'line' or 'boundaries' between groups together demonstrate the complexity and diversity in how different individuals' perceptions of their own positions within a social setting can be shaped by their different self-positioning strategies, namely identities, subjectivities, and ontologies. While the identity perspective states a fixed position characterized by 'nationalities' or 'ethnicities', a perspective on subjectivity reflects more our abilities to form our own 'being' as well as the constraints that prevent us from 'becoming' (knowledge expertise, social skills, and open-mindedness), and the ontological approach to understanding our 'self and being' may create either a protection from or a pathway to the 'other'. While the intrinsic links between these perspectives are not to be ignored, differences are illustrated through the narratives: how each perspective deals with conference space is different, and may lead individuals to different situations. An example is shown below:

I do tend to stick with Chinese people at international conferences … a big reason is that the possibility to collaborate with them is much higher. You gain more from that. I'm very pragmatic. I have limited time, so I'm not there just to socialize with random people … Also, I think it's not really fair if it's only me adjusting myself to others, approaching them, and 'becoming' like them. It's not necessary. (Participant 14)

Interestingly, while Participants 14's 'pragmatic attitude' suggests a rather logical and natural result of 'staying on this side of the line', the last sentence also shows his active and conscious decision to not cross the line, because 'it's not really fair' that effort is made from only one side. Another participant digs a bit deeper into the lack of effort from 'the other side', describing:

I feel it's easier to talk to non-English-speaking people at international conferences … because we make an effort to make ourselves understood. With people whose mother tongue is English, sometimes I might not understand them perfectly, such as the slang and references they use. But the thing is … they just assume that you get all the jokes and cultural references. I don't think I'm that Western yet. (Participant 16)

Again, how attendees 'end up' in a particular group is dependent not only on their own willingness and attitudes, but also on those of their counterparts. Besides the repeatedly mentioned 'line' between the Chinese and Western scholars, another line between 'small potatoes' (early-career and young scholars) and 'big names' (daniu, 大牛, senior and established scholars) is also mentioned as a barrier by several early-career scholars. For instance, one participant proposes that *'most of the negative feelings we get at international conferences are because we're small potatoes, not because we're Chinese'* (Participant 2), while another also talks about being *'transparent'* to others because he is just a *'small potato'* (Participant 6). That early-career scholars often face difficulties in gaining access to the 'academic tribes' or in simply being seen has been well discussed in the literature within critical tourism studies (Ateljevic et al., 2005; Khoo-Lattimore, 2018; Tribe, 2010; Winter, 2009). And, indeed, according to Participant 2 above this is universal. However, as several participants mention, if you are at an early career stage and are from a non-Western country, you are often confronted with the dilemma that, if you are working with Western scholars you will be in their shadow, and if you are not working with them you will be seen as not measuring up to the international standard and people will not remember you. Participant 15 summarizes his observations of the dynamic between non-Western and Western scholars at international conference over the years, suggesting that there has been a change: in the early years, non-Western and Chinese tourism scholars had low self-confidence due to language issues and the dominant Western paradigm, and this could be observed in the common scene of Chinese and young scholars circling around Western and successful scholars. However, it is a different situation today:

Today, many English speakers are more willing to listen to non-English speakers' presentations, which was rare back in the 1990s. They didn't bother to listen to us then. It's not great now, but much better. Interestingly, nowadays it's often the non-English-speaking scholars who don't want to listen to our own people (ziji ren 自己人). (Participant 15)

Another participant describes being mistaken for a student by other Western attendees only because she took a photograph with a famous tourism scholar, though she has worked in the field for some time. She reflects on this, saying:

It feels like many people cannot accept that Chinese scholars are getting better. Especially when you're a woman ... Chinese scholars going out into the world is still a process. (Participant 9)

These narratives on the hierarchic line, and even the gender line, certainly reflect a process of mapping and positioning one's place at a conference, or even a process of positioning China in the world. They also take us back to the discussions of including non-Western knowledge in tourism studies, and evoke many questions: What does it mean to 'include'? Who is including whom? Further, how can we judge these Chinese scholars who do not listen to 'their own people'? Perhaps they are trying to become more 'relevant' and thus 'included'? Moreover, an underlying question lurking beneath these narratives is: Who is going to make non-Western knowledge, and how?

Who makes non-western knowledge, and how?

Scholars advocating for a 'decolonization of methodologies and knowledge' encourage non-Western scholars to relearn the 'mandates of knowing' (Hollinshead & Suleman, 2018) and to provide alternative discourses for the social sciences (Alatas, 2006). As a number of participants reflect on, it is more complicated than simply moving from a to b:

The question of changing the ideology of knowledge-making is not only one way, but two ways. On the one hand we're quite loud in Asian society that we need our own characteristics and localizing knowledge-making system, while on the other we still rely on university ranking, journal ranking, etc., trying to join the internationalization. I'm not saying that the international and the local cannot coexist; it's just complicated to balance and negotiate. For instance, studying the same question but with culturally characterized methodology is important but difficult. (Participant 15)

Another participant also mentioned the paradox between local and global systems, expressing that although he believes publishing in Chinese journals will ensure more knowledge transfer and have more impact, the university promotion system only recognizes English publications (Participant 14). He also emphasizes the 'two-way knowledge exchange', saying:

It's not only about us joining them ... joining or integrating is just a means, not the goal. If the goal is to have people understand us, then I don't really need to ingratiate myself with them. It's not about whether I like it or not; it's about self-confidence. (Participant 14)

Such voices indeed challenge us to think if the 'knowledge' in general discussion is referred to 'Western knowledge' by default. The pressure of publishing in English journals is also tied to global universities' ranking, for which only internationally recognized journal papers count. Hence knowledge exchange has become limited and what happens at the micro-level is only a small reflection of the global trend of pursuing English journals. SSCI, for example, is ranked based purely on impact factor rather than any significant social impact. Conferences are therefore positioned here as a means of demonstrating one's identity as a scholar in a certain area that is defined by h index and journal impact scores.

Such pressure affects how academics develop scholarly collaboration. When asked whether conferences provide space for their future career development, several participants start evaluating the pros and cons of finding collaboration with either Chinese or Western scholars at conferences. One participant describes his dilemma:

> It's easier to form collaborations with Chinese scholars nowadays. But the problem is also this: that if you only worked with other Chinese scholars you'd only have one perspective, so your paper would be easily rejected or be subject to major revisions. Because they (reviewers) don't understand our perspectives. I need to work with more Western scholars. But it's hard to collaborate with them. (Participant 8)

Another participant also expresses the frustration of working with Western scholars: 'We have very different perceptions of the world and society' (Participant 3). It seems that, for these participants, 'the integration process' becomes a goal and is equal to 'collaborating with Western scholars', while another participant points out that 'Younger scholars feel they're just coming to learn, but not in a more advanced position' (Participant 9). Again, as presented earlier, Participant 14's argument that the integration of knowledge systems is not the final goal brings us back to the current debate within tourism studies on alternative epistemologies and non-Western knowledge. What is the whole point of talking about Western/non-Western knowledge? Is it simply about increasing the co-authorship between Western and non-Western scholars? Or is it about the subordinates rejecting and resisting the masters? As one participant reflects:

> Of course, we shouldn't produce knowledge that reinforces the norms, but I don't think we should be extreme either. For example, saying that the Western paradigm doesn't work in Chinese contexts is wrong; it does work in some cases, really well. What we should focus on is getting our message across in everyday conversations with our Western colleagues. It's hard for people to accept challenges, not only Westerners but Chinese as well. Even the constructive criticisms are difficult to take. I think we should be moderate in doing this. (Participant 17)

An important point in Participant 17's comment is the everyday discourse of 'how to get our message across' is also part of the knowledge-making process. Rather than the radical attitude expressed in anti-colonial and decolonizing literature, again a pragmatic, rational and analytical attitude is seen here. As another scholar reasons:

> There are indeed power relations affecting the phenomenon that some knowledge systems are not acknowledged. But many existing knowledge systems have been formed over long periods of time; it's not possible that they would collapse at once. The core is about communicating with the same language, and then seeing the difference, back and forth, to view your own research subjectively and objectively. (Participant 18)

One can make the point, from the decolonization literature, that 'using the same language' is the exact danger of submitting to the other's mindset and ways of knowing (Çakmak & Isaac, 2017). However, as Diversi and Moreira (2009) state in their work, a 'betweenness' in knowledge production is inevitable when one spontaneously exists in multiple spaces and crosses borders of different territories. And it is challenging – though possible – to work 'between and among different spaces, and to '[unify] differences (bodies) to make a more inclusionary movement' (p. 174). However, according to one participant, the reason some Chinese scholars (especially young ones) are asked why they do not use Chinese perspectives in their research is that 'they hold an assumption that we know every Chinese classic, and this is totally because of the binary of Western/non-Western they have in their mind' (Participant 16).

Go (2016), in his search for a solution for going beyond Eurocentric influence in the social sciences, argues that it is the 'law of division' and consequently 'analytic bifurcation' that perpetuates the persistent Orientalism of social theory. From a postcolonial perspective, what to do is ridiculously simple in a way: "if the imperial episteme's law of division cuts the world up into separate entities, a postcolonial approach would start by reconnecting the separated parts" (Go, 2016, p.111). As one participant finds connection with her non-Western colleagues, she suggests that we stop framing knowledge-making as an *'Asian/Western issue'* and instead see it as *'balancing these different experiences and having a consciousness towards people who are different'* (Participant 17).

The never-ending dynamics between the 'self' and the 'Other' entail, as Said (1993) has pointed out, how cultural identities are formed through contrapuntal ensembles; that 'no identity can ever exist by itself and without an array of opposites, negatives, oppositions' (p.52). The reflections and discussions that emerge from the narratives indeed demonstrate that international conferences are a space where 'putatively separate and opposed cultures or identities are actually contaminated by each other' (Go, 2016, p.113), regardless of whether one finds it easy or hard to work with others from different backgrounds. The discussion on 'who will make non-Western knowledge and how' indicates that it is the 'connections' and equal 'dialogues' we need in order to bring newer perspectives into the existing knowledge systems. What we need to further ask is perhaps what counts as 'connection' and how to determine this. As Go (2016) recognizes, we cannot avoid such questions by simply claiming that everything is connected. We leave this inquiry for future studies.

Conclusion

Within tourism studies, the 'critical turn' has evoked growing reflective and critical perspectives on the role of the researcher in conducting research and producing knowledge (Hollinshead et al., 2009). This has also led to calls for building an inclusive research community, particularly through including non-Western and non-positivist methodologies (Hollinshead & Ivanova, 2013; Mura & Khoo-Lattimore, 2018a; Pritchard & Morgan, 2007; Winter, 2009). While it is noted that non-Western scholarship has gained more visibility in the international tourism research community through publications in prestigious academic journals and attendance at international conferences, few studies discuss non-Western scholars' responses to the calls for non-Western knowledge. In this paper, we have attempted to contribute to this discussion in the context of international conferences and investigated how Chinese tourism researchers' conference experiences shape their positionality in knowledge-making process.

As Zhang (2018) points out, Chinese tourism scholars' positioning shows that Chinese tourism scholarship is 'something in between' the global and the local, while binary positions (such as local/global, China/West) are still dominant in tourism studies and limit our understandings of our own identities and subjectivities. To this point, Tucker and Zhang (2016) contend that knowledge-making is always a process of inter-reacting; we are therefore always simultaneously distributing and receiving knowledge. Thus, the process of positioning is not simply to find a 'spot' for oneself, but to reconnect the ties one was previously unaware of or had lost. In a way, to position oneself in the conference space is to find one's way through the 'overlapped territories' and 'intertwined histories' (Said, 1993), and to play an active role in the complex interchange of representations and meanings. In a similar vein, the narratives collected from semi-structured interviews in this study show a process of mingling and mapping, where conference attendees attempt to find meanings of being at an academic conference and to understand the relations embedded in the conference space. In this process of mingling and mapping, reflexivity on conference dynamics is seen in terms of how one goes around certain constructed groups, be it the group of 'Western scholars' or 'male and senior scholars'. And clearly such a process of mingling and mapping affects the participants views on who and how will make the non-Western knowledge. While the qualitative nature of this study means that the narratives collected do not present the whole group of 'Chinese tourism scholars', they do shed light on voices that go beyond the normal rationale of motivations for attending, addressing the significance of positionality for knowledge production in all academic settings. For instance, uncertainty, ambiguity and ambivalence are common in the narratives, redirecting the focus from the end results of knowledge production (network, publication and university ranking) to the making process, where emotional and ethical encounters are often neglected but are essential in influencing attendees' decision-making. The narratives may seem descriptive at times, due to this point and more questions are raised than answered or concluded.

Several points emerge from the materials that deserve future investigation. Firstly, the concept of the 'relevance of knowledge' should be explored more in broader contexts. We believe that it is important to invite further discussion on how to make knowledge relate to not only different groups of scholars, but also the wider society, building communicative bridges. We also suggest that future studies should continue the line of this inquiry to dig deeply into the complex mechanism through which conference experiences impact on knowledge-making, and the often taken-for-granted but vague conceptions of 'connections' and 'relations' between different individuals, groups and ideas. A recent study on slow conferencing through camping together is one example of exploring unusual forms (silence and fiction) of communication in moments of meeting (Veijola et al., 2019). Furthermore, with increasing awareness and efforts coming from the 'non-Western' scholars' side to understand the dynamics of knowledge production in today's scientific community, there is little study done on a more general population. This makes us wonder whether the 'line' between the Western and non-Western scholars mostly exists or is imagined as a hurdle for the non-Western scholars. Finally, research ethics or the question of what roles intellectuals have in today's society should be addressed more in the calling for non-Western scholarship in tourism studies. While this study focuses on individual scholars' internal processes of conference attending and knowledge making, future studies could be designed to understand such personal experiences in relation to broader arenas. For instance, a time perspective may shed light on how Chinese scholars' experiences at international conferences have changed in relation to how China as a nation is positioned in the world. The geopolitical perspective, as Rowen (2019) suggests, is also needed to reflect on tourism slogans regarding promoting peace and sustainability. It should be noted that innovative and creative methodologies may be required to conduct research on this area of inquiry due to the sensitivity and responsibility involved.

Disclosure statement

No potential conflict of interest was reported by the author(s).

References

Alatas, F. (2006). *Alternative discourses in Asian social science: Responses to Eurocentrism*. Sage.

Ateljevic, I., Harris, C., Wilson, E., & Collins, F. L. (2005). Getting 'entangled': Reflexivity and the 'critical turn'in tourism studies. *Tourism Recreation Research*, *30*(2), 9–21. doi:10.1080/02508281.2005.11081469

Bakas, F. E. (2017). A beautiful mess': Reciprocity and positionality in gender and tourism research. *Journal of Hospitality and Tourism Management*, *33*, 126–133. doi:10.1016/j.jhtm.2017.09.009

Bao, J., Chen, G., & Ma, L. (2014). Tourism research in China: Insights from insiders. *Annals of Tourism Research*, *45*, 167–181. doi:10.1016/j.annals.2013.11.006

Bell, E., & King, D. (2010). The elephant in the room: Critical management studies conferences as a site of body pedagogics. *Management Learning*, *41*(4), 429–442. doi:10.1177/1350507609348851

Çakmak, E., & Isaac, R. K. (2017). A future perspective about tourism and power: A polyphonic dialogue in the agora. *Tourism Culture & Communication*, *17*(1), 75–77. doi:10.3727/109830417X14837314056979

Chambers, D., & Buzinde, C. (2015). Tourism and decolonisation: Locating research and self. *Annals of Tourism Research*, *51*, 1–16. doi:10.1016/j.annals.2014.12.002

Chang, C. (1962). *Wang Yang-ming: Idealist philosopher of sixteenth-century China*. University Press.

Czarniawska, B. (2004). *Narratives in social science research*. Sage.

Delanty, G. (2001). *Challenging knowledge: The university in the knowledge society*. Open University Press.

Denzin, N. K. (2001). The reflexive interview and a performative social science. *Qualitative Research*, *1*(1), 23–46. doi:10.1177/146879410100100102

Diversi, M., & Moreira, C. u. (2009). *Betweener talk: decolonizing knowledge production, pedagogy, and praxis*. Left Coast Press.

Edelheim, J. R., Thomas, K., Åberg, K. G., & Phi, G. (2018). What do conferences do? What is academics' intangible return on investment (ROI) from attending an academic tourism conference? *Journal of Teaching in Travel & Tourism*, *18*(1), 94–107. doi:10.1080/15313220.2017.1407517

Ford, J., & Harding, N. (2008). Fear and loathing in Harrogate, or the study of a conference. *Organization*, *15*(2), 233–250. doi:10.1177/1350508407086582

Ford, J., & Harding, N. (2010). Get back into that kitchen, woman: management coferences and the making of the female professional worker. *Gender, Work & Organization*, *17*(5), 503–520. doi:10.1111/j.1468-0432.2009.00476.x

Foucault, M. (1972). *The archaeology of knoweldge and the discourse on language*. Pantheon.

Go, J. (2016). *Postcolonial thought and social theory*. Oxford University Press.

Halford, S., & Leonard, P. (2006). Place, space and time: Contextualizing workplace subjectivities. *Organization Studies*, *27*(5), 657–676. doi:10.1177/0170840605059453

Hall, C. (Ed.) (2010). *Fieldwork in tourism: Methods, issues and reflections*. Routledge.

Haraway, D. J. (1991). *Simians, cyborgs, and women: The reinvention of nature*. Free Association Books.

Henderson, E. F. (2015). Academic conferences: Representative and resistant sites for higher education research. *Higher Education Research & Development*, *34*(5), 914–925. doi:10.1080/07294360.2015.1011093

Hollinshead, K. (2010). Tourism studies and confined understanding: The call for a "new sense" post-disciplinary imaginary. *Tourism Analysis*, *15*(4), 499–512. doi:10.3727/108354210X12864727693669

Hollinshead, K., Ateljevic, I., & Ali, N. (2009). Worldmaking agency-worldmaking authority: The soverign constitutive role of tourism. *Tourism Geographies*, *11*(4), 427–443. doi:10.1080/14616680903262562

Hollinshead, K., & Ivanova, M. (2013). The multilogical imagination: Tourism studies and the imperative for postdisciplinary knowing. In M. Smith & G. Richards (Eds.), *The routledge handbook of cultural tourism* (pp. 53–62). Routledge.

Hollinshead, K., & Suleman, R. (2018). Tourism studies and the lost mandates of knowing: Matters of epistemology for the inscriptive/projective industry. In P. B. Mura & C. Khoo-Lattimore (Eds.), *Asian qualitative research in tourism: ontologies, epistemologies, methodologies, and methods* (pp. 51–80). Springer.

Huang, S. S., & Chen, G. (2016). Current state of tourism research in China. *Tourism Management Perspectives*, *20*, 10–18. doi:10.1016/j.tmp.2016.06.002

Khoo-Lattimore, C. (2018). The ethics of excellence in tourism research: A reflexive analysis and implications for early career researchers. *Tourism Analysis*, *23*(2), 239–248. doi:10.3727/108354218X15210313504580

King, L., MacKenzie, L., Tadaki, M., Cannon, S., McFarlane, K., Reid, D., & Koppes, M. (2018). Diversity in geoscience: Participation, behaviour, and the division of scientific labour at a Canadian geoscience conference. *FACETS*, *3*(1), 415–440. doi:10.1139/facets-2017-0111

Labov, W., & Waletzky, J. (1997). Narrative analysis: Oral versions of personal experience. *Journal of Narrative and Life History*, *7*(1-4), 3–38. doi:10.1075/jnlh.7.02nar

Latour, B. (1988). *The pasteurization of France*. Harvard University Press.

Latour, B. (1993). *We have never been modern*. Harvester Wheatsheaf.

Liao, L. F. (2008). Impact of manager's social power on R&D employees' knowledge-sharing behaviour. *International Journal of Technology Management*, *41*(1/2), 169. doi:10.1504/IJTM.2008.015990

Lloyd, P. (2015). The making of a conference. *Constructivist Foundations*, *11*(1), 30–31.

Maher, F., & Tetreault, M. (1993). Frames of positionality: Constructing meaningful dialogues about gender and race. *Anthropological Quarterly*, *66*(3), 118–126. doi:10.2307/3317515

Mair, J. (2015). *Conference and conventions: A research perspective*. Routledge.

Mair, J., & Frew, E. (2016). Academic conferences: A female duo-ethnography. *Current Issues in Tourism*, *19*(1), 1–21. doi:10.1080/13683500.2016.1248909

Mair, J., & Thompson, K. (2009). The UK association conference attendance decision-making process. *Tourism Management*, *30*(3), 400–409. doi:10.1016/j.tourman.2008.08.002

Morgan, N., Pritchard, A., Causevic, S., & Minnaert, L. (2018). Ten years of critical tourism studies: Reflections on the road less traveled. *Tourism Analysis*, *23*(2), 183–187. doi:10.3727/108354218X15210313504517

Munar, A.M., Biran,A., Budeanu,A., Caton, K., Chambers, D., Dredge, D., Gyimóthy, S., Jamal, T., Larson, M., Lindström, K. N. and Nygaard, L. (2015). *The gender gap in the tourism academy: Statistics and indicators of gender equality.*

Mura, P., & Sharif, S. P. (2017). Narrative analysis in tourism: A critical review. *Scandinavian Journal of Hospitality and Tourism, 17*(2), 194–207. doi:10.1080/15022250.2016.1227276

Mura, P. B., & Khoo-Lattimore, C. (2018a). *Asian qualitative research in tourism: Ontologies, epistemologies, methodologies, and methods.* Springer.

Mura, P. B., & Khoo-Lattimore, C. (2018b). Locating Asian Research and Selves in Qualitative Tourism Research. In P. B. Mura & C. Khoo-Lattimore (Eds.), *Asian qualitative research in tourism: ontologies, epistemologies, methodologies, and methods* (pp. 1–22). Springer.

Mura, P. B., Mognard, E., & Sharif, S. P. (2017). Tourism research in non-English-speaking academic systems. *Tourism Recreation Research, 42*(4), 436–445. doi:10.1080/02508281.2017.1283472

Oppermann, M., & Chon, K. S. (1995). Factors influencing professional conference participation by association members: A pilot study of convention tourism. *Travel and Tourism Research Association Journal, 26*(1), 254–259. doi:10.1177/0047287507312421

Parker, I. (1991). *Discourse dynamics: Critical analysis for social and individual psychology.* Sage.

Parker, I. (1998). *Social constructionism, discourse, and realism.* Sage.

Polkinghorne, D. E. (1995). Narrative configuration in qualitative analysis. *International Journal of Qualitative Studies in Education, 8*(1), 5–23. doi:10.1080/0951839950080103

Pritchard, A., & Morgan, N. (2007). De-centering tourism's intellectual universe, or traversing the dialogue between change and tradition. In I. Ateljevic, A. Pritchard, & N. Morgan (Eds.), *The critical turn in tourism studies: Innovative research methodologies* (pp. 11–28). Elsevier.

Ren, C., Johannesson, G. T., & Duim, R. v. d. (Eds.). (2017). *Co-creating tourism research: Towards collaborative ways of knowing.* Routledge.

Riessman, C. K. (2012). Analysis of personal narratives. In J. F. Gubrium, J. A. Holstein, A. B. Marvasti, & K. D. McKinney (Eds.), *The SAGE handbook of interview research: The complexity of the craft* (2nd ed., pp. 367–380). Sage.

Rittichainuwat, B. N., Beck, J. A., & Lalopa, J. (2001). Understanding motivations, inhibitors, and facilitators of association members in attending international conferences. *Journal of Convention & Exhibition Management, 3*(3), 45–62. doi:10.1300/J143v03n03_04

Rowen, I. (2019). Tourism studies is a geopolitical instrument:Conferences, Confucius Institutes, and 'the Chinese Dream. *Tourism Geographies*, 1–20. doi:10.1080/14616688.2019.1666912

Said, E. W. (1993). *Culture & imperialism* (1st ed.). Random House.

Sweeting, B., & Holh, M. (2015). Exploring alternatives to the traditional conference format: introduction to the special issue on composing conferences. *Constructivist Foundations, 11*(1), 1–7.

Tribe, J. (2010). Tribes, territories and networks in the tourism academy. *Annals of Tourism Research, 37*(1), 7–33. doi:10.1016/j.annals.2009.05.001

Tucker, H. (2009). Recognizing emotion and its postcolonial potentialities: Discomfort and shame in a tourism encounter in Turkey. *Tourism Geographies, 11*(4), 444–461. doi:10.1080/14616680903262612

Tucker, H., & Zhang, J. (2016). On Western-centrism and" Chineseness" in tourism studies. *Annals of Tourism Research, 61*, 250–252. doi:10.1016/j.annals.2016.09.007

Veijola, S., Höckert, E., Carlin, D., Light, A., & Säynäjäkangas, J. (2019). The Conference Reimagined. Postcards, Letters, and Camping Together in Undressed Place. *Digithum*, (24), 21–35. doi:10.7238/d.v0i24.3168

Walters, T. (2018). Gender equality in academic tourism, hospitality, leisure and events conferences. *Journal of Policy Research in Tourism, Leisure and Events, 10*(1), 17–32. doi:10.1080/19407963.2018.1403165

Wijesinghe, S. N. R., Mura, P., & Culala, H. J. (2019). Eurocentrism, capitalism and tourism knowledge. *Tourism Management, 70*, 178–187. doi:10.1016/j.tourman.2018.07.016

Winter, T. (2009). Asian tourism and the retreat of anglo-western centrism in tourism theory. *Current Issues in Tourism, 12*(1), 21–31. doi:10.1080/13683500802220695

Wolcott, H. (1994). *Transforming qualitative data: description, analysis and interpretation.* Sage.

Xiao, H. (2016). A Humanities Wanderer "Lost" in Tourism Studies: A Critical Reflection. *Journal of China Tourism Research*, *12*(1), 144–154. doi:10.1080/19388160.2016.1162760

Yu, F.-L T., & Kwan, D. S. M. (2008). Social construction of national identity: Taiwanese versus Chinese consciousness. *Social Identities*, *14*(1), 33–52. doi:10.1080/13504630701848515

Zhang, J. (2017). The irreducible ethics in reflexivity: Rethinking reflexivity in conducting ethnography in Shangri-La, Southwest China. *Tourism Culture & Communication*, *17*(1), 19–30. doi:10.3727/109830417X14837314056816

Zhang, J. J. (2018). How could we be non-western? Some ontological and epsitemological ponderings on Chinese tourism research. In P. B. Mura & C. Khoo-Lattimore (Eds.), *Asian qualitative research in tourism: ontologies, epistemologies, methodologies, and methods* (pp. 117–138). Springer.

Tourism studies is a geopolitical instrument: Conferences, Confucius Institutes, and 'the Chinese Dream'

Ian Rowen ⓘD

ABSTRACT

Tourism scholarship can advance the multifarious geopolitical projects of state actors and aligned commercial entities. Such effects are achieved not only through tourism itself, but through the production and circulation of politically-inflected forms of knowledge. Such work is conducted by tourism scholars and allied industry and state actors. A first-person account of a 2017 tourism studies conference held at an Australian university demonstrates the argument by examining the ways in which scholars, industry, and state actors navigated and facilitated the geopolitical and geoeconomic agendas of not only domestic but potentially contentious international regimes. The conference received financial and administrative support from state, industry, and academic agencies from both Australia and China. Hosted by the Griffith Institute for Tourism (GIFT) and the Griffith University's Tourism Confucius Institute (TCI), the conference was the third in a series of 'East-West Dialogues on Tourism and the Chinese Dream'. By actively positioning the international collaboration within the rhetorical bounds of the 'Chinese Dream', and by conducting the conference in collaboration with the Tourism Confucius Institute, a quasi-educational operation directly managed by the Chinese party-state, the Australian and international tourism academy implicitly supported the geopolitical designs of the Chinese Communist Party. Renewed attention to academic ethics and increased areal expertise are a necessary response, especially in a time of global geopolitical instability and structural economic transformation in the academy.

摘要

旅游学术可以促进国家行为者和结盟的商业实体的多种多样的地缘政治项目。这种影响不仅通过旅游业本身，而且通过具有政治影响的知识生产和流通来实现。这些工作是由旅游学者、相关产业和国家行动者运作的。本文以第一人称叙述了澳大利亚一所大学举行的2017年旅游研究会议，表明学者、行业人士和国家行动者在引导和推动地缘政治和地缘经济议程方面的方式，不仅包括国内的，还包括潜在的有争议的国际制度。会议得到了来自澳大利亚和中国的国家、行业和学术机构的财政和行政支持。此次会议由格里菲斯旅游学院(GIFT)和格里菲斯大学旅游孔子学院(TCI)共同主办，是'旅游与中国梦的东西方对话'系列活动的第三次会议。通过积极定位修辞色彩的国际合作的'中国梦'，并与旅游孔子学院合作运作这个旅游会议，直接由中国党政,澳大利亚及国际旅游研究院运作准教育运作暗中支持中国共产党的地缘政治设计。重新

重视学术道德和增加领域专业知识是一种必要的反应，特别是在
该学院全球地缘政治不稳定和经济结构转型之际。

Introduction

Tourism scholarship can advance the multifarious geopolitical projects of state actors and aligned commercial entities. This owes not only to tourism's intrinsic political potency, but to its development and promotion by tourism scholars and allied industry and state actors via the production and circulation of knowledge. To theorize this point, this article builds on past research that treats tourism as a technology of state territorialization—i.e. as a mode of social and spatial ordering that produces tourists and state territory as effects of power—to take a further reflexive step to argue that the transnational knowledge production and circulation conducted by tourism *scholars* plays a key part in such practices. In so doing, scholars, intentionally or not, can implicitly and explicitly advance the various territorial and 'soft power' projects of state actors and aligned commercial entities.

As a case study, I analyze a tourism studies conference held in 2017 by an Australian university in collaboration with Chinese researchers and funding agencies. This conference received financial and administrative support from state, industry, and academic agencies from both countries. Hosted by the Griffith Institute for Tourism (GIFT) and the Griffith University's Tourism Confucius Institute (TCI), this conference was the third in a series of 'East-West Dialogues on Tourism and the Chinese Dream'. The first round was held in Griffith University in 2014 and the second at Shanghai Normal University in 2016. In addition to providing a platform research presentations and panels, these occasions were, like many academic conferences, also an opportunity for business and political networking. To gather thick ethnographic data on this event, I participated as a regular attendee. Needless to say, like all ethnographically-informed articles, this is an account of one particular person's experience of an event, among many other possible interpretive approaches. In addition to presenting my own paper on a conference-appropriate topic, I conducted both formal and informal interviews with 16 organizers, scholars, industry representatives, and government officials, as well as volunteers, staff, and other attendees. In line with the research protocol reviewed in advance by my university ethics board, I disclosed my interest in researching the conference itself as an example of 'international collaboration' to informants as part of my interview process. Depending on their personal preferences and the relative publicness of the spaces in which data was collected, some informants are quoted by name, while others are anonymized.

The article begins with a discussion of the geopolitical instrumentality of tourism studies in general, before providing a genealogy of the 'Chinese Dream' discourse and its embedding in tourism studies. It then turns to an analysis of the conference itself and of its organizing bodies, before considering the broader political implications and questions raised by this and similar international collaborations. Ultimately, I argue that by actively positioning the research collaboration within the rhetorical bounds of the 'Chinese Dream' and by conducting the conference together with a quasi-educational organization directly overseen by Chinese party-state organizations, the international tourism academy's participation directly supported the geopolitical designs of the Chinese Communist Party (CCP). This raises broader intellectual, political, and ethical questions about researcher reflexivity and positionality that have yet to be adequately addressed by the global tourism academy, and that are likely to assume increasing importance as the CCP accelerates its involvement in and funding of international academic research.

The geopolitics of tourism studies and the 'Chinese Dream'

Geopolitics is a small but growing focus of scholarship conducted under the flag of 'tourism studies'. Over the years, it has received increasing attention in both tourism journal and book publications (Gillen & Mostafanezhad, 2019; Hall, 1994; Hannam, 2013; Kim, Prideaux, & Timothy, 2016; Rowen, 2014, 2016; Timothy, 2004). The geopolitics of tourism has also been addressed in adjacent social science and area studies forums (Gillen, 2014; Mathews, 1975; Norum, Mostafanezhad, & Sebro, 2016; Park, 2005; Richter, 1983). Such work has directly addressed the impact of violence (or peace-making) and borders on tourist flows and vice versa (Becken & Carmignani, 2016; D'Amore, 1988; Farmaki, 2017; Jafari, 1989; Litvin, 1998), as well as the international political economy of tourism more generally (Britton, 1991; Salazar, 2005). Although these are useful contributions, what has received little attention is *the geopolitical work conducted by tourism scholars* themselves, whether intentionally, implicitly, or unwittingly, and *the geopolitical effects of tourism scholarship* in general.

The work of tourism scholarship—including its capacity to conjure, consolidate, and propagate territorial demarcations and social, cultural, and political categories—has proven useful to a wide array of state and market actors, as attested to by the proliferation of academic tourism research centers, public-private partnerships, and corporate consultancies. Some tourism scholars have periodically engaged in disciplinary self-evaluations and indulged in programmatic calls for, among other things, 'post-disciplinary' or postcolonial inquiry (Ateljevic, Pritchard, & Morgan, 2007; Coles, Hall, & Duval, 2006; Michael Hall & Tucker, 2004), but there has been little systematic research on the geopolitical implications of some of tourism studies' most fundamental geographical assumptions and practices, let alone the broader political commitments and practices of particular scholars or institutions.

Although tourism's potential social and environmental risks have certainly received scholarly attention, prompting the emergence of various cautionary approaches (Jafari, 2001) to promoting well-being or 'sustainability', tourism scholars have had relatively little to say about the geopolitical implications of their work. Indeed, some of tourism studies' most basic, yet unavoidably geopolitical, practices include the reification of divisions between domestic and international spaces, and the segmentation of markets based on sending or receiving states or ethnic groups which are themselves often contested categories.

Tourism studies' general lack of geopolitical reflexivity, with a few notable exceptions, is all the more striking when the field is compared to other social science and area studies communities, which have grappled with such issues for decades. For example, within anthropology and geography, particularly contentious and widely debated issues have included ideological and technical support for colonial projects and other forms of geopolitical violence. Many participating scholars, both past and contemporary, have received intense scrutiny and even censure from within their own disciplinary communities. Such criticism and soul-searching, part of what has been more broadly called the 'critical' or 'reflexive' turns, attends to both conceptual and ethical dimensions.

A full interrogation of the geopolitical problematics of tourism theory is beyond the conceptual scope of this article, but their salience should signal caution especially for scholars attempting to do not only theoretical work, but well-informed empirical work about regions with which they are unfamiliar. It is impossible for a single scholar to have a firm grasp of not only a body of theory and method, but of their possible applications across a world of cultural, social, and political difference. Therefore, research collaborations between area specialists and funding bodies, and theoretically or methodologically sophisticated (or simply well-placed and widely-published) tourism scholars are an understandably common way to compensate. However, these diverse actors' personal, political, and intellectual agendas may not be entirely visible, let alone convergent, which demands careful attention to the ethical and intellectual implications and trade-offs of such collaborations. For example, a senior Chinese tourism scholar once told me that advocacy for the political unification of mainland China and Taiwan was a 'precondition' to our pursuing any collaborative work on tourism between the two polities. However, I am uncertain if this scholar's many previous co-authors had been explicitly informed of such an agenda, especially those who were not area specialists and therefore possibly inattentive to the institutional or ideological imperatives that can drive such research projects.

With this in mind, I now turn to the history, geopolitical underpinnings and agenda of the 'Tourism and the Chinese Dream' project, as a case study conjoining tourism and areal studies, before presenting an ethnographically-informed analysis of the conference, its geopolitical effects, and its wider context. I will begin first by examining the broader history and geopolitical instrumentality of the 'Chinese Dream' itself, before tracing its transit into tourism studies. As such, it is necessary to explore not only the ambiguities of the 'Chinese Dream' and the soft power potential of tourism and tourism studies as constituting geopolitical resources in their own right, but also to consider the particular actors facilitating their cross-fertilization, who in this case include international scholars and the Confucius Institute.

The geopolitical instrumentality of the 'Chinese Dream'

International Anglophone tourism scholarship effectively incorporated the language of the 'Chinese Dream' via publications spawned by several conferences that specifically preceded and enabled the conference spotlighted in this article's case study (Weaver, 2015; Weaver et al., 2015). These publications would have both benefitted from and been problematized by greater attention to the historical conditions and political resources that Chinese leader Xi Jinping has drawn upon in the proclamation of the

Dream as a 'master narrative' to pursue his policy agenda both within and beyond China (Mahoney, 2014).

'The Chinese Dream' was first announced by Xi during his first month as the chairman of the Communist Chinese Party, in November 2012. Although the phrase had already been formulated by Chinese intellectuals and policy makers in the years prior, its embrace by the head of state repositioned it as a discursive resource with major implications for national redefinition, and domestic and foreign policy (Callahan, 2015b, p. 222). Xi's Chinese Dream, as he elaborated later, calls for the 'great rejuvenation of the Chinese nation', which requires China and its partners to 'to make the country prosperous and strong, rejuvenate the nation, and make the people happy' (Xi, 2014, p. 6).

While 'The Chinese Dream' points positively to prosperity and greatness, international relations scholar William Callahan has argued that it also has a negative aspect implicitly intended to criticize the countervailing ideals of other sorts of dreams: 'The point of China Dream policy ... is not only to tell people what they can dream, but more importantly, what they cannot dream: the negative soft power strategy ... serves to exclude many individual dreams, the constitutional dream, the American dream and so on' (Callahan, 2015b, p. 224). This dual quality is evident from Xi's choice of sites for the announcement of the 'Dream': the 'Road to Rejuvenation' exhibit of China's National History Museum, a site which showcases China's purported, '5,000 years of glorious civilization, but also its 170 years of humiliation where 'capitalist imperialist powers invaded and plundered China', and where Xi said he 'learned deep historical lessons'. Xi's interpretation consistently emphasizes the role and destiny of the nation over that of the individual: 'History informs us that each person's future and destiny are closely linked to those of their country and nation. One can do well only when one's country and nation do well' (Xi, 2014, p. 4). Alternative visions of the China Dream—for example, the Guangzhou-based Southern Weekly newspaper 2013 New Year editorial 'dream for freedom and constitutional government'— were forcibly censored and re-written into calls for greater party-state centralization (Yuen, 2013).

The 'Chinese Dream' draws both on the reconstruction of ancient Chinese traditions of 'harmony' favored by his predecessor, Hu Jintao, as well as the more modern nationalist calls of Mao Zedong for China to surpass the US in economic strength. Although its imaginative scope is vast, and its recognizably parallel structure with a globally-recognized confection like the 'American Dream' gives it an uncanny appeal, historical reviews suggest that the 'Chinese Dream' is but one more slogan in a series of 'master narratives' designed to legitimize and promote the Chinese Communist Party (CCP). As put by political scientist Zheng Wang, 'Xi's Chinese Dream narrative is like an old wine in a new bottle with the dream's name replacing [past leaders] Jiang and Hu's national rejuvenation, Deng's invigoration of China, and Mao's realization of socialism and communism'. While the exact wording has changed, there is continuity in these iterations: '... the most important message of these narratives from different periods is the same: the Party wants its people to believe that only under the leadership of the CCP can the dream of a better life be realized' (Wang, 2014, p. 7).

This uneasy transnational blend of old and new slogans, in parallel with the assumed continuity of party-state rule, has considerable ideological implications for both domestic and foreign policy, seen clearly through the international promotion of the 'Chinese Dream'. Repeated at home and abroad, this phrase has become a signature slogan of his administration. Xi's official book on the Chinese Dream suggests that it 'not only enriches the Chinese people, but also benefits the people of the world'. Foreign Minister Wang Yi presented the China Dream as Xi Jinping's 'key conceptual innovation in foreign affairs, which led to a successful year for Chinese diplomacy in 2013'. Influential Chinese public intellectuals, some of whom had already advocated for similar stance, have also been quick to embrace this formulation, calling for a 'China dream/world dream', with 'China leading the rise of the Global South against the West' in which it will '"leap forward" to overtake the United States in a "great reversal" of power in which "American hegemony" will be replaced by a World of Great Harmony controlled by the Global South' (Callahan, 2015a, pp. 15–16). As a 'master narrative', it articulates both domestic and global ambitions and subsumes them under an ethno-national flag that can be marshaled by the Chinese Communist Party and its organs. China's former ambassador to the United Kingdom, Ma Zhengang, declared, 'China's dream is the world's dream', a claim that was echoed uncannily throughout the Tourism Studies and the Chinese Dream conference.

The [tourism] Confucius Institute as vehicle of 'the Chinese Dream'

Strikingly, 'The Chinese Dream' has been introduced to international tourism scholarship not by Chinese scholars or China area specialists, but by international scholars who have pursued collaborations with China-based scholars and funding agencies. These scholars have also collaborated with the Tourism Confucius Institute (TCI) at Griffith University, one of a growing network of PRC-administered language training and cultural influence centers based in universities across the globe. Together, Griffith University scholars and the TCI hosted the conference of this case study.

Although Griffith University's Tourism Confucius Institute is the first tourism-themed Confucius Institute, it is but one of many arms of the Confucius Institutes (CI), the 'brightest brand of China's soft power', in the words of its founding director-general, Vice-minister Xu Lin (Callahan, 2015b, p. 225). First established in 2004, the CIs have been depicted by anthropologist Marshall Sahlins as 'an instrument of the party state operating as an international pedagogical organization' (Sahlins, 2013). They have since been further rearticulated as key sites for the promotion of the 'Chinese Dream'. In words of CI council chairperson, Liu Yandong, a Politburo member and Vice Premier of the State Council of China, 'The Confucius Institute, as an effective vehicle for cultural exchanges and important platform for reinforcing international friendships, is where the China Dream, the dreams of all countries and the world dream come to convergence' (quoted in Lahtinen, 2015, p. 212).

Nominally a cultural and educational promotion agency, the CI is supervised by party officials, however variably implemented in practice. Although the CI parent organization, Guojia Hanban (state management of Chinese), is affiliated with the Ministry of Education, it is managed by a council comprised of officials from the Ministry of Foreign Affairs, the State Council Information Office (also known as the Office of Overseas Propaganda), and ten other ministries and commissions (Lahtinen,

2015). This forms a major cultural arm for the implementation of political initiatives, is backed with a significant financial budget (US$189 million as early as 2009), and is aimed at a variety of targets (Link, 2017). In the words of CCP Politburo Member and Minister of Propaganda, Liu Yunshan, to his staff in a talk on 'cultural battlegrounds' in 2010, 'We should actively carry out international propaganda battles on issues such as Tibet, Xinjiang, Taiwan, human rights, and Falun Gong. Our strategy is to proactively take our culture abroad … We should do well in establishing and operating overseas cultural centers and Confucius Institutes' (Sahlins, 2015, p. 6).

Distinct from the cultural promotion strategies of the UK's British Council or Germany's Goethe Institute, which set up outposts in urban centers outside of universities, the CIs are almost always incorporated within foreign educational institutions, where they run classes (sometimes for university course credit), and conduct a variety of activities in collaboration with staff and faculty. The CIs' placements within international public and private universities have turned them into lightning rods for controversy about China's soft power campaigns. Critics have expressed fears that CIs may be used for direct surveillance or other explicitly nefarious activities. Evidence for this remains largely anecdotal, but self-censorship of sensitive topics is a much likelier outcome for affiliated or affected scholars—as implied by Minister Liu's quote above, discussions of challenging issues like Tibet, Taiwan, or 'democracy' and 'human rights' is spun in a 'politically correct' fashion, if not forbidden outright (Dirlik, 2014). As part of its declared activities, CIs 'monitor content in everything from language textbooks to lectures and cultural programs' and 'shape the views of foreigners, especially young foreigners, by introducing them to China—its language, its traditional civilization, and, seamlessly, a version of its modern life that the Communist Party wishes to put forward' (Link, 2017, pp. 166–167).

Despite the concerns and criticisms of some prominent scholars, Hanban's offer of free Chinese language tuition and university access to other Chinese political and commercial networks and resources has proven appealing to many international institutions. Chinese universities have likewise been enticed by the opportunity to set up a foothold and send faculty and staff to run CIs in international universities, which has led to a long application queue that can be expedited by demonstrating the strategic value of a particular partnership arrangement. This has led to the creation of themed CIs, including the Tourism Confucius Institute, based at the Griffith's Gold Coast campus, where tourism research was promoted as one of the university's strengths.

According to interviews with several senior current and past Griffith professors, the university's establishment of its TCI was neither smooth nor obvious. Although Sun Yat-Sen University in Guangzhou had a longstanding relationship with Griffith, its efforts in the 2000s to establish a Confucius Institute had been quashed by Griffith's then vice-chancellor, who was concerned about the reputational risk of what was already viewed as a politically-suspect initiative. According to interviews with several involved faculty, conditions changed after Ian O'Connor was appointed as vice-Chancellor and President and set greater engagement with China as a university-wide strategic direction.[1] Enrollment of Chinese students soared, as happened in many Australian universities.

As part of its China-focused development strategy, Griffith looked to establish a Confucius Institute, but had difficulty finding a candidate Chinese partner university that would not have to wait in a long queue for approval from Chinese officials. The impasse was solved by the personal involvement of the director of a major China-focused consulting company which provides services to Griffith and many other Australian universities. The company director, who had a strong personal connection to the China University of Mining and Technology (CUMT), reasoned that Australia's significant tourism assets and Griffith's tourism studies strengths would provide unique appeal to the leader of Hanban, the ministry that oversees the Confucius Institute, and approached CUMT's president. The tactic proved successful and the TCI was set up soon after, despite the unlikely pairing of mining and tourism faculties. The TCI's strategic importance was highlighted at its groundlaying ceremony by the participation of an unusually high-level political visitor, Jia Qingling, a member of the CCP's Standing Committee. The TCI's inaugural director was Colin Mackerras, a scholar of Chinese theater who was personally thanked in Australia's Parliament by Xi Jinping for his many years supporting Australia-China relations. He was succeeded after one year by Leong Liew, a political economist originally from Malaysia. CUMT has sent two of its own professors as partner directors, while the acting head of the institute, tourism scholar Ding Peiyi, has been involved since its inception.

The opening of the TCI was followed by the establishment of GIFT, also at Griffith's Gold Coast Campus, which was established in 2014 with Susanne Becken as its inaugural director. Featuring its own roster of researchers as well as affiliated faculty from other departments, GIFT drew additional staff from an older tourism-focused unit at the university. Outbound Chinese tourism was a research focus from GIFT's inception, while the rest of its strategic orientation was partially determined by an advisory board chaired by 'tourism industry heavyweight' and Griffith University lecturer, Don Morris (Ausleisure, 2014). On its homepage, in addition to claiming a commitment to the 'triple bottom line' concept of economic, social, and environmental sustainability, GIFT also listed 'Soft power, travel and the Chinese Dream' as an 'Area of Expertise' throughout 2017 and 2018.

Based on interviews with several participating scholars, the initial idea to link tourism with the 'Chinese Dream' arose either independently or during conversations between GIFT scholars, TCI deputy director Ding Peiyi, and David Weaver, a full professor in Griffith's Department of Tourism, Hotel, Leisure, and Sports Management, which is located down the hall from GIFT's office. According to an emailed communication from Weaver, 'In 2013 or 2014 I started to think about the obvious connections between the 'Chinese Dream' and tourism, and came to realise that no one had linked them before. It therefore seemed to be a good theme for GIFT's first international tourism conference. That event, and an early article on the subject I wrote, attracted attention in China, and led to a number of speaking invitations. I think that many of my Chinese colleagues find it interesting that it was a Westerner who made this link. Since then I've switching my research focus increasingly to China, where I've been developing a good working relationship with colleagues who are eager to publish in top tier English language journals to improve their international academic standing. They can benefit and learn from my long years of experience as a tourism researcher, while I am gaining a lot of knowledge about the Chinese research and cultural environments. In 2017 I was very fortunate to be awarded a Yangtze River Scholarship, China's top academic award, and this has allowed me to become even more embedded as a researcher of Chinese tourism. This is especially relevant since

President Xi's keynote speech at the recent 19th Congress of the Communist Party of China late last year, which prioritises a better quality of life for the Chinese people through poverty alleviation and development of an "eco-civilization", all of which can be facilitated by sustainable tourism'.

In November 2014, GIFT and the TCI hosted the 'G20 First East-West Dialogue on Tourism and the Chinese Dream', which coincide with the G20 Summit in Brisbane. Based on this, TCI organized Weaver and several collaborators at Griffith soon published reports on the conferences and advanced the 'Chinese Dream' as a 'framework for engagement' in two flagship journals, the *Annals of Tourism Research* (Weaver, 2015) and the *Journal of Travel Research* (Weaver et al., 2015). The following conference round, in September 2016, was titled 'Second East-West Dialogue on Tourism and Chinese Dream'. It was hosted by Shanghai Normal University in collaboration with Griffith, TCI, the China Tourism Academy, and the China Academy of Natural Resources, among other agencies. The Call for Papers began by asserting that, 'The initiation of 'Chinese dream' by Chinese President Xi Jinping accelerated the construction of an ecologically sustainable society'. These two conferences and their published outcomes enabled the 3rd conference in November 2017, on which the remainder of this paper will focus.

Articulating tourism with the 'Chinese Dream': analysis of the conference

The 2017 'Third East West-Dialogue on Tourism and the Chinese Dream', subtitled 'Eco-civilisation: Managing tourism for sustainable growth', was announced by conference co-convenors, GIFT Director Susanne Becken and Tourism Confucius Institute Deputy Director Ding Peiyi. The organizers' welcome letter, which was released well in advance of the conference, positioned the event's themes in line with national and international discourses of sustainability and harmony, the temporal imaginaries of non-governmental organizations such as the United Nations, and uniquely state-driven imaginaries such as 'eco-civilization' and the 'China Dream'. Building on the work of the previous two conferences, it aimed specifically to link 'Sustainable Tourism for Development with the Chinese Government's efforts towards becoming an "Eco-Civilisation"'. The Call for Papers framed the conference themes and location as follows:

> ... This year's conference is especially important for two reasons. Firstly, the United Nations have declared 2017 as the International Year of Sustainable Tourism for Development. Tourism is one of the largest and fastest-growing sectors in the world, and has the potential to stimulate economic growth, create decent jobs, help preserve ecosystems, contribute to protecting cultural heritage, and support the peace building process. Secondly, 2017 is the China-Australia Year of Tourism, celebrating the close relationship between China and Australia and recognising the importance of the China market for Australia's tourism economy.

> The conference addresses the challenging topic of 'Managing Tourism for Sustainable Growth'. President Xi Jinping introduced the concept of the 'China Dream' to highlight the aspirations of the Chinese people, and 'Eco-civilisation' as a form of development, an ethic, or blueprint for an environmentally harmonious compatible society. Given the increasing importance of Chinese outbound tourism around the world, the success of sustainable tourism is shaped by Chinese tourism trends. Thus, conference dialogue will link Sustainable Tourism for Development with the Chinese Government's efforts towards becoming an 'Eco-Civilisation'.

This conference provides a platform for tackling challenges and opportunities that arise from rapid tourism growth with a particular focus on the Asia Pacific region, and the China-Australia relationship. Building on the first two successful conferences, this event will further strengthen and broaden the connections between the academic and business communities in Australia and China, and within the region, through creating opportunities for intellectual, business, and cultural dialogues. This will enhance mutual understanding and engagement between the two countries at a time when geopolitical interests are creating divisions.

The Gold Coast is the perfect location for this conference. It receives almost half a million Chinese visitors per year. In a recent Ctrip survey, the Gold Coast was voted as the favourite Australian holiday destination by Chinese travellers, and the only Western destination to make the global top ten list. Eco-civilization, sustainable tourism development and China-Australia relations are topics of great importance to Gold Coast, Queensland, Australia and China ...

Such narrative framing directed the intellectual production of participating scholars, businesses, and institutions, partially determining the permissible language and political position of potential publications and other outcomes. The effectiveness of this framing is demonstrated in the following sub-sections, which examine the specific and contingent ways in which institutional and personal actors came together to reinforce old narratives and produce new ones.

These narratives were produced over took place over two days, largely inside the Gold Coast Surfers Paradise Marriott Hotel. The event featured presentations from leading Australian industry representatives, including the trade groups Tourism Australia, Gold Coast Tourism, Tourism Transport Forum, the Pacific Asia Travel Association and the Mantra Group; Chinese state actors including the PRC vice-consul; and a variety of academics and other actors that blur these distinctions, such as the head of the China National Tourism Academy, which is a party-state research agency.

To further trace these interactions and connections, the following account is divided into two headings: conference structure, and narratives and representations. The focus on conference structure, including a discussion of state-industry-academic interaction, is meant to identify and illuminate the specific actors and networks involved in the production of knowledge at this event. The succeeding focus on representation draws from the field of critical geopolitics, an approach to scholarship which casts a 'critical perspective on the force of fusions of geographical knowledge and systems of power' (Dalby & Tuathail, 1996, p. 452). This approach attends not only to representations of China as a nation-state or of Chinese people as an imagined community, but of specific state and party organs, such as the TCI. As geopolitical narratives are inherently unstable and shifting formations, and can take years to consolidate and materialize, the conclusions are somewhat tentative due to the timeline of the analysis, which was conducted shortly after the conference. This timeliness is all the more reason for further reflection on the tourism academy's role in the framing of such high-stake political narratives.

Conference structure

The conference was held at the Marriott Hotel in Surfer's Paradise, itself a popular destination for Gold Coast holidaymakers, including inbound Chinese tourists. Close collaboration between state, industry, and academic actors was evident from the start.

Informal proceedings began with an opening party on the patio of the hotel's sprawling pool, hosted by Paul Donovan, the chair of Gold Coast Tourism, who declared that 'China is the size of the prize'. Praising the Tourism Confucius 'Center' [sic] as a 'great development', he said, 'We need to do more in Australia to promote China. People here don't understand the history, the culture, the warmth. We need to do more to promote China. That's one of the missions I have in life'. His opening toast, including an admission that he had studied Chinese language for 3 months before giving up, was capably translated into Mandarin by his office staff.

The conference's formal program began the next morning in the hotel's meeting rooms. Griffith University Vice-Chancellor Ian O'Connor, wearing a bright red tie, introduced 2017 as the year of Australia-China tourism and noted the large and growing financial sums circulated in the bilateral tourism trade. He observed that Griffith hosts the world's only Tourism Confucius Institute and that GIFT is a 'world-leading institution'. Following this introduction, Bob East, CEO of the Mantra Group and the new chair of trade association, Tourism Australia, used his keynote speech to explore ways of 'unlocking this amazing potential that is the Chinese market'.

The keynote was followed by an expression of thanks from the PRC Vice-Consul, who noted that that 2017 was not only the year of Australia-China tourism but of 'sustainability' and also of the 19th Congress of the Chinese Communist Party. The Vice-Consul was followed by greetings from Dai Bin, the head of the China National Tourism Academy, a central party think tank and policy institute, who said, 'Welcome, Australians, to a beautiful and strong China. The Chinese dream will not just be our dream. It will be everyone's dream. This idea does not belong to China. It belongs to all of the world'. This marked the end of the conference introductions.

The program continued with a panel on tourist dispersal, chaired by GIFT director Suzanne Becken, before splitting into separate tracks on business and sustainability, with the former proving far more popular for Chinese attendees. The full conference then reconvened for a session on Geopolitics, moderated by TCI head and Griffith Professor of political economy, Leong Liew, and featuring talks by Dai Bin, Griffith Professor and the inaugural head of the TCI, Colin Mackerras, and representatives from the Tourism Transport Forum and Pacific Asia Travel Association. The afternoon's academic sessions ended with a summary round-up from David Simmons, a professor at Lincoln University in Christchurch, New Zealand, and also Suzanne Becken's past PhD advisor.

Day turned to night and climaxed with a hotel banquet dinner, resplendent with red lanterns and Chinese knot decorations and dancing, *pipa* (Chinese lute) performances and other live entertainment provided by the TCI. The dinner also featured a talk by Tourism Australia stalwart Don Morris, who had been involved in the initial ADS negotiations and was credited for the hugely successful television campaign that featured actor 'Crocodile Dundee' actor Paul Hogan. Morris began by flattering China's 'Five thousand years of history and 230 years of modernity' before asking 'What is tourism?' and how can it be marketed to China, and included lively slide titles like, 'The killer fact about the Chinese middle class', 'An army of Chinese millennials is reshaping global travel', and 'Australian Tourism Tsunami'. He then noted, 'I've never once met a Chinese who wasn't an absolutely lovely person. They make lovely

Australians'. The event then featured a Chinese-style lucky draw and awards event, with gifts from China Southern Airlines and other sponsors.

The conference's second and final day began with a talk to the full group about the new era of Chinese theme parks by Bao Jigang, the dean of the tourism program at Sun Yat-Sen University in Guangzhou, and the only other researcher besides David Weaver to have received a Yangtse River Scholarship. The conference then split into various sessions on everything from sustainability to destination marketing. My own paper, a critical look at the 'sustainability' of bilateral tourism and the conference itself, using the examples of China's tourism cuts to Taiwan and South Korea, was delivered during this period. Despite being the only paper to explicitly address geopolitical risk, it was scheduled in the 'sustainability' session alongside papers on land use planning. The conference ended with another summary session by David Simmons, who further rearticulated the Chinese Dream language into both indigenous New Zealand and global registers: 'We've all heard about the Chinese dream, but that shouldn't just be for China, the dream should be for the world. We should all be part of some idea of an ecocivilization. Can we borrow it and work on it as well?' He suggested that the idea complements a Maori term, 'kaitiake', which he glossed as 'stewardship'. He continued by talking about shared cultural affinities for food and 'bread-breaking', and concluded with a quote from former US Vice President and prominent environmentalist Al Gore.

The afternoon involved an optional field trip, at additional charge, to the Currumbin Wildlife Park, where participants could cavort with koalas and kangaroos. The group included myself and mostly Chinese academic delegates, and afforded an opportunity to conduct further informal interviews.

Narratives and representations

From start to finish, the conference recapitulated the conceptual connections between tourism, tourism studies and the 'Chinese Dream' narrative. By emphasizing the important links between tourism practices and tourism research, it consolidated discursive and social links that had been initiated by previous rounds of gatherings, research projects, and publications. This effect was primed and reinforced through the ideological framing of the 'Chinese Dream' discourse in the call for papers, the transnational circulation of the social, intellectual, and financial capital of conference participants and backers, and the creation of an affective atmosphere that celebrated China as a major outbound tourism market and unique 'eco-civilization'.

Such effects were achieved despite, or perhaps enhanced due to the fact that the majority of participating international scholars neither spoke Chinese nor had any significant research experience within China prior to this and previous rounds of 'Chinese Dream'-driven collaborations. As noted by political geographers Dalby and Tuathail, '[geopolitical] devices need not be very sophisticated to function in political discourse... repetition is an important facet of rendering particular understandings "common sense". The ideological production and reproduction of societies can, in part, be understood as the mundane repetition of particular geopolitical tropes which constrain the political imaginary' (Dalby & Tuathail, 1996, p. 452). Indeed, the mundane repetition of tropes, not just those of the 'Chinese Dream', but also of Australian industry desire, such as 'China is the size of the prize', preceded and pervaded this conference. This latter phrase was repeated not only by Paul Donovan, of Gold Coast Tourism, but also by conference keynote speaker, Mantra CEO Bob East's, who observed asked, 'The macro tailwinds mean the size of the prize keeps growing... We in Australia are sitting in the heart of the world's new GDP center. We're in geographically fortunate place... We at Tourism Australia know China will be the biggest market for Australia. We know it already is in terms of spending. We know that the greatest migration in the world's history, China's urbanization, has unlocked the consumer class... This region, Gold Coast, is at the forefront of opening up the China market'.

Although some international tourism scholars had already advocated for the 'Chinese Dream' in well-regarded journals and subsequently received major fellowships in China, several Chinese scholars hesitated to frame their research in terms of the 'Chinese Dream', expressing privately to me that such rhetoric was unduly political. The prominent exception to this rule was Dai Bin, the head of the China Tourism Academy, a party official, who adopted the Chinese Dream rhetoric wholeheartedly. Dai's conference presentations followed the intellectual path blazed by David Weaver, the international scholar who had provided the most impactful intellectual advocacy for this position and had already received significant official accolades and research support in China following his previous work. Dai's approach makes sense, given his unique position as a scholar-official who presides over a government research and policy agency. This political sensibility was also reflected in the official register and cadence of his speech. In his introductory remarks, he said, 'We have to thank China, as tourism has become an important part of the Chinese dream. China's already become New Zealand's biggest market, and I hope one day it will also be the biggest for Australia. Welcome, Australians, to a beautiful and strong China. The Chinese dream will not just be our dream. It will be everyone's dream. This idea does not belong to China. It belongs to all of the world'.

Dai expanded on this theme later in the day during a session on Geopolitics, moderated by TCI head and Griffith Professor of political economy, Leong Liew, and also featuring presentations by Griffith Professor and the inaugural head of the TCI, Colin Mackerras, and representatives from the Tourism Transport Forum and Pacific Asia Travel Association. His talk was delivered in spoken Mandarin with English translations on projected slides, and began with a direct statement of the geopolitical foundations of bilateral tourism: 'China-Australia tourism cooperation is always delivered under the framework of comprehensive strategic partnership between the two Countries, and practical and in-depth tourism cooperation effectively stimulates the joint efforts for maintaining the partnership... the partnership has been strengthening in mutual respect and parallel advance while overcoming the differences in national situation and political system'.

Dai continued by pointing explicitly to the role of government officials, including himself, in directing tourism towards particular destinations: 'Chinese tourists' choice of destination would be immediately and detectably transformed when the Chinese government expressed the attitude toward the relevant event, or the People's Daily, CCTV, or the Xinhua News Agency had voices on the event... But the information search and demand would be quickly stimulated when the diplomatic relation between China and the Destination Country or Region was improved, or when a high-level visit was effectively made. From this point, the most influential factor in Chinese tourist's choice of a destination is the strategic position of China's diplomatic relation with a destination Country or region: and the high-level mutual visit is the most profitable intellectual property (IP), the Key Opinion leader (KOL), or the 'Internet celebrity'. At this point, Dai went off-script (or at least off-slide) and said with a smile, 'So, with my coming here, people say China-Australia relations will improve'.

Dai then reiterated the language of the Chinese Dream: 'In less than five years, China will achieve the first 100th anniversary goal of the Chinese dream. This is to say that China will realize the dream of building a well-off society in all-round way. This is a great dream that will be recorded in history. And it means that Chinese people will pursuit [sic] better life and have more budget and time for enjoying high-quality tourism, including outbound tourism. The latest forecast of China Tourism Academy suggests in the coming five years China will generate as many as 700 million outbound tourists who will have 800 billion US dollar spending in the outbound tourist destinations. Looking into the future, the joint efforts will advance the comprehensive strategic partnership, people at both sides will have better impression of each other, and the degree of popularity of tourist attractions will be increased. Those always make me have an optimistic prediction on China-Australia tourism exchanges ... China has become the world's second-largest economy, the quality of economic growth and the level of social development and people's life are getting better, and we would also like to attract more tourists from all over the world. Chinese dream has become the most powerful impetus to all-round recovery and growth of Chinese inbound tourism; and to international tourists, it is also more attractive than the natural historical, and cultural resources ...' Dai gestured towards potential challenges before breezily citing a quote from a 2014 speech at the Australian Parliament by Xi Jinping, There is an Australian saying, 'Keep your eyes on the sun and you will not see the shadows', a proverb with which few Australians actually seemed familiar (Stokes, 2014).

Dai continued, 'Australia has become a model or sample of Chinese outbound tourist destination. And Australia has frequent interactions with China tourism industry and would not have direction that is unfavorable for mutual tourism exchanges. It is my hope that in near future our two sides could make strategic communication and technical negotiation as soon as possible on relevant topics such as the destination status agreement plus (ADS+) and free trading area (FTA)'. He went on to mention particular enterprises, including travel booking engine Ctrip, and China Southern Airlines (also a sponsor of the conference), before closing by reaffirming his agency's cooperations with Griffith, Tourism Australia, and Tourism Research Australia.

The subsequent speakers from the Tourism Transport Forum and the Pacific Asia Travel Association talked triumphantly about the boom in bilateral tourism. The mood turned more severe when Colin Mackerras addressed bilateral relations more broadly: 'I agree that our tourism relationship with China is good. [But] this year China-Australia relations are actually not good. Yesterday a [Australian government ministry] white paper gave a bleak assessment of China, suggested it's a threat. This idea is not gone forever ... Maybe I'm too bleak but I think recent [Prime Minister] Turnbull decisions over the South China Sea ... have not been helpful to our China relationship ...

I think this FTA is good and it's the future but I'm worried in the short term. Also I want to raise the question of Australian impressions of China, of fears of a rising China. [There was a] report that Chinese students are being incited by their government to undermine Australian values… Most students I know are just trying to get on. Will tourism solve them [the problems]? I hope and think so. First, tourism influences images. Australians go to China get impression of China and vice versa. On holiday, people usually get good impressions. I think it helps. China has a wonderful culture and scenery and friendly people…. But I find it extraordinary that things are getting less politically favorable. I don't dispute aviation growth [but] I see tensions getting worse, nationalisms competing, I wish I could be more optimistic. The US president seems to be very inconsistent… As for politics, will tourism improve it? I hope so. Tourism is economic. It's also social. I'm optimistic but don't think that tourism has been directly involved in improving images'.

The following Q&A session with the panelists provided one of the entire conference's only mention of the potential risk of overreliance on Chinese tourism under shifting geopolitical conditions. This consideration was prompted by a question from moderator Leong Liew, who noted the same recent foreign policy paper, as well as China's ongoing politically-motivated tourism cuts to South Korea and Taiwan. The panelists were generally sanguine about this possibility, although Mackerras suggested that Australia 'shouldn't put all its eggs in one basket… I think politics does matter. In the 60 s, we had bad relations and no tourists. Now it has momentum and would be hard to change. The basic central fact about this country: China matters for us'.

Remarks from other participants clarified that Chinese tourism not only matters for Australia's geopolitical and geoconomic opportunities, but for its domestic electoral politics, as well. For example, after Mantra CEO Bob East's mentioned economic benefits from China, he observed the political might of the tourism industry. 'There's an election tomorrow [in Queensland]. Every single district has at least 5000 tourism workers, so we can bring our weight to the process'. Likewise, the PRC vice-consul's noted in his welcome remarks that, 'The Chinese dream is the aspiration of Chinese people to a better life… Travelling around is part of that dream. Tomorrow, there will be an election here. I can see that the Chinese dream is shared not only by Australian but by Queenslanders. The Chinese dream closely connects with the dream of the Australian people. The dream is beautiful but we can't take it for granted. We have to work hard and work together. That's why I think today's conference so vital'.

The conference was also occasion for PRC officials to link the Chinese Dream and notions of eco-civilization with the rule of the Chinese Communist Party. For example, after the PRC vice-consul noted that 2017 was not only the year of Australia-China tourism but of 'sustainability' and also of the 19th Congress of the Chinese Communist Party. With the 19th Party Congress, said the vice-consul, 'Socialism with Chinese characteristics has entered a new phase and will bring new benefits to the China Australia relationship…'

The productive polysemy of the Chinese Dream was demonstrated by the comments of a representative of Tourism Australia, a government agency which has been targeting China since 1999. He noted that Australia was one of the first countries to which China granted Approved Destination Status, which accelerated the development of group tourism, and said, 'I'd say the Chinese dream is now a reality'. This revealed a rather different understanding of the 'Chinese Dream'—one based more on the value of renminbi than of Xi Jinping Thought.

The first day's summary session, provided by David Simmons, provided yet another take on the 'Chinese Dream'. Simmons projected a few preliminary schematic charts about the role of industry and academia in promoting 'The Chinese Dream' and 'Eco-Civilization' in China, Australia, and New Zealand. He then noted that New Zealand scholars were particularly interested in reducing carbon outputs to mitigate climate change, although this issue was getting less attention elsewhere. He then adopted the Chinese Dream rhetoric, asking on his slide, 'Are these the first steps on the pathway to the Chinese Dream? Chinese green growth … and understanding of Chinese themselves as 'global citizens'?' Such a conclusion was remarkable given how much more active Chinese conference participation was in the business than in the sustainability sessions. The second day's summary session, also by Simmons, further rearticulated the Chinese Dream language into both indigenous New Zealand and global registers, citing a Maori term for co-existence, and then exclaiming, 'We've all heard about the Chinese dream, but that shouldn't just be for China, the dream should be for the world. We should all be part of some idea of an ecocivilization. Can we borrow it and work on it as well?'

Considering these various accounts together, 'The Chinese Dream' served as a poly-semic rhetorical device in multiple registers that articulated in various ways depending on the speaker's positionality. For PRC officials such as the pro-consul and Dai Bin, it pointed to the continued rule of the Chinese Communist Party and an advantageous relationship with the Australian state. For sustainability-minded scholars such as David Simmons, it was an occasion to invoke ecological notions. For Australian state officials or industry representatives, it gestured more towards dreams of industry growth, and may be facilitated by supportive voters. Although likely only the first of these groups is cognizant of the overriding political imperatives of 'Chinese Dream' language, spo-ken and written repetition served to further consolidate and extend its presence and potency in all registers.

This event was the culmination of two previous rounds of international conferences, as well as several jointly-authored papers. As such, major shifts in the Australia-China relationship notwithstanding, it is likely to continue shaping industry, state, and aca-demic collaborations. Although the Chinese Dream language was wielded uncritically by a number of international scholars, as well as Chinese political officials, it earned mixed reviews from Chinese scholars. For example, during the bus ride and walks to Currumbin Wildlife Sanacuary, I individually queried several participants about their feelings about the Chinese Dream framing of the conference. Said one, 'The Chinese dream is not really used in academic discourse, at least not for us. But hearing it from a foreigner, sure it sounds familiar and comfortable'. Another said, 'Chinese Dream sounds kind of American to me and my friends in Beijing. Not too academic, really. I don't know if it's been picked up in Chinese tourism studies'. A senior scholar from southern China observed, 'We don't talk about the Chinese dream. We're academics and that's political talk'. However, he did quickly add that political considerations do constrain the bounds of acceptable academic discourse within China.

Beyond the anticipated outcomes of academic knowledge production and circula-tion, the conference may have yielded other fruit. On the final night, according to an attendee, several of the high-level Chinese national and ethnic Chinese Australian delegates and organizers joined a small dinner hosted by a major Chinese real estate developer with multi-million dollar projects in Gold Coast, including a theme park. Several months later, this theme park was listed as part of China's Belt and Road Initiative, Xi Jinping's signature geopolitical and geoeconomic program (Walsh & Xiao, 2018). While it is impossible to conclude that this particular meeting directly resulted in the listing, the timing is likely not insignificant.

Conclusion

Tourism in, from, and to China is one the great stories of 21st century mobility. It is a story that encompasses not only millions of hosts and guests, but their storytellers as well. By promoting tourism as a component of the 'Chinese Dream', a discursive instrument devised and deployed to support the rule of the Chinese Communist Party, prominent international tourism scholars enrolled themselves into an ambiguous and ambitious geopolitical program that deserved more nuanced and critical examination. This account should serve as a cautionary tale for other credulous tourism scholars or geographers researchers building collaborations in regions beyond their initial expertise. Given the involvement of industry and government officials, it is not surprising that the presentations and publications of were framed in a celebratory rather than impartial fashion. Still, it would have behooved participants to take pause at such efforts and carefully evaluate their intellectual, ideological, and ethical implications.

As tourism scholars, industry practitioners, and government administrators in China, Australia, and elsewhere continue consolidating this 'meta-narrative' in the midst of shrinking university budgets, it is worth reflecting on the rationality of such behavior. China's soft power campaigns have not been uniformly successful (Callahan, 2015b). Yet, it is remarkable that at the very moment that China's soft power was being publicly questioned in a Australian government white paper (Australian Government Department of Foreign Affairs and Trade, 2017), its own tourism academy was channeling such soft power in collaboration with the Tourism Confucius Institute. One merely needs to trace the line between the claims of PRC diplomats and conference scholars that the Chinese Dream should be the world's dream to find evidence that such soft power campaigns had borne fruit, however belatedly, through as unlikely an academic realm as international tourism studies. This demonstrates the sticky power (Mead, 2009) of the institutional links and shared discourses developed through the previous rounds of tourism studies conferences and publications, and the likely value of this iteration to future such projects.

If the lessons of reflexive scholarship have taught us anything, it is that the conceptual (and, of course, national) flags under which we conduct our own research bear intellectual and ethical consequences. Indeed, this article is itself a kind of political intervention, even if it is one that calls for caution and critique. While its empirical target is cheerleading for the 'Chinese Dream', it should not be misread as a simple or single-minded attack on Chinese state aspirations, or an argument for other nations, whether competitive or collaborative, to get a free pass. Nationalist or jingoistic stances are questionable options for scholars who care to engage critically with their or other nations or the world at large. Indeed, all states engage in violence. As a discursive device, 'the Chinese Dream' obscures its own potential for incipient violence no more or less than does 'the American Dream', a phrase which has been marshaled towards a variety of chauvinistic and imperialistic projects. While scholars have few if any straightforward or universally-accepted formulas for the determination of ethical conduct, what should be clear in our scholarship, especially in a time of global geopolitical uncertainty and financial and structural transformation in the academic industry, is the need for intellectual and ethical reflection that transcends the temporary allures of entrepreneurial outreach and national cheerleading.

If claims that tourism can engender peace and sustainable development and other laudable results are to be taken seriously, such possibilities owe precisely to the geopolitical instrumentality of tourism and tourism studies. Therefore, converse possibilities—that tourism can threaten economies and environments or aggravate geopolitical contention– must be considered as well. Such possibilities are better evaluated not by indulging in unsustainable if temporarily lucrative pipe dreams, but by waking up to a far more complicated geopolitical and intellectual reality.

Note

1. In the Australian university management structure, "Vice-Chancellor" is the chief executive position, with the title of "Chancellor" reserved as an honorary and symbolic designation. Unusually, O'Connor was given the additional title of "President," largely to clarify his executive role during personal negotiations with Chinese institutions, according to an interview with a Griffith faculty member who preferred to remain anonymous.

Disclosure statement

No potential conflict of interest was reported by the authors.

Funding

Nanyang Technological UniversityInstitutional Review BoardThis research was supported in part through Nanyang Technological University, Singapore start-up grant and travel funds. The research protocol was approved in advance by the university's Institutional Review Board (reference number IRB-2017-10-049).

ORCID

Ian Rowen (iD) http://orcid.org/0000-0002-9674-5669

References

Ateljevic, I., Pritchard, A., & Morgan, N. (Eds.). (2007). *The critical turn in tourism studies: Innovative research methodologies.* Oxford: Elsevier Ltd.

Ausleisure. (2014). Griffith University tourism research institute launched. Retrieved from https://www.ausleisure.com.au/news/griffith-university-tourism-research-institute-launched

Australian Government Department of Foreign Affairs and Trade. (2017). 2017 foreign policy white paper: Opportunity security strength. Retrieved from https://www.fpwhitepaper.gov.au/

Becken, S., & Carmignani, F. (2016). Does tourism lead to peace? *Annals of Tourism Research, 61,* 63–79. doi:10.1016/j.annals.2016.09.002

Britton, S. (1991). Tourism, capital, and place towards a critical geography of tourism. *Environment and Planning D: Society and Space, 9*(4), 451–478. doi:10.1068/d090451

Callahan, W. A. (2015a). History, tradition and the China dream: Socialist modernization in the World of Great Harmony. *Journal of Contemporary China, 24*(96), 983–1001. doi:10.1080/10670564.2015.1030915

Callahan, W. A. (2015b). Identity and security in China: The negative soft power of the China dream. *Politics, 35*(3-4), 216–229. doi:10.1111/1467-9256.12088

Coles, T., Hall, C. M., & Duval, D. T. (2006). Tourism and post-disciplinary enquiry. *Current Issues in Tourism, 9*(4-5), 293–319. doi:10.2167/cit327.0

D'Amore, L. J. (1988). Tourism: A vital force for peace. *Annals of Tourism Research, 15,* 269–283.

Dalby, S., & Tuathail, G. Ó. (1996). The critical geopolitics constellation: Problematizing fusions of geographical knowledge and power. *Political Geography, 15*(6-7), 451–456. doi:10.1016/0962-6298(96)00026-1

Dirlik, A. (2014). June Fourth at 25: Forget Tiananmen, you don't want to hurt the Chinese people's feelings – And miss out on the business of the new "New China"! *International Journal of China Studies, 5*(2), 295–330.

Farmaki, A. (2017). The tourism and peace nexus. *Tourism Management, 59,* 528–540. doi:10.1016/j.tourman.2016.09.012

Gillen, J. (2014). Tourism and nation building at the War Remnants Museum in Ho Chi Minh City, Vietnam. *Annals of the Association of American Geographers, 104*(6), 1307–1321. doi:10.1080/00045608.2014.944459

Gillen, J., & Mostafanezhad, M. (2019). Geopolitical encounters of tourism: A conceptual approach. *Annals of Tourism Research, 75,* 70–78. doi:10.1016/j.annals.2018.12.015

Hall, C. M. (1994). *Tourism and politics: Policy, power and place.* Chichester: John Wiley & Sons.

Hannam, K. (2013). "Shangri-La" and the new "great game": Exploring tourism geopolitics between China and India. *Tourism Planning & Development, 10*(2), 178–186. doi:10.1080/21568316.2013.789655

Jafari, J. (1989). Tourism and peace. *Annals of Tourism Research, 16*(3), 439–443. 10.1016/0160-7383(89)90059-5 doi:10.1016/0160-7383(89)90059-5

Jafari, J. (2001). The scientification of tourism. In V. L. Smith & M. Brent (Eds.), *Hosts and guests revisited: Tourism issues of the 21st century* (pp. 28–41). Elmsford: Cognizant Communication Corporation.

Kim, S. S., Prideaux, B., & Timothy, D. (2016). Factors affecting bilateral Chinese and Japanese travel. *Annals of Tourism Research, 61,* 80–95. doi:10.1016/j.annals.2016.08.001

Lahtinen, A. (2015). China's soft power: Challenges of Confucianism and Confucius Institutes. *Journal of Comparative Asian Development, 14*(2), 200–226. doi:10.1080/15339114.2015.1059055

Link, P. (2017). Confucius murders squirrels. *Harvard Journal of Asiatic Studies, 77*(1), 163–173. doi:10.1353/jas.2017.0011

Litvin, S. (1998). Tourism: The world's peace industry? *Journal of Travel Research, 37*(1), 63–66. doi:10.1177/004728759803700108

Mahoney, J. G., (2014). Interpreting the Chinese Dream: An exercise of political hermeneutics. *Journal of Chinese Political Science, 19*(1), 15–34. doi:10.1007/s11366-013-9273-z

Mathews, H. G., (1975). International tourism and political science research. *Annals of Tourism Research, 2*(4), 195–203. doi:10.1016/0160-7383(75)90032-8

Mead, W. R. (2009, October). America's sticky power. Foreign Policy. Retrieved from https://foreignpolicy.com/2009/10/29/americas-sticky-power/

Michael Hall, C., & Tucker, H. (Eds.). (2004). *Tourism and postcolonialism*. London and New York: Routledge

Norum, R., Mostafanezhad, M., & Sebro, T., (2016). The chronopolitics of exile: Hope, heterotemporality and NGO economics along the Thai–Burma border. *Critique of Anthropology, 36*(1), 61–83. doi:10.1177/0308275X15617305

Park, C. J., (2005). Politics of Geumgansan Tourism: Sovereignty in contestation. *Korean Studies, 8*(3), 113–135.

Richter, L. K., (1983). Tourism politics and political science: A case of not so benign neglect. *Annals of Tourism Research, 10*(3), 313–335. 10.1016/0160-7383(83)90060-9 doi:10.1016/0160-7383(83)90060-9

Rowen, I., (2014). Tourism as a territorial strategy: The case of China and Taiwan. *Annals of Tourism Research, 46*, 62–74. doi:10.1016/j.annals.2014.02.006

Rowen, I., (2016). The geopolitics of tourism: Mobilities, territory, and protest in China, Taiwan, and Hong Kong. *Annals of the American Association of Geographers, 106*(2), 385–393. doi:10.1080/00045608.2015.1113115

Sahlins, M. D. (2013, October). China U. The Nation. Retrieved from http://thenation.com/article/china-u/

Sahlins, M. D. (2015). *Confucius Institutes: Academic malware*. Chicago: Prickly Paradigm Press.

Salazar, N. B., (2005). Tourism and glocalization: "Local" tour guiding. *Annals of Tourism Research, 32*(3), 628–646. doi:10.1016/j.annals.2004.10.012

Stokes, A., (2014, November 19). Tony Abbott and Xi Jinping take the proverbial. Sydney Morning Herald. Retrieved from http://search.proquest.com.ezproxy.une.edu.au/docview/1625529364?accountid=17227%5Cnhttp://gr6md6ku7c.search.serialssolutions.com/?ctx_ver=Z39.88-2004&ctx_enc=info:Ofi/enc:UTF-8&rfr_id=info:Sid/ProQ%3Aanznews&rft_val_fmt=info:Ofi/fmt: Kev:Mtx:Journal&rft.g

Timothy, D. J. (2004). *Tourism and political boundaries*. London: Routledge.

Walsh, M., & Xiao, B., (2018, March 5). One belt one road: China lists $400m gold coast theme park as "key project" of global initiative. *Australian Broadcasting Corporation*. Retrieved from http://www.abc.net.au/news/2018-03-05/china-lists-planned-gold-coast-theme-park-as-a-key-project/9508904

Wang, Z., (2014). The Chinese Dream: Concept and context. *Journal of Chinese Political Science, 19*(1), 1–13. doi:10.1007/s11366-013-9272-0

Weaver, D. B., (2015). Tourism and the Chinese Dream: Framework for engagement. *Annals of Tourism Research, 51*, 54–63. doi:10.1016/j.annals.2015.01.001

Weaver, D. B., Becken, S., Ding, P., Mackerras, C., Perdue, R., Scott, N., & Wang, Y., (2015). Research agenda for tourism and the Chinese Dream: Dialogues and open doors. *Journal of Travel Research, 54*(5), 578–583. doi:10.1177/0047287515588594

Xi, J. (2014). *The Chinese Dream of the great rejuvenation of the Chinese nation*. Beijing: Foreign Languages Press.

Yuen, S., (2013). Debating constitutionalism in China: Dreaming of a liberal turn? *China Perspectives, 2013*(4), 67–72.

(Post-) pandemic tourism resiliency: Southeast Asian lives and livelihoods in limbo

Kathleen M. Adams (ID), Jaeyeon Choe, Mary Mostafanezhad and Giang Thi Phi

ABSTRACT

While tourism scholars have sought to problematize the unevenly distributed impacts of the COVID-19 pandemic, we know much less about how resilience is cultivated among tourism practitioners and communities whose lives and livelihoods are have been placed in limbo. Drawing on literature at the intersection of critical tourism studies and resilience theory as well as interviews with local tourism practitioners and academics, four historically situated and place-based trends in Southeast Asia that are reshaping tourism in the region are outlined: livelihood diversification, ecosystem regeneration, cultural revitalization, and domestic tourism development. These trends highlight how the political economy of tourism in the region has both challenged and facilitated opportunities for reshaping the industry in (post-) pandemic times. These interconnected trends should not be understood in silo but rather as historically rooted and place-based experiences. The examples of resilience among Southeast Asian residents presented in the article demonstrate that local individuals and communities are active agents in resilience. While the concept of resilience has been applied widely by scholars from multiple disciplines during the COVID-19 pandemic, a critical tourism studies approach to resilience theory accounts for the historically situated nuances of local scale dynamics and their relationship to macro-level processes. Rather than simply focusing on the pandemic's sudden transformative effects, practices of resilience in Southeast Asia reflect ongoing political-economic and cultural shifts that have often been underway in the region for several decades. The conclusion identifies several policy implications and future directions for tourism research in (post-) pandemic times.

摘要

虽然旅游学者试图对新冠肺炎产生影响的不均匀提出了问题，但对如何培养旅游业从业人员的恢复力以及如何培养那些生活与生计都处于不确定状态的社区旅游的恢复力，我们尚知之甚少。本文借鉴批判性旅游研究与恢复力理论的交叉学科文献，在访谈当地旅游从业者和学者的基础上概括出四个重塑东南亚旅游业的、适应该地区历史情势与地方特色的发展趋势，即生计多样化、生态系统恢复、文化振兴与发展国内旅游。这些趋势突出表明，在后新冠疫情时期，该区域旅游业的政治经济现状如何既挑战又促进重塑该地区旅游业的机会。这些相互关联的趋势不应被孤立地理解，而应被视为植根于历史和基于地方的经验。本文中列举的东南亚居民恢复力的例子表明，当地个人和社区是区域旅游恢复力的积极推动者。尽管在新冠肺炎爆发期间，恢复力的概念已被来自多个学科的学者广泛应用，但是恢复力理论的批判性旅游研究方法解释了当地尺度旅游发展动态的历史差异及其与宏观层面过程的关系。东南亚旅游恢复力的实践反映了该地区几十年来经常发生的政治、经济和文化变化，而不是仅仅强调新冠疫情突发的变革性影响。该结论确定了后疫情时代旅游研究的若干政策影响和未来方向。

Introduction

In Summer 2017, the lead author found herself chatting with Indonesia's Minister of Communications and Information Technology, Rudiantara, on a flight from Makassar to the Toraja highlands. Although the popular Toraja tourist destination had an airport perched on a levelled mountain peak, the small planes, short runway, and unpredictable flights obliged most visitors to weather an exhausting 8-hour bus ride traversing steamy lowland plains and stomachchurning, sinuous mountain roads. While in flight, Minister Rudiantara shared his hopes that improved transport and communications systems would "better serve not only locals but tourists." Over the next few days, the author's discussion with the Minister continued during chance encounters at a Toraja ritual and village tourist site where the author bases her long-term research. By then, the Minister had become enchanted with Toraja vistas and cultural riches and in the spirit of "if you build it, they will come," he shared plans to build a new, modern airport capable of receiving tourist flights directly from Bali. Earlier that year, the government had declared the Toraja highlands Indonesia's eleventh "emerging tourism location" and announced preparations for a new airport to accelerate international and domestic tourist arrivals (Susanty, 2017).

Now, three years later, during a pandemic that has strangled international tourism, the lead author's cell phone pings almost daily with social media messages from Toraja friends documenting construction of the region's new Buntu Kuni Airport and, more recently, footage of trial flights and an airport "tour" for local guides (Figure 1). While the gleaming new airport has generated a hopeful buzz amongst tourism-tethered Torajans, at present, the planes' primary passengers are government officials and wealthy Torajans: tourists constitute just a handful of those boarding the flights. Yet, a Toraja guide's recent text conveys cautious optimism about the new airport, "In the future, I think it will be better...domestic tourists will be first to come on the planes." Certainly, even if the larger planes lack the hoped-for tourists, new flights facilitate visits from government officials, who may subsequently sponsor relief aid for pandemicimpacted Toraja tourism sector workers.

The COVID-19 pandemic has brought tourism to a grinding halt throughout the world. While tourism scholars have highlighted the impacts of the COVID-19 pandemic, this scholarship is often based on theoretical rather than empirical experience.

Figure 1. The new Toraja Airport, September 2020. (Photo courtesy of Daud Tangjong).

Additionally, we still know much less about what factors contribute to resilience among tourism practitioners, businesses, and communities whose livelihoods are now on hold. Thus, we begin with this story of a newlyopened airport (operational since September 2020) envisioned as a tourism panacea because it touches on several of the themes advanced in this article.

Drawing on literature at the intersection of critical tourism studies and resilience theory, as well as conversations with local tourism practitioners and academics, this paper accounts for how the fallout of the COVID-19 pandemic has pushed millions of tourism practitioners' lives and livelihoods into limbo as they wait for tourists to return, and yet, despite being in a state of limbo, many are demonstrating extraordinary resilience. We home in on this resilience and identify four ongoing tourism resiliency trends in Southeast Asia: livelihood diversification, ecosystem regeneration, cultural revitalization, and domestic tourism. Via this examination, we argue that resilience theory would benefit from an accounting of the historically situated nuances of local scale dynamics and macro-level processes. Rather than simply spotlighting the pandemic's sudden transformative effects, we argue that current practices of resilience in Southeast Asia reflect ongoing political-economic and cultural shifts that have often been underway in the region for several decades. These shifts, we further contend, will be critical to the reshaping of Southeast Asia's (post-) pandemic tourism industry. After a brief review of resilience theory in tourism studies, we examine the COVID-19 crisis vis-á-vis tourism in Southeast Asia and outline our methodology. Subsequent sections highlight current regional trends and their historical and socio-economic contexts. We conclude with a series of placebased policy implications and suggestions for future research addressing the intersection of tourism, COVID-19, and socio-economic change in Southeast Asia.

Resilience theory in tourism studies

Originating from the Latin term, *resilire*, 'to leap back' (Bec et al., 2015), resilience is commonly understood as the ability to build capacity (Gallopin, 2006) and the capacity to rebound from adverse events (Ledesma, 2014). With deep roots in medicine, psychology, engineering and education (Masten & Obradovic, 2006), resilience theory was developed by an ecologist, C. S. Holling (1973), and introduced into ecological literature to explain the non-linear dynamics observed in ecological systems. Resiliency theory has since been linked to socialecological systems which recognize the role of human action in resiliency and accounts for social contexts such as a community (Bec et al. 2015). Due to its diverse academic origins, multiple definitions of resilience exist, yet all share a view of systems as dynamic and constantly adapting to changes (Masten & Obradovic, 2006). Resilience is defined here as "the capacity of a system to absorb disturbance and reorganize while undergoing change so as to retain essentially the same function, structure, identity and feedbacks" (Walker et al. 2004, p. 6).

Within tourism studies, resilience research has largely focused on economic resilience (Lew, 2014), short-term disasters and hazards (Bec et al.2015), highly vulnerable systems (Coaffee & Wood, 2006; Larsen, et al. 2011), or long-term climate change (Dogru et al. 2019). Groundbreaking research on tourism resiliency tends to be either predominantly conceptual (Bec et al, 2015) or case study-oriented, while broader theoretical constructs are rarer (Lew, 2014). Additional significant areas of resilience-oriented research include communities, policy and planning, and sustainable development (Hall, 2018). More recently, scholars have sought to understand how community resilience can be used as a tool for responding to and managing long-term structural change and environmental change, through changes in regulations, policy, and laws (Moyle et al. 2010). Its intuitive appeal suggests the urgent need for new frameworks and applications to meet immediate life-saving and sustaining needs (Hall, 2018; Lew, 2017).

Given its relevance for analyzing how systems deal with and overcome crisis, the concept of resilience has been applied widely by scholars from multiple disciplines during this global pandemic. In their systematic review of 35 recently published papers about tourism in the wake of this pandemic, Sharma et al., (2021) identified resilience as one of the most prominent themes. Their analysis and subsequent proposal of a resilience-based framework for the global tourism industry post-COVID 19 has a broader focus on business/industry resilience within the context of organizational studies. Building on this scholarship, we propose a framework that accounts for the nuances of experiences of resilience among local tourism practitioners. Our approach is rooted in grounded qualitative research and envisions local individuals and communities, not as passive victims, but as active agents in resilience.

Methods

This article draws from critical tourism studies (CTS), which distinguishes itself from positivistic tourism research by highlighting tourism's entanglement with neoliberal development, socio-political inequities, and classic paradigms' privileging of some voices over others (Atelievic et al., 2007; Bianchi, 2009; Tribe, 2007). Scholars of CTS work to decolonize tourism scholarship and foster social justice. We also combined several qualitative research methods of data collection. As Bernard (2006) notes, participant observation research entails spending lengthy periods of time conducting on-site research, engaging in daily activities, observing, and documenting mundane and extraordinary events, all while taking extensive research notes. While the pandemic hinders our ability to plant ourselves in our respective field sites in Indonesia, Thailand, and Vietnam, we incorporated our long-term research experience and drew on our ongoing personal communications with collaborators in once heavily-toured Southeast Asian communities.

Collectively, the authors have conducted over 66 years of field research in various Southeast Asian nations. Each of us draws from earlier participant observation and in-field interviews on tourism-related themes to inform our understanding of long-term trends that extend well beyond the current COVID-19 era. As qualitative researchers widely note, ethnographic methods (participant observation, casual spontaneous conversations, etc.) require a level of rapport and trust typically cultivated over years rather than weeks. In this sense, the authors are particularly well-positioned to follow-up with personal contacts—many of whom we consider to be friends—to elicit candid perspectives on pandemic-induced challenges and changes. Such insights might not be possible with short-term or survey-based methods (Adams, 2012; Cole, 2004; Pelto, 2017). While all "data" are subjective, in qualitative research in the CTS tradition, positionality and reflexivity mediate both the data collection and analysis processes. Thus, we use the term, "friends" to indicate our long-term, close relationships with our research collaborators.

Our primary data collection included informal conversations and semi-structured interviews with 16 purposefully chosen research participants between June and September of 2020. We primarily used video and audio calls via WhatsApp, Zoom, and Facebook Messenger to communicate with our participants who are represented here using pseudonyms to protect privacy. Our research participants include tour guides, tour operators, artisans, homestay owners, local hotel managers, tourism consultants, NGO practitioners, and tourism academics in Indonesia, Myanmar, Thailand, Vietnam, the Philippines, and Singapore. We asked participants a series of similar, semi-structured questions such as "what are the long and short-term implications of COVID-19 on tourism?" and "what are some of the ways people in your community are dealing with the loss of tourism revenue?" This method enabled us to learn about local community-level initiatives. Significantly, these local initiatives are not typically covered by regional, national, or international media nor visible on news media beyond those communities. The primary data were further triangulated with secondary data including news media, NGOs reports, webinars, and academic publications. We analyzed materials collected

COVID-19 PANDEMIC-ACCENTUATED PLACE-BASED TRENDS IN SOUTHEAST ASIAN TOURISM			
Livelihood Diversification	Ecosystem Regeneration	Cultural Revitalization	Domestic Tourism
Souvenir crafters reorient to local markets	Catalyzation of political support for new limits on carrying capacity at National Parks	Development of minority cultural products and performances	Government-funded domestic tourism stimulus plans
Tourism businesses shift develop e-commerce	Development of environmental policy that is more appropriate to local religious and spiritual beliefs	Tourism labor migrants' return to rural homelands and reinvigorate ancestral subsistence patterns and cultural practices	Local community enjoyment of previously foreign-oriented tourism facilities
Development of virtual tours and experiences	Public recognition of ecological limits of overemphasis on growth-oriented tourism	Revitalization of grassroots cultural initiatives	Rural tourism sites serve citizens' needs for socially-distanced leisure

Figure 2. COVID-19 pandemic-accentuated place-based trends in Southeast Asian tourism. (source: authors).

using thematic analysis, identifying recurrent and notable themes in our data (Bernard, 2006). Through this process, we identified livelihood diversification, ecosystem regeneration, cultural revitalization, and domestic tourism development as core themes (Figure 2).

COVID-19, tourism, and crises in Southeast Asia

Home to some of the most popular tourism destinations in the world, the Southeast Asian region welcomed 129 million arrivals in 2018 and the tourism sector constituted 12.6% of GDP and approximately 1 out of every 10 jobs in the region (UNESCO, 2020). By early 2020, the COVID-19 pandemic slowed tourism to a trickle, resulting in numerous challenges for those whose livelihoods depend on the industry. While in 2019, the industry's 4.6 percent annual growth rate outpaced the global average of 3.5 percent (WTTC, 2020), today, restricted movement, quarantines, and travel anxiety have paralyzed the industry. Scholars have documented some of the specific health and community impacts of COVID in the region (Foo et. al., 2020; Schmidt-San et. al. 2020; Yuniti et. al. 2020), and a remarkable number of articles have addressed the potential for COVID-19 to reset tourism for a more sustainable future (e.g. Brouder, 2020; Galvani et al., 2020; Ioannides & Gyimóthy, 2020; Nepal, 2020; Romagosa, 2020). Yet, in Southeast Asia, there is still much to be learned by attending to how place-based trends influence post-pandemic tourism (Fauzi & Paiman, 2020).

Southeast Asia boasts numerous celebrated tourism destinations. Bangkok, with over 22 million international tourists per year, was recognized as the most visited city in the world for the last four consecutive years. Bali's fame as the quintessential exotic island destination dates back to the Dutch colonial era (Kodhyat, 1996) and the island

Figure 3. Empty streets at tourist hot spot, a Tuk Tuk driver is waiting for customers. If he is lucky, he could earn some money for the day, Chiang Mai, Thailand, March 23, 2020. (Photo courtesy of Jittrapon Kiacome).

has enjoyed such magnetism as a vacation destination that, by the 1970s, consultants warned of mass tourism's threats to local cultural "vitality" (Adams, 2018a; Picard, 1996). Likewise, the region's profusion of UNESCO World Heritage Sites (i.e. Cambodia's Angkor, Java's Borobudur, and the rice terraces of the Philippine Cordilleras) have drawn ever-growing numbers of domestic and international tourists, boosting revenue possibilities for local entrepreneurs. Today, tourism plays a significant role in the economies of most countries in the region, presenting numerous challenges as well as opportunities for post-crisis recovery.

Southeast Asia is no stranger to disasters. In the past two decades alone, it has been affected by several epidemics such as the 2002-2004 SARS outbreak and the 2014-2015 Avian influenza (Chan & King, 2019). Similarly, disasters including the 2004 Indian Ocean Tsunami, 2017s eruptions of Mount Agung on Bali and the 2008 Cyclone Nargis that struck Myanmar's Irrawaddy delta region have disrupted local lives. Despite some inroads in containing the current virus in certain Southeast Asian nations (Beech & Dean, 2020), the COVID-19 pandemic has been an unprecedented crisis for the region's tourism industry with an estimated loss of US$34.6 billion in 2020 (PATA, 2020). (Figure 3)

While the COVID-19 pandemic situation is rapidly changing, as of October 2020, a few (though not all) countries in Southeast Asia such as Vietnam (1,177 cases) and Thailand (3,775 cases) were described as COVID-19 successes (Beech & Dean, 2020; Jones, 2020). However, this success was tentative and the potential for an additional wave of transmissions has created fear and anxiety, both of contagion and travel (Paddock, 2020). Notably, testing capacities vary by countries and population sizes are relatively disparate (rendering the case totals difficult to compare). For example, the World Health Organization and medical experts in Indonesia noted that low-cost test kits and false rapid testing used to screen domestic tourists have surged COVID cases in Bali (Aljazeera, 2020).

Thus, while the accuracy of testing numbers may be questionable, currently, Myanmar has identified 1,518 cases, Cambodia has 274, Brunei has under 150 and East Timor and Laos each reported under 30 cases (Dong et al., 2020). However, despite these relatively low official counts, Malaysia (9,459 cases) and Singapore (57,044 cases) have fared less well and Indonesia and the Philippines (with 197,000 and 239,000 cases respectively) (Dong et al., 2020), have been hard hit by the pandemic. In what follows, we address several popular tourism destinations in Southeast Asia. We recognize that given the region's extraordinary diversity, ongoing shifts in COVID-19 situations, uneven testing abilities and reporting, and variations in respondents' abilities to speak candidly, there are inherent gaps in our portrait of current dynamics in the region. Nevertheless, we offer an initial review of the relationship between historically rooted trends in Southeast Asia and the experience of disaster as well as resilience (Dayley, 2019).

Findings and discussion

Livelihood diversification

The concept of 'diversified livelihoods' is frequently used by development studies scholars to emphasize the complex economic realities of people living in rural areas. Ellis identifies livelihood diversification "as the process by which rural families construct a diverse portfolio of activities and social support capabilities in order to survive and to improve their standards of living" (1998, p.1). For many rural communities throughout Southeast Asia, tourism is a key activity in this portfolio, as it encourages the creation of diverse inter-sectoral linkages such as agriculture and artisan production to support the tourism value chain (Phi & Whitford, 2017). In Southeast Asia, individuals as well as communities with histories of diversified livelihood strategies demonstrate resilience relative to their less-diversified counterparts. Never having become fully dependent on tourism (Phi, 2020) or reorienting tourism-honed skills to non-tourist markets (Adams 2018b), some individuals and communities across the region are tapping into alternative income streams such as cattle rearing, weaving, and field plantations (Ha, 2020).

In the cultural tourism destination of Toraja (Sulawesi), Indonesia, where one of the author's research has been based for several decades, souvenir-makers faced with the COVID-19 induced evaporation of tourist customers have shifted to crafting goods for local markets. Returning to an adaptive strategy initiated in the early 2000s when political instability, Avian influenza, and SARs outbreaks prompted dramatic decreases in tourism, carvers who previously earned livelihoods sculpting souvenir statues and decorative trinkets are now crafting new-genre coffins incised with Toraja designs (Adams 2018b). Likewise, weavers, painters and batikers who once sold their products in tourist shops are now producing protective masks embellished with Toraja motifs: their chief clients are other Torajans (Figure 4).

Significantly, some of these livelihood diversification strategies entail new non-face-toface approaches to business practices. Along with the rise of technological

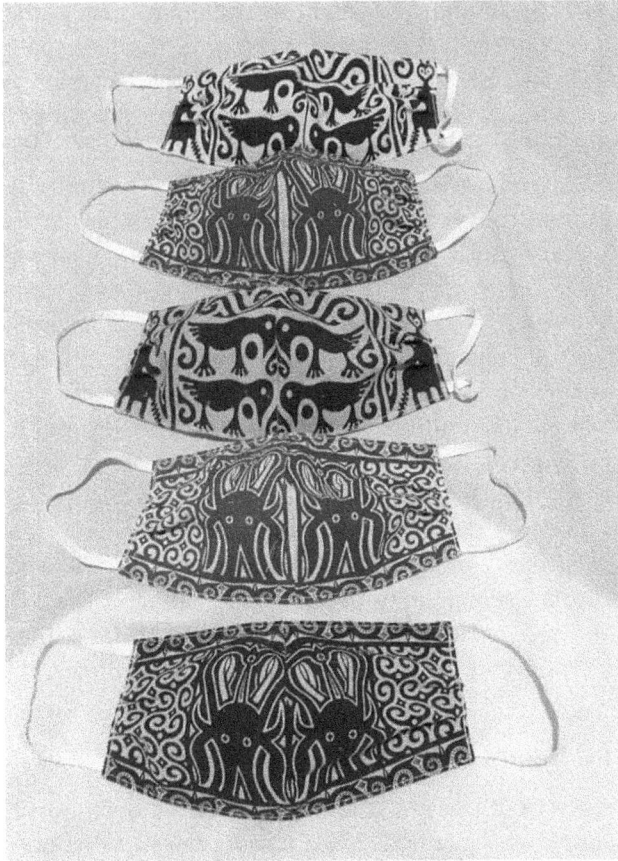

Figure 4. Masks designed by F. Pongsamma, with Toraja motifs. (Photo courtesy of F. Pongsamma).

innovations and the widespread adoption of internet/mobile phone services, many rural communities have embraced e-Commerce to sell local products both to others residing locally as well as to cyber-tourists in distant countries. For instance, in Myanmar, Gewa, a 26-year-old Karen community-based tourism consultant explained to one of the authors how, since COVID-19, many former tourism practitioners have left the industry and founded new small businesses such as translation services and social media marketing firms in Yangon. Additionally, many struggling tourism entrepreneurs and Airbnb hosts now make use of AirBnB's Southeast Asian "online/virtual experiences" platform. Virtual tourists can experience batik painting with Malaysian artisans, explore the "trail of *Crazy Rich Asians*" in Singapore, enjoy "peaceful temple life" in Bangkok, undergo "spiritual awakening" via a Balinese blessing ritual, and gain a "cultural appreciation" of Vietnam via coffee-making lessons. Soon after COVID-19's arrival, the Indonesian travel company Jakarta Good Guide began staging virtual tours of central Javanese cities (Wira, 2020). Likewise, the Singapore-based tour operator Monster Day Tours shifted its focus to interactive virtual gaming tours for Singaporeans and foreign groups. As an employee explained, "we knew young people were interested in gaming and tours, and were exploring this... [COVID-19] could be a new beginning...where less privileged, elderly and disabled people from all over the world can visit Singapore virtually". Online and virtual experiences are partially orchestrated by government and tourism authorities. For instance, the Singapore Tourism Board curates and/or funds an array of online tourism experiences featuring local sites and characters. Some of these were initiated in pre-pandemic times but have since blossomed into stand-alone online experiences.

Beyond the private and governmental tourism sectors, online experiences are also utilized by social enterprises to support their missions. For instance, Friends in Bali (a Balinese-based tourism social enterprise that uses tourism revenue to cross-subsidy local charities) now offers an assortment of online Balinese experiences (e.g., cooking classes and batik making), and donates US$2 from each purchase to local families thereby supporting the collective resilience of Balinese communities.

Thus, in the wake of the COVID-19 pandemic, diverse economic practices have contributed to resilience among former tourism actors. This includes many established yet largely overlooked practices that, in pre-COVID-19 times, formed part of the emerging array of economies of the region (especially in indigenous communities). These practices highlight a diversity of income creation beyond the capitalist, industrialised structures of direct tourismbased employment (Cave & Dredge, 2020). They also highlight the robust community development networks that have operated for several decades in most areas of the region. In times of crisis, these organizations have served as an important buoy in an otherwise turbulent sea.

Ecosystem regeneration

Historically, numerous Southeast Asian destinations have been hotspots of 'overtourism' such as Thailand's now infamous Maya Bay, Indonesia's Komodo Island, and the Philippines' Boracay (Erb, 2015; Koh & Fakfare, 2019). Journalists and academics alike have widely noted how the COVID-19 triggered tourism pause has provided opportunities for ecosystems to recuperate from decades of uninterrupted tourist flows (Crossley, 2020). For instance, unhindered by tourist boats, marine ecosystems are flourishing in Phuket, Thailand where rarely seen species of sharks, dolphins and whales are now being spotted with increasing frequency. Inland national parks have also reported the return of tigers and leopards, while elephants are beginning to thrive in destinations like Khao Yai National Park (Regan & Olarn, 2020). In many places, the pandemic catalyzed political support for new policies that reduce 'overtourism' and its corollary ecological impacts. Yet, in numerous countries this support was already growing for decades (Forsyth, 2002). The Thai Minister of National Resources and Environment, for example, had announced the closure of national parks for two months each year, a decision largely derived from evidence of ecological regeneration following park closures.

In Buddhist regions of Southeast Asia (Thailand, Myanmar, Cambodia, Laos and Vietnam), religious philosophy reminds populations of the importance of living mindfully in alignment with Nature and developing compassion for human and non-human beings (Gross, 1997). Similarly, in predominantly-Hindu Bali, efforts are underway to reimagine tourism not as "an industrial production line but a living, networked system embedded in a natural system called Nature and subject to Nature's operating rules and principles" (Pollock, 2019, p. 7).[1] Hinduism's key belief

'Ahimsa' ('the principle of non-violence') encourages humans to respect the natural world as all life forms are sacred parts of God, while 'Karma' ('consequences resulting from one's actions) further encourages individual responsibilities towards ethical conduct and environmental protection (Mittal & Thursby, 2009). Bali's recently-proposed $10 tax on foreign tourists to fund environmental and cultural programs (Regan & Olarn, 2020) embraces local perspectives of the destinations' ecosystems (Cheer, 2020). In this way,

Crossley's contention that the COVID-19 pandemic may reveal "expressions of environmental hope" (2020, p. 542) materializes in a decidedly Southeast Asian form.

Yet, tourism site closures often go hand-in-hand with layoffs and, as desperation mounts, closures may trigger a return to historically reliable livelihoods such as illegal logging and, fishing thereby threatening the environmental gains seen in the initial months of the pandemic

(Fabro 2020; Poole, 2020). In the Philippines, the pandemic has "hit the reset button" for Palawan, impacting thousands of families who had shifted from fishing and farming to tourism sector work (Fabro, 2020). Boat-operators who once carried tourists to the island's UNESCO World Heritage sites have experienced some of the greatest revenue losses forcing many to return to other resource-depleting income generating activities. Additionally, national parks throughout the region now face significant funding gaps because of lost ecotourism revenue which in many areas has contributed to illicit activities such as animal poaching and logging in protected areas. These controversial impacts of COVID-19 on Southeast Asian ecosystems reflect both the ecological limits of growth-oriented tourism (Cheer et al., 2019, p. 554) and the risks of tourism-financed conservation (Fletcher et al., 2020). Moving from conventional support for "trickle-down environmentalism," where conservation follows income generation, the 'political ecology of tourism' paradigm demonstrates local communities' tacit knowledge of traditional livelihoods enabling them live symbiotically with the ecosystem (Broad & Cavanagh, 2015; Mostafanezhad et al, 2016). These examples illustrate how pandemic-triggered ecosystem regeneration exists within a broader historical and place-based context of livelihood diversification and bubbling political support for more sustainable forms of tourism recreation.

Further, many communities have long practiced a range of indigenous environmental conservation strategies that are now being reprioritized in the face of the current pause on tourism flows.

Cultural revitalization

Cultural revitalization is on the rise throughout Southeast Asia. In many cases, this growth was spearheaded by tourism practitioners who returned to their rural homelands and, in others, cultural tourism enterprises have been replaced—at least temporarily—by creative grassroot initiatives. This section highlights three dynamics pertaining to cultural revitalization in the COVID-19 era. First, throughout the region, cultural revitalization goes hand-in-hand with domestic cultural tourism and is often prompted by direct government interventions. In

Myanmar, the first phase of the COVID-19 Tourism Relief Plan is strongly focused on reopening the country's pagodas and cultural sites for domestic tourists (UNESCO,

2020). While for Western tourists these sites may simply be "attractions," domestic tourists tend to view these as pilgrimage destinations and sites for reconnecting with heritage (Singh, 2009). These contrasting visions of are noteworthy, underscoring the continuation and revitalization of historic sociospiritual mobility patterns.

In Bali, cultural revitalization is being encouraged by Governor Wayan Koster, whose plans for economic diversification entail developing new traditional and creative products. One product on the roster is *arak*, Bali's traditional liquor. Although the sale and marketing of *arak* within Bali was recently legalized, a local community group had worked for several years the tradition to be consumed and respected. As a representative of this group lamented to one of the authors, "[*arak*] is our Balinese traditional drink, yet these Bali hotels only serve foreign alcoholic beverages like Heineken and Vodka, and that is wrong." COVID-19-accelerated government efforts to jump-start the economy by promoting heritage products are apparently reviving pride in Balinese products that once held second-class status vis á vis foreign prestige brands.

Second, the pandemic has reversed historical trends of rural-to-urban migration. The shuttering of tourism businesses prompted thousands of tourism migrant workers to return to their ancestral farms and fishing villages. In tandem with this reverse migration, we observe a notable revitalization of cultural practices, particularly those related to subsistence activities (Laula & Paddock, 2020). In Indonesia's Toraja highlands, for example, former tourism-sector workers now not only grow and harvest traditional foods on ancestral plots, but post social media of themselves learning to paint and batik Toraja scenes and symbols in their leisure time.

Southeast Asian countries are ethnically diverse with a myriad of languages and cultures

(e.g., Vietnam, Thailand and Myanmar have 54, 70 and 135 different groups respectively, and Indonesia and the Philippines officially recognize 633 and 175 ethnic groups, respectively) (Wijeyewardene, 1990). However, most ethnic minority groups with homelands in remote areas, have long experienced cultural domination and been targeted for cultural assimilation via schooling and economic development projects in their homelands (Croissant & Trinn, 2009; Winichakul, 1994). This reverse migration pattern is particularly important in the Southeast Asian context because it facilitates new pathways for revitalizing minority ethnic cultures.

Third, many grassroot cultural revitalization initiatives have emerged, albeit with contemporary twists. In Malaysia, a theatre group that enjoyed past support from the nation's Ministry of Tourism, Arts and Culture is now producing online traditional shadow puppet performances to warn Malaysians about COVID-19 (Maganathan, 2020). Likewise, a small business representative in Ubud, Bali reports that the parking lot of Ubud's Monkey Forest (a major tourist attraction) has been partially re-purposed to host entertainment events for locals, such as the 'Bali Revival Festival.' As he explained to one of the authors, "this is good for keeping livelihoods going and to keep people motivated. Even small initiatives and events all count!"

These examples reveal the potential for local cultural revitalization through new practices that challenge centralized planning norms and defy neoliberal, corporate models (Carr, 2020). Southeast Asia's (except Thailand) broader historical context entails a long history of colonialism, through which tourism was initially introduced and local identities and ethnic relations were constructed (Hitchcock, 2009). Contemporary tourism continues to see both national and ethnic minority cultures negotiated via globalization processes, often commodified by powerful actors ranging from transnational tourism corporations and the media to statedirected policies (Picard & Wood, 1997). As such, these sprouting grassroot cultural initiatives should be viewed as part of a long history of local efforts to reclaim and protect cultural identities and lifeways. As various scholars have demonstrated, such efforts by minority cultures to rearticulate and reframe touristic representations of their cultural identities pre-date COVID19 (Adams 2006; Kahn 2011), yet the pandemic has provided a new framework for these efforts.

Domestic tourism development

While domestic tourists have long outnumbered international tourists in Southeast Asia (c.f. Adams, 1998), governments have disproportionately focused on international tourism and tourist facilities are often geared toward foreign tourists (Singh, 2009). Today, many Southeast Asian governments have refocused their efforts on growing domestic tourism. Hilda (a German tourism NGO founder and long-time Yangon resident) notes that some popular destinations are already experiencing domestic tourist fueled 'overtourism', especially during holidays and long weekends. Similarly, Gewa, a community-based tourism consultant, describes how "There is now domestic tourism and a lot of local people are enjoying it at the moment surprisingly!! …

Tourism now seems to focus and rely on domestic ones so far". Dayu, (a homestay association representative in Ubud, Bali) echoed these sentiments noting how many hotels and resorts that previously catered to tourists now promote dining and entertainment (e.g. pool use) programs for Balinese residents. Another Balinese respondent who is enjoying staycations said, "To be honest, I love this situation so much. There are no traffic jams, and things are much cheaper now!"

These observations reflect the rapid impacts of multiple pandemic-prompted state policies and initiatives to support domestic tourism. For instance, while Indonesia reopened popular tourist destinations such as Bali for domestic tourists in July, 2020 (Juniarta, 2020), its doors are now closed to foreign tourists. Similarly, the Singapore Tourism Board, invested S$45 million to encourage residents to take local holidays through the 'Singapoliday campaign' (Min, 2020), while the Thai government committed 22.4 billion-baht (US$723 million) to a similar domestic tourism stimulus plan. These funds subsidize accommodations, transport, food, and attractions to support domestic tourist hotspots.

Indeed, many areas of Southeast Asia are experiencing a rise in domestic rural and community-based tourism as operators shift towards catering to demands for socially distanced leisure and recovery from pandemic anxiety (Glusac, 2020; Saengmanee, 2020). A Vietnamese respondent noted that even Hop Thanh, a remote commune in Vietnam's Northern mountains, still receives steady flows of tourists during the pandemic from the adjunct city of Lao Cai. He believed that "there is unlimited potential to develop rural and community-based tourism targeting the urban citizens of Vietnam's 87 cities." Domestic tourism has long been envisioned as both an antidote to dependence on foreign tourists and a nation-building strategy (Picard & Wood, 1997; Werry, 2011). Indonesia's focus on domestic tourism dates back to the 1970s when

President Suharto promoted it as an avenue for cultivating citizen's love of country (Adams, 1998) and critiques arose regarding the exclusive focus on foreigners as Bali's economic lifeblood (Picard, 1997). Similarly, James Guild recently critiqued news reports bemoaning COVID19's potentially devastating impact on Bali as foreign tourists vanished. Guild notes that, "this idea that Bali will die without tourists comes uncomfortably close to a White Saviour narrative, implying that local people have no choice but to hunker down and endure this crisis until foreigners start showing up again to rescue them" (2020). As he argues, this framing "strips Indonesians of their agency in rising to meet this challenge, something they are quite capable of doing and have done many times before" (2020). Indeed, previous studies have shown that during the global financial crisis of 2007-2008, domestic tourism in Southeast Asia grew faster than in North America and Europe (Singh, 2011). This reflects the broader shift in leisure mobility due to the expansion of the middle-class in the region. Relatedly, proximity tourism encourages people to become 'tourists' in their own 'backyards,' exploring or reconnecting with places closer to home (Romagosa, 2020). Such 'staycations' can help both reduce tourists' risks and channel pent-up travel demands towards supporting tourism businesses at home (Phi, 2020).

While domestic tourism is seen as a viable path to partial recovery in many countries, Singaporean tourism academic, Tian, explained to one of the authors that "For a country with a resident population of only 5.8 million, it would be hard pressed to imagine that domestic tourism can fill the gap in Singapore for what used to be 19.11 million international visitor arrivals in 2019." On the other hand, Tian further notes how "Perhaps never before have this many middle and upper income Singaporeans remained within the country for such an extended period of time": COVID-19 has challenged what was once a long-held perception that leisure is something you do outside of Singapore (Cheong & Sin, 2019). In short, while domestic tourism has a long history in Southeast Asia, dating to at least the 1970s (Hitchcock, 2009), the pandemic has prompted some Southeast Asian governments to refocus their efforts on growing this sector as an antidote to the lost revenues from international tourism.

Overall, this section has demonstrated a wide range of resilience strategies utilised by Southeast Asian tourism practitioners, communities, and governments throughout the COVID-19 pandemic. Whilst some are grassroot and organic, others are top-down and led by the governments' funding and policies. Regardless of their nature, these resilience strategies flourished because of (1) stakeholders' capability to recognise opportunities within crises, (2) comprehensive existing/emerging networks that enable the flow of knowledge exchange and (3) coordination/collaboration among the diverse stakeholders involved (Sharma et al., 2021). These factors reflect the protective factor model of resilience which contributes to fostering positive outcomes for individuals and/or systems despite averse external circumstances (Ledesma, 2014).

The emergence of these resilience strategies also reflects the ongoing political-economic and cultural shifts that have been underway in the region for several decades. While the COVID19 pandemic has shed light on vulnerabilities, we recognize that local scale-communities and individuals are active agents in resilience. To better understand the broader issues and multiple dimensions such as livelihood diversification, ecosystem regeneration, cultural revitalization, and domestic tourism development, as well as small-scale resilience, our findings call for a placebased resilience framework. We believe such an approach expands current understandings of the interactions between local scale dynamics and macro level processes.

Place-based policy implications of COVID-19 for Southeast Asia

Significantly, the four trends outlined above reveal how the seemingly-novel resilience responses to COVID-19 are, in actuality, historically rooted and place-based. By attending to the nuances of how resilience to the pandemic is enacted and experienced, we can develop more appropriate and effective post-disaster tourism recovery plans. For example, while tourism is often prescribed as a panacea for local/regional development and poverty alleviation, local communities often face challenges integrating tourism markets and depend on maintaining diversified livelihoods practices (Phi, 2020). The rise of social entrepreneurship in the region may offer opportunities to strengthen links with organizations that have long supported local livelihoods (Biddulph, 2018; Laeis & Lemke, 2016). When further supported by government policies, these types of initiatives can help foster the emerging generation of tourism social entrepreneurs (Phi & Whitford, 2017).

The pandemic has reopened debates about the meaning and politics of sustainable tourism. As Southeast Asian governments and tourism authorities reassess industry priorities, local voices may come to the fore demonstrating the potential of domestic tourism to foster social entrepreneurship and diversified livelihoods for rural residents (Phi & Dredge, 2019). Frameworks for these policies can be drawn from emerging literature on inclusive tourism, which focuses on supporting economic and social inclusions of marginalized populations through practices such as improving access to tourism as producers and consumers, facilitating self-presentations, and challenging dominant power relations (Scheyvens & Biddulph, 2018).

The replacement of voluntary measures with governmental environmental policies that mitigate overtourism can dramatically improve the region's ecosystems (Cheer, 2020). Similarly, these policies should limit reliance on tourism as the key source to fund environmental conservation. Instead, redistributive mechanisms like a conservation basic income can be funded through direct taxation of extractive economic activities, including tourism (Fletcher et al., 2020). For example, long before the pandemic, the government of Bhutan introduced the 'high value, low impact' policy in which tourists are charged US $250 daily while in the country, $65 of which funds education, health and environmental care (Phi, 2019).

Finally, in the (post)-pandemic period, many destinations would benefit from developing domestic tourism in order to enhance community resilience in the face of future disasters and border closures. By adjusting pricing strategies and diversifying tourism products, these policies will not only support destinations' resilience but also contribute towards reducing the social inequity inherent in the current global political economy of tourism. In a similar vein, these efforts may help mitigate the further development of neo- or post-colonial tourism experiences (Hitchcock, 2009). Such moves articulate with recent trends towards local governance in some

Southeast Asian countries. For instance, Indonesia's efforts to decentralize governance (including aspects of tourism planning) by giving outlying provinces and regencies more autonomous decision-making authority means that community members can potentially secure new opportunities to shape tourism planning in their own locales.

Conclusion: future research directions on (post-) pandemic tourism in Southeast Asia

Over five decades ago, American anthropologist Marshal Sahlins (1972) described how crisis tends to reveal the structural contradictions of the modes of production. In this article, we outlined four trends that have proliferated throughout Southeast Asia in the wake of the COVID19 crisis including livelihood diversification, ecosystem regeneration, cultural revitalization, and domestic tourism development. We contend that these trends highlight how the political economy/ecology of tourism in the region has both challenged as well as facilitated opportunities for reshaping tourism in (post-) pandemic times. These interconnected trends—we further argue—should not be understood in silo but rather as historically rooted and place-based experiences. For instance, shifting to emphasize domestic tourism can potentially support the revitalization of minority ethnic cultures and solidarity movements. It may also enable future cultural exchanges between domestic ethnic groups.

In a region that just a year ago welcomed 129 million international arrivals, the COVID19 pandemic has unquestionably triggered catastrophic impacts for millions of residents. In our communications with tourism practitioners in the region, many described being laid off, filing bankruptcy, and ongoing searches for new livelihoods amid economic crises. Yet, we caution that reportage of these effects may inadvertently condone a "theatre or pornography of violence" (Scheper-Hughes & Bourgois, 2004, p 1), or what we term pandemic porn. That is, by focussing exclusively on clinically chronicling the COVID-19 era's economic violence to the tourism sector, we may obscure the numerous examples of agency that are also present. In this vein, alongside tragedy, we have highlighted examples of resilience among Southeast Asian residents and argued that understandings of the implications of COVID-19 on tourism, lives and livelihoods requires a place-based account of local experiences and contexts. Though resilience theory in tourism studies is still in its infancy, our article helps validate resilience theory's application to tourism studies.

To conclude, we offer several departure points for future research concerning COVID19's implications for increasingly precarious Southeast Asian tourism sector lives and livelihoods. First, scholars may consider how COVID-19 prompted urban to rural migration contributes to the revitalization of cultural practices as well as how it reshapes rural demographics. Scholars may investigate creative initiatives undertaken by these return-migrants, whose prior involvement in tourism lends them insights into potential new products with regional, national, and global market appeal, should the pandemic travel-dampening effects linger. Also, given demand for social distancing and psychological recuperation, scholars may consider the role of rural villages in tourism recovery efforts. Additionally, scholars may address how the heightened focus on domestic rather than international tourism departs from but also echoes dynamics chronicled in pre-pandemic ethnographies of tourism destinations, a theme largely overlooked by tourism officials, planners and with some notable exceptions, academics. The socio-economic implications of livelihood diversification in the region and the extent to which it is a mechanism for economic and ecological resilience in disaster times is another topic of critical importance. Scholars may also investigate how former tourism sector workers returning to rural homelands may carry tourism knowledge that can foster successful new domestic tourism ventures in rural areas, ranging from palm wine brewing tours to indigenous dance lessons. Additionally, more research is needed on residents' perspectives on rejuvenated ecosystems and how they may challenge and/or reinforce their understandings of the ecological limits of tourism development. Finally, future research on COVID-19 would benefit from diversification of scholarship and literature beyond the Western-centric interpretations and methods (Adams, 2020; Chang, 2019; Sin et al, 2020; Tucker & Hayes, 2019). In the midst and aftermath of the COVID-19 pandemic, this will provide more culturally relevant tourism knowledge(s) and practices for Southeast Asia in what is now widely dubbed the Asian Century.

Note

1. Although spirituality's political deployment merits further scrutiny (Roth & Sedana, 2015).

Acknowledgements

The authors would like to thank the interviewees for sharing their insights. In addition, we thank the following individuals for allowing us to reproduce their photographs in this article: Jittrapon Kaicome, Franz Pongsamma, Daud Tangjong. Finally, we thank the anonymous reviewers and journal editors for their valuable feedback on earlier versions of this article.

ORCID

Kathleen M. Adams (iD) http://orcid.org/0000-0001-5041-076X

References

Adams, K. M. (1998). Domestic tourism and nation-building in South Sulawesi. *Indonesia and the Malay World, 26*(75), 77–97. https://doi.org/10.1080/13639819808729913

Adams, K. M. (2006). *Art as politics: Recrafting identities, tourism and power in Tana Toraja, Indonesia.* University of Hawai'i Press.

Adams, K. M. (2012). Ethnographic methods. In L. Dwyer, A. Gill, & N. Seertaram (Eds.), *Handbook of research methods in tourism: Qualitative and quantitative methods* (pp. 339–351). Edward Elgar.

Adams, K. M. (2018a). Revisiting "Wonderful Indonesia": Tourism, economy and society. In R. Hefner (Ed.), *Routledge Handbook of Contemporary Indonesia* (pp. 197–207). Routledge.

Adams, K. M. (2018b). Local strategies for economic survival in touristically volatile times: An Indonesian case study of microvendors, gendered cultural practices, and resilience. *Tourism Culture & Communication, 18*(4), 287–301. https://doi.org/10.3727/109830418X15369281878422

Adams, K. M. (2020). What western tourism concepts obscure: Intersections of migration and tourism in Indonesia. *Tourism Geographies.* https://doi.org/10.1080/14616688.2020.1765010

Aljazeera. (2020, September 7). Bali COVID-19 surge blamed on inaccurate rapid tests for visitors. *Aljazeera.* https://www.aljazeera.com/news/2020/09/bali-covid-19-surge-blamed-inaccurate-rapid-tests-visitors-200907052734606.html

Bec, A., McLennan, C., & Moyle, B. D. (2015). Community resilience to long-term tourism decline and rejuvenation: A literature review and conceptual model. *Current Issues in Tourism, 19*(5), 431–457. https://doi.org/10.1080/13683500.2015.1083538

Bernard, H. R. (2006). *Handbook of methods in cultural anthropology: Qualitative and quantitative methods.* Alta Mira Press.

Bianchi, R. (2009). The 'critical turn' in tourism studies: A radical critique. *Tourism Geographies, 11*(4), 484–504. https://doi.org/10.1080/14616680903262653

Biddulph, R. (2018). Social enterprise and inclusive tourism. Five cases in Siem Reap, Cambodia. *Tourism Geographies, 20*(4), 610–629. https://doi.org/10.1080/14616688.2017.1417471

Broad, R., & Cavanagh, J. (2015). Poorer countries and the environment: Friends or foes? *World Development, 72*, 419–431. https://doi.org/10.1016/j.worlddev.2015.03.007

Brouder, P. (2020). Reset redux: Possible evolutionary pathways towards the transformation of tourism in a COVID-19 world. *Tourism Geographies, 22*(3), 1–7. https://doi.org/10.1080/14616688.2020.1760928

Carr, A. (2020). COVID-19, Indigenous peoples and tourism: A view from New Zealand. *Tourism Geographies, 22*(3), 491–502. https://doi.org/10.1080/14616688.2020.1768433

Cave, J., & Dredge, D. (2020). Regenerative tourism needs diverse economic practices. *Tourism Geographies, 22*(3), 1–11. https://doi.org/10.1080/14616688.2020.1768434

Chang, T. C. (2019). 'Asianizing the field': Questioning critical tourism studies in Asia. *Tourism Geographies,* 3–17. https://doi.org/10.1080/14616688.2019.1674370

Cheer, J. M. (2020). Human flourishing, tourism transformation and COVID-19: A conceptual touchstone. *Tourism Geographies, 22*, 1–11. https://doi.org/10.1080/14616688.2020.1765016

Cheer, J. M., Milano, C., & Novelli, M. (2019). Tourism and community resilience in the Anthropocene: Accentuating temporal overtourism. *Journal of Sustainable Tourism, 27*(4), 554–572. https://doi.org/10.1080/09669582.2019.1578363

Cheong, Y. S., & Sin, H. L. (2019). Going on holiday only to come home: Making happy families in Singapore. *Tourism Geographies,* 1–22. https://doi.org/10.1080/14616688.2019.1669069

Coaffee, J., & Wood, D. (2006). Security is coming home: Rethinking scale and constructing resilience in the global urban response to terrorist risk. *International Relations, 20*(4), 503–517. https://doi.org/10.1177/0047117806069416

Cole, S. (2004). Shared benefits: Longitudinal research in Eastern Indonesia. In J. Phillimore & L. Goodson (Eds.), *Qualitative research in tourism: Ontologies, epistemologies and methodologies* (pp. 292–310). Routledge.

Croissant, A., & Trinn, C. (2009). Culture, identity and conflict in Asia and Southeast Asia. *Asien, 110*(S), 13–43. https://doi.org/10.1.1.531.5748

Crossley, É. (2020). Ecological grief generates desire for environmental healing in tourism after COVID-19. *Tourism Geographies, 22*, 1–10. https://doi.org/10.1080/14616688.2020.1759133

Dayley, R. (2019). *Southeast Asia in the new international era.* Routledge.

Dogru, T., Marchio, E. A., Bulut, U., & Suess, C. (2019). Climate change: Vulnerability and resilience of tourism and the entire economy. *Tourism Management, 72*, 292–305. https://doi.org/10.1016/j.tourman.2018.12.010

Dong, E., Du, H., & Gardner, L. (2020). An interactive web-based dashboard to track COVID-19 in real time. *The Lancet Infectious Diseases, 20*(5), 533–534. https://doi.org/10.1016/S1473-3099(20)30120-1

Ellis, F. (1998). Household strategies and rural livelihood diversification. *Journal of Development Studies, 35*(1), 1–38. https://doi.org/10.1080/00220389808422553

Erb, M. (2015). Sailing to Komodo: Contradictions of tourism and development in Eastern Indonesia. *Austrian Journal of South-East Asian Studies, 8*(2), 143–164. https://doi.org/14764/10.ASEAS-2015.2-3

Fabro, K. (2020). *No tourism income, but this Philippine community still guards its environment.* Mongabay. https://news.mongabay.com/2020/04/no-tourism-income-but-this-philippinecommunity-still-guards-its-environment/

Fauzi, M. A., & Paiman, N. (2020). COVID-19 pandemic in Southeast Asia: Intervention and mitigation efforts. *Asian Education and Development Studies, 10*(2), 176–184. https://doi.org/10.1108/AEDS-04-2020-0064

Fletcher, R., Büscher, B. E., Koot, S. P., & Massarella, K. (2020). Ecotourism and conservation under COVID-19 and beyond. *ATLAS Tourism and Leisure Review, 2*, 42–50.

Forsyth, T. (2002). What happened on "The Beach"? Social movements and governance of tourism in Thailand. *International Journal of Sustainable Development, 5*(3), 326–337. https://doi.org/10.1504/IJSD.2002.003756

Gallopín, G. C. (2006). Linkages between vulnerability, resilience, and adaptive capacity. *Global Environmental Change, 16*(3), 293–303.

Galvani, A., Lew, A., & Perez, M. (2020). COVID-19 is expanding global consciousness and the sustainability of travel and tourism. *Tourism Geographies, 22*(3), 1–10. https://doi.org/10.1080/14616688.2020.1760924

Glusac, E. (2020, June 24). The New Escapism: Isolationist Travel. *New York Times.* https://www.nytimes.com/2020/06/24/travel/camping-outdoors-resorts-viru.html

Gross, R. M. (1997). Toward a Buddhist environmental ethic. *Journal of the American Academy of Religion, 65*(2), 333–353. https://doi.org/10.1093/jaarel/65.2.333

Guild, J. (2020, May 26). Have reports of Bali's death been greatly exaggerated? New Mandala: New Perspectives on Southeast Asia. https://www.newmandala.org/have-reports-of-balis-deathbeen-greatly-exaggerated/

Ha, N. (2020). *Homestay CBT remains resilience during the pandemic COVID-19.*https://theleader.vn/cac-homestay-cbt-dung-vung-giua-dai-dich-COVID-19-1587804923373.htm

Hall, M. (2018). Resilience in Tourism: Development, theory, and application. In M. J., Cheer & A. Lew (Eds.), *Tourism, resilience, and sustainability: Adapting to social, political and economic change* (pp. 18–33). Routledge.

Hitchcock, M. (Ed.). (2009). *Tourism in Southeast Asia: Challenges and new directions.* Nias Press.

Holling, C. S. (1973). Resilience and stability of ecological systems. *Annual Review of Ecology and Systematic, 4*, 1–23.

Ioannides, D., & Gyimóthy, S. (2020). The COVID-19 crisis as an opportunity for escaping the unsustainable global tourism path. *Tourism Geographies, 22*, 1–9. https://doi.org/10.1080/14616688.2020.1763445

Jones, A. (2020, May 15). Coronavirus: How 'overreaction' made Vietnam a virus success. *BBC.* https://www.bbc.com/news/world-asia-52628283

Juniarta, W. (2020, August 2). Bali reopens for domestic tourists with modest ceremony. *Jakarta Post.* https://www.thejakartapost.com/news/2020/07/31/bali-reopens-for-domestic-tourists-withmodest-ceremony.html

Kahn, M. (2011). *Tahiti beyond the postcard: Power, place, and everyday life*. University of Washington Press.

Kodhyat, H. (1996). *Sejarah Pariwisata dan Perkembangan di Indonesia*. Gramedia Lembaga Studi Pariwisata Indonesia.

Koh, E., & Fakfare, P. (2019). Overcoming "over-tourism": The closure of Maya Bay. *International Journal of Tourism Cities, 6*(2), 279–296. https://doi.org/10.1108/IJTC-02-2019-0023

Laeis, G. C. M., & Lemke, S. (2016). Social entrepreneurship in tourism: Applying sustainable livelihoods approaches. *International Journal of Contemporary Hospitality Management, 28*(6), 1076–1093. https://doi.org/10.1108/IJCHM-05-2014-0235

Larsen, R. K., Calgaro, E., & Thomalla, F. (2011). Governing resilience building in Thailand's tourism-dependent coastal communities: Conceptualising stakeholder agency in social– ecological systems. *Global Environmental Change, 21*(2), 481–491. https://doi.org/10.1016/j.gloenvcha.2010.12.009

Laula, N., & Paddock, R. (2020, July 20). With tourists gone, Bali workers return to farms and fishing. New York Times. https://www.nytimes.com/2020/07/20/world/asia/bali-tourismcoronavirus.html

Ledesma, J. (2014). Conceptual frameworks and research models on resilience in leadership. *Sage Open, 4*(3), 1–8. DOI: 2158244014545464. https://doi.org/10.1177/2158244014545464

Ledesma, J. (2014). Conceptual frameworks and research models on resilience in leadership. *Sage Open, 4*(3), 1–8. https://doi.org/10.1177/2158244014545464

Lew, A. A. (2014). Scale, change and resilience in community tourism planning. *Tourism Geographies, 16*(1), 14–22. https://doi.org/10.1080/14616688.2013.864325

Lew, A. A. (2017). Tourism planning and place making: place-making or placemaking?. *Tourism Geographies, 19*(3), 448–466. https://doi.org/10.1080/14616688.2017.1282007

Maganathan, D. (2020, May 8). Theatre veteran uses wayang kulit to warn people about lingering dangers of COVID-19. *The Star*. https://www.thestar.com.my/lifestyle/culture/2020/05/08/theatre-veteran-uses-wayang-kulit-towarn-people-about-lingering-dangers-of-COVID-19

Masten, A. S., & Obradovic, J. (2006). Competence and resilience in development. *Annals of the New York Academy of Sciences, 1094*(1), 13–27. https://doi.org/10.1196/annals.1376.003

Min, C. (2020). *S$45 million tourism campaign launched urging locals to explore Singapore*. https://www.channelnewsasia.com/news/singapore/singaporediscovers-45-million-tourismcampaign-stb-singapoliday-12952932.

Mittal, S., & Thursby, G. (Eds.). (2009). *Studying Hinduism: key concepts and methods*. Routledge.

Mostafanezhad, M. (2020). Covid-19 is an unnatural disaster: Hope in revelatory moments of crisis. *Tourism Geographies, 22*, 1–7. https://doi.org/10.1080/14616688.2020.1763446

Mostafanezhad, M., Norum, R., Shelton, E. J., & Thompson-Carr, A. (Eds.). (2016). *Political ecology of tourism: Community, power and the environment*. Routledge.

Moyle, B. D., Croy, W. G., & Weiler, B. (2010). Community perceptions of tourism: Bruny and magnetic Islands. *Asia Pacific Journal of Tourism Research, 15*(3), 353–366. https://doi.org/10.1080/10941665.2010.503625

Nepal, S. K. (2020). Travel and tourism after COVID-19–business as usual or opportunity to reset?*Tourism Geographies, 22*, 1–5. Online First. https://doi.org/10.1080/14616688.2020.1760926

Paddock, R. (2020, July 25). Vietnam, lauded in coronavirus fight, has first local case in 100 days. *New York Times*. https://www.nytimes.com/2020/07/25/world/asia/coronavirusvietnam.html

PATA. (2020). *PATA does the virus impact numbers and they're not pretty to read*.https://www.travelweekly-asia.com/Travel-News/Travel-Trends/PATA-does-the-virus-impactnumbers-and-theyre-not-pretty-to-read

Pelto, P. (2017). *Applied ethnography: Guidelines for field research*. Routledge.

Phi, G. T. (2020). Development-tourism and poverty alleviation: Towards an integrative framework. In F. Brandão, Z. Breda, R. Costa & C. Costa (Eds.), *Handbook of research on role of tourism in achieving the sustainable development goals* (pp. 20–41). IGI Global.

Phi, G. T., & Dredge, D. (2019). Collaborative tourism-making: An interdisciplinary review of co-creation and a future research agenda. *Tourism Recreation Research, 44*(3), 284–299. https://doi.org/10.1080/02508281.2019.1640491

Phi, G. T., & Whitford, M. (2017). Social entrepreneurship in Viet Nam. In *Managing growth and sustainable tourism governance in Asia and the Pacific* (pp. 132–139). UNWTO.

Phi, G. T. (2019). Tourism is not sustainable. Conscious travel may be? *Sydsvenskan.* https://www.sydsvenskan.se/2019-06-02/fragan-ar-om-medvetet-resande-kan-gora-skillnadeller-om-det-bara-ar-den-senaste-modenycken

Phi, G. T. (2020). *Staycation & backyard tourism: Silver linings for tourism recovery in the COVID-19 pandemic?*https://destination-review.com/en/staycation-backyard-tourism-silverlinings-for-tourism-recovery-in-the-COVID-19-pandemic/

Picard, M. (1996). *Bali: Cultural tourism and touristic culture.* Archipelago Press.

Picard, M., & Wood, R. (Ed.). (1997). *Tourism, ethnicity, and the state in Asian and Pacific societies.* University of Hawaii Press.

Pollock, A. (2019, Feb 6). Flourishing beyond sustainability the promise of a regenerative tourism. *Presentation to ETC workshop in Krakow.* https://etccorporate.org/uploads/2019/02/06022019_Anna_Pollock_ETCKrakow_Keynote.pdf

Regan, H., & Olarn, K. (2020). Before the virus, Asia's ecosystems were buckling under overtourism. When the tourists return, it has to be different. https://www.cnn.com/travel/article/southeast-asia-overtourism-coronavirus-intl-hnk/index.html

Romagosa, F. (2020). The COVID-19 crisis: Opportunities for sustainable and proximity tourism. *Tourism Geographies, 22*, 1–5. https://doi.org/10.1080/14616688.2020.1763447

Roth, D., & Sedana, G. (2015). Reframing *Tri Hita Karana*: From 'Balinese culture' to politics. *The Asia Pacific Journal of Anthropology, 16*(2), 157–175. https://doi.org/10.1080/14442213.2014.994674

Saengmanee, P. (2020, June 8). Repainting the tourism landscape. *Bangkok Post.* https://www.bangkokpost.com/life/social-and-lifestyle/1931116/repainting-the-tourismlandscape

Scheper-Hughes, N., & Bourgois, P. (Eds.). (2004). *Violence in war and peace: An anthology.* Mountaineers Books.

Scheyvens, R., & Biddulph, R. (2018). Inclusive tourism development. *Tourism Geographies, 20*(4), 589–609. https://doi.org/10.1080/14616688.2017.1381985

Schmidt-San, M., Ripoll, S., & Wilkerson, A. (2020). *Key considerations for COVID-19 management in marginalised populations in Southeast Asia: Transnational migrants, informal workers, and people living in informal settlements.*https://opendocs.ids.ac.uk/opendocs/bitstream/handle/20.500.12413/15324/SSHAP%20COVID19%20Key%20considerations%20Southeast%20Asia.pdf?sequence=1

Sharma, G. D., Thomas, A., & Paul, J. (2021). Reviving tourism industry post-COVID-19: A resilience-based framework. *Tourism Management Perspectives, 37*, https://doi.org/10.1016/j.tmp.2020.100786

Sin, H. L., Mostafanezhad, M., & Cheer, M. J. (2020). Recentering tourism geographies in the "Asian Century". *Tourism Geographies*, Online First.

Singh, S. (Ed.). (2009). *Domestic tourism in Asia: Diversity and divergence.* Earthscan.

Susanty, F. (2017, February 8). Toraja to be next top-priority tourist destination. *Jakarta Post.* https://www.thejakartapost.com/news/2017/02/08/toraja-to-be-next-top-priority-touristdestination.html

Tribe, J. (2007). Critical tourism: Rules and resistance. In I. Ateljevic, A. Pritchard, & N. Morgan (Eds.), *The critical turn in tourism studies* (pp. 51–62). Routledge.

Tucker, H., & Hayes, S. (2019). Decentring scholarship through learning with/from each 'other'. *Tourism Geographies*, 1–21. https://doi.org/10.1080/14616688.2019.1625070

UNESCO. (2020). Culture & COVID-19: Impact and response tracker. https://en.unesco.org/sites/default/files/issue_11_en_culture_COVID-19_tracker-6.pdf

Walker, B., Holling, C. S., Carpenter, S. R., & Kinzig, A. (2004). Resilience, adaptability and transformability in social–ecological systems. *Ecology and Society, 9*(2), 5. https://doi.org/10.5751/ES-00650-090205

Werry, M. (2011). *The tourist state: Performing leisure, liberalism, and race in New Zealand*. University of Minnesota Press.

World Travel and Tourism Council. "Economic Impact Reports." World Travel and Tourism Council, 6, August 2020, https://wttc.org/Research/Economic-Impact.

Wijeyewardene, G. (Ed.). (1990). *Ethnic groups across national boundaries in mainland Southeast Asia*. Institute of Southeast Asian Studies.

Winichakul, T. (1994). *Siam Mapped: A history of the geo-body*. University of Hawaii Press.

Wira, N. N. (2020, May 24). Indonesian travel organizers tap into virtual tours amid pandemic. *Jakarta Post*, https://www.thejakartapost.com/travel/2020/05/23/indonesian-travel-organizerstap-into-virtual-tours-amid-pandemic.html

Yuniti, I., Samsita, N., Komara, L., Purba, J., & Pandawani, N. (2020). The impact of COVID-19 on community life in the province of Bali, Indonesia. *International Journal of Psychosocial Rehabilitation*, *24*(10), 1918–1929.

Afterword: a critical reckoning with the 'Asian Century' in the shadow of the anthropocene

Tim Oakes

ABSTRACT
What does the advent of an 'Asian Century' portend for critical tourism geographies? This commentary argues that two recent developments have made a critical reckoning with the Asian Century more pressing than ever. First, the Asian Century is in danger of being eclipsed by the deteriorating relationship between China and the United States. And second, the COVID-19 pandemic has created unprecedented challenges for tourism throughout the world and threatens to fundamentally upend travel as we've known it for the foreseeable future. These two developments suggest a pressing need not just for a decentering of Eurocentric approaches in tourism scholarship - as has long been argued by critical tourism scholars - but for a more thorough unraveling of the politics of knowledge in tourism scholarship.

What does the advent of an 'Asian Century' portend for critical tourism geographies? While the extent to which we are indeed seeing the rise of such a Century is still debated, the empirical reality of global tourism's center of gravity shifting toward Asia is undeniable. Much of the work in tourism studies over the past decade has also shifted toward a more Asian orientation, reflecting Asia's increasing dominance in tourism markets, and the increasing presence of Asia-based scholars in tourism studies. What are the critical geographies of this shift? In what ways are we compelled to address Asia's prominence in the tourism landscape theoretically, analytically, and methodologically?

If an analytical reckoning with 'Asia on Tour' was an important and necessary step a decade ago (Winter et al., 2009), two developments have made a further *critical* reckoning more pressing than the authors of the current collection could have imagined when they were writing. First, the 'Asian Century' is in danger of being eclipsed by the deteriorating relationship between China and the United States (Loong, 2020). While the papers in this collection were all written prior to what can now be confidently viewed as the abandonment of the notion of constructive engagement that has defined the US-China relationship since the late-1970s (see Schell, 2020), they still point us toward what have emerged as key questions of concern: How does critical tourism scholarship navigate the broader geopolitical context within which the Asian Century is now being imagined and debated? To what extent will tourism thrive in an Asia increasingly polarized by an antagonistic US-China relationship? To what extent will tourism also become polarized into two separate yet competing spheres, one dominated by China, the other by the US?

The second development is, of course, the COVID-19 pandemic, which has created unprecedented challenges for tourism throughout the world and threatens to fundamentally upend travel as we've known it for the foreseeable future. The emergence of COVID-19 as a global pandemic has reshaped, but also sharpened, many of the critical issues raised in these papers. The binaries by which we have conveniently organized our scholarship – among them Western and non-Western, private and public, or human and non-human – have been scrambled by a pathogen that refuses to be contained by social organization of any kind. The ways that the pandemic has revealed, and taken advantage of, inequalities, absences of leadership, habits of discrimination, uneven patterns of connectivity, political-economic dysfunctions, and the detrimental impacts of bankrupt ideologies, provides ample evidence of the need for critical geographical perspectives on tourism as we enter a new and uncertain era.

I will argue, in this brief commentary, that these two developments suggest a pressing need for a critique of the politics of knowledge in tourism scholarship.

Flattening the hierarchies of binary thinking

Most of the papers in this collection grapple with the critical imperative of transcending the binaries and dualisms which still drive conventional approaches within tourism studies. The papers provide a collective testament to the growing realization that an Asian orientation to critical tourism scholarship cannot be satisfied by simply replacing Eurocentric theorizations and analytical approaches with Asia-centric ones. While it is still important to develop 'Asia-based' theories that derive from Asian social contexts, the more difficult critical work involves a radical decentering in which binary thinking itself is challenged on a more fundamental level. Such a decentering would unsettle and flatten the hierarchies – many of which derive from imperialist and colonial legacies – that hide within and energize binary thinking. And this flattening is, in turn, a crucial step in reckoning with a world upended by the SARS-CoV-2 virus.

Chris Gibson's (2019) call for an "integrative update" of critical tourism studies suggests some productive steps in this direction. His observation of a world "haunted by the spectre of catastrophic ecological futures, while grappling with unparalleled mobilities, new disruptive technologies, perverse combinations of financialization and austerity, and economic and geopolitical uncertainty" (p. 3) has only grown more acute with the emergence of COVID-19 and increasing geopolitical tensions between the US and China. Noting "the exceeding of planetary boundaries by humans across a range of spheres," Gibson proposes a reorganization around the cross-cutting themes of excess and volatility. This pushes the conversation about criticality beyond dualisms *per se* to a more direct confrontation with the abstract categories by which our knowledges and analyses are ordered, such as the 'economic,' the 'environmental,' or the 'cultural'. Gibson's focus on climate change and the Anthropocene thus raises more fundamental issues that speak directly to the kind of radical decentering that I have in mind and that these papers are inching toward. On one level, this is also a question of transcending binaries since the binary between humans and non-humans is perhaps the most basic of all. But on another level, our attention is drawn to the ways this binary is obliterated by the reality of the Anthropocene. Gibson's paper thus demonstrates the need for an enhanced concern with materiality, and with the ways a 'more-than-human' approach might help flatten the hierarchies underlying binary thinking. Mostafanezhad's (2020) framing of haze in northern Thailand in terms of more-than-human sociality offers one provocative example of this.

The provocations offered by Gibson and Mostafanezhad point to an overdue reckoning with 'the material turn' among critical tourism geographers. While Chang's (2019) incisive analysis of 'Asianizing the field' makes clear that an embrace of 'new materialism' (Coole & Frost, 2010) is not necessary to critically interrogate the binary thinking inherent in much tourism studies, I am particularly interested here in what purchase the material turn *does* have on a critical reckoning with the 'Asian Century.' What does a non-binary, non-dualistic criticality look like in a world turned upside down by a non-human virus? What sorts of spaces might such a criticality create and allow us to inhabit? I want to suggest that the critical imperative of transcending binaries and decentering hegemonic theory – imperatives central to 'Asianizing the field' – represent not the end goal of a critical tourism studies agenda, but merely its starting point. In the remainder of this commentary I will focus on two key issues – the COVID-19 pandemic and the increasingly confrontational relationship between the US and China – in order to demonstrate this point.

Tourism, the anthropocene, and COVID-19

It is still too early to say whether and how the tourism industry will rebound and what future tourism practices will look like in a post-COVID-19 world. But tourism is directly implicated in increasing rates of zoonosis, of which SARS-CoV-2 is but one example in an increasing number of viruses that jump from their non-human reservoirs to human hosts. Just one consideration here is the development of large-scale transportation infrastructures that make more and more of the world accessible to tourists. These infrastructures feed the industry's appetite for convenient and comfortable access to new frontiers, 'non-touristy' places, pristine and exotic locations. A recent study estimated that by 2050, 25 million kilometers of new roads would be built and that 90% of these roads would be built in developing countries, particularly in areas with exceptional biodiversity and vital ecosystem services (Laurance et al., 2014).

While the 'material turn' in human geography has been characterized by a more sustained focus on the socio-technical dimensions of infrastructure projects, highlighting for instance the 'distributive agency' of infrastructural materials (Bennett, 2010; Easterling, 2014), the COVID pandemic has brought into sharper focus how the "matter of nature" is implicated in this kind of road building (Coole & Frost, 2010, 6). Roads drive habitat loss, fragmentation, and other environmental degradations, often with irreversible impacts on ecosystems (O'Callaghan-Gordo & Anto, 2020). Zoonosis emerges as a key result of this kind of habitat loss and fragmentation, while tourism offers up an excessively mobile collection of human hosts ready to temporarily inhabit the expanding boundary regions where viruses make the jump. Economic and commercial practices that destroy natural habitats and diverse animal populations – and we can include much tourism development within these practices – enhance the frequency of zoonosis. A critical tourism geography must, then, account for the likelihood that tourism development and viral disease outbreaks will continue to accompany each other.

But the key issue here is not simply how tourism is implicated in the kinds of ecosystem disturbances that encourage zoonosis, but in how COVID-19 forces a broader reckoning in critical scholarship with the need for entirely different perspectives on what constitutes criticality in the first place. "COVID-19," O'Callaghan-Gordo and Anto (2020) tell us, "is a paradigmatic example of an Anthropocene disease." They suggest that we consider the virus "from the perspective of Planetary Health, that is, to understand that the response to the pandemic must not only be the right one for humans but also the right one for the Planet." Environmental activists have been saying this for years, of course, but what is being suggested here is something more: that 'the environment' within which humans interact does not exist on the other side of some boundary separating humans and non-humans. We are the planet, and the planet is us, and its viruses are increasingly ours as well. A 'more-than-human' planetary approach could thus be made more explicit as an agenda for critical tourism research and practice (Gibson, 2019). In these terms, tourism can no longer be thought of, if it ever was, as a merely human activity, a social activity. The critical issues must no longer be framed in human terms alone (e.g. discrimination, inequality, geopolitics, Eurocentrism) but in non-human terms as well. Meanwhile, Asia is one of the world's key regions where ecosystem disturbance, urban expansion, infrastructural development, and rapidly rising rates of consumption and leisure travel come together to make these critical issues especially important.

A focus on Asia raises another particularly thorny issue for critical tourism scholarship, and that has to do with the need to rethink the standard categories and concepts through which criticality has typically been expressed. Recent debates over how to interrogate the Asian origins of the SARS-CoV-2 virus have, for example, been mired in the human-centered framings of cultural and identity politics, with accusations of racism, essentialism, and colonialist mentalities dominating much of the critical discourse (see e.g. Carrioco, 2020; Pan, 2020; Rodenbiker, 2020; Yuen & Lung, 2020). But a more complete critical reckoning with the pandemic may require a decentering of these framings. Coole and Frost (2010, 25) argued that "it is ideological naiveté to believe that significant social change can be engendered solely by reconstructing subjectivities, discourses, ethics, and identities." And as Whatmore (2006) argued some time ago, the shift to a more-than-human mode of inquiry also entails a shift from a focus on the politics of identity to the *politics of knowledge*. Recent claims that certain cultural practices (e.g. consumption of wild animals) are to blame for the COVID-19 pandemic – and the inevitable responses wielding accusations of racism and essentialism – suggest Whatmore's shift remains overdue.

Critical inquiry into the politics of knowledge might draw our attention, then, to the socio-technical, infrastructural, and scientific dimensions of how we understand viral spread, and of how our approaches to tourism development do or do not take these factors into account. How will critical tourism scholars position themselves within the knowledge controversies swirling around viral origins, diffusions, and mitigations? How are these knowledge controversies played out within (particularly Asian) states, where tourism has been a prominent development mechanism? These are questions not simply of critical discourse or representation, but of how we engage the matter of nature in our critical analyses of the social world. More to the point, a planetary critical tourism studies needs to engage in a conversation about what constitute the elements of life, resilience, and human distinctiveness in tourism practices.

What constitutes criticality in a new era of superpower confrontation?

Another set of questions emerge when we consider the practice of critical scholarship in an era of heightened geopolitical tensions. The deteriorating relationship between the US and China – the world's two great tourism superpowers – has significant implications for critical tourism scholarship. And the COVID-19 pandemic has only intensified and exacerbated the situation. As Mostafanezhad et al. (2020) have noted, the dramatic gap in effectiveness between Asian and Euro-American responses to the pandemic seem to have heralded the arrival, finally, of the Asian Century. Indeed, the spectacular failure of the United States to contain the virus has emboldened China's claims that the world may be witnessing the emergence of a new global order. But if COVID-19 has rewritten the "geopolitical script" (Mostafanezhad et al., 2020), then what are the implications of this for critical tourism studies?

Here I would argue that we are seeing how superpower geopolitics have a tendency to narrow the parameters of critical inquiry, such that key questions get refracted through an increasingly limited set of issues. In 2020, superpower confrontation and pandemic have worked in concert to close borders and this is giving rise to 'bordered-thinking,' a new and expanded version of methodological nationalism. Just as 'pandemopolitics' (Mionel et al., 2020) closes down channels of engagement, so too can it shut down productive avenues of criticism. The dominant discourse of criticality framing the current US-China relationship, for example, is being articulated as 'academic freedom' vs 'self-censorship.' Critical thought in China, in these terms, is framed primarily around the ways scholars must watch out for certain 'invisible lines' that should not be crossed. One such line, on the geographical origins of COVID-19, has already been drawn quite clearly.

But there is no single definition of criticality, and one of the casualties of a decentered critical tourism scholarship surely must be the hegemonic 'Western' definition of what constitutes 'being critical' in the first place. Critical tourism scholarship has thus far largely been framed around a postcolonial or anticolonial positioning of 'non-Westerners' reclaiming tourism studies from its hegemonic home in the 'Western' academy. Yet that framing is threatened by an embrace in the Western academy of what I would call a much more confrontational 'politics of academic freedom,' part of which involves a new willingness to adopt a more bluntly critical stance toward the Chinese state and its increasingly aggressive practices (in, for example, Hong Kong, Xinjiang, and the South China Sea). Of course, critical scholarship has always called out state practices of violence and discrimination wherever they appear. But a geopolitical Cold War channels critical inquiry into a binary – and thus in many ways contradictory – positionality that can be difficult to overcome.

This makes it imperative for critical tourism scholarship to develop a non-binary perspective on criticality. How might criticality be conceived in a context outside of the US-China binary? How might tourism studies move beyond a binary of 'free speech' and 'self-censorship'? Answering these questions compels us to build constructive academic spaces where alternative ways of being critical can interact and thrive. Such spaces include the Critical Tourism Studies Asia-Pacific network that organized the papers for this collection, and where binary and 'bordered' thinking in tourism studies are being actively challenged. It seems imperative that such transnational networks of tourism scholars cultivate a critical politics that pushes back against nation-based framings of criticality. This means, I believe, cultivating a politics centered on the planetary dimensions of our rapidly transforming and increasingly unstable relationship with the non-human realm.

To conclude, then, the task facing critical tourism studies involves not just decentering Eurocentric academic hegemony, or transcending modernist binaries, but a broader interrogation of the politics of knowledge by which the boundaries between the human and non-human are policed and challenged. At the same time, it is crucial to recognize how attention to the politics of knowledge, in turn, reveals the need for a reset on how we conceive criticality itself.

Acknowledgement

Many thanks to Dr. Harng Luh Sin and anonymous reviewers for helpful comments on an earlier version of this paper.

Disclosure statement

No potential conflict of interest was reported by the author(s).

References

Bennett, J. (2010). *Vibrant matter: A political ecology of things*. Duke University Press.
Carrioco, K. (2020, March 25). Ming Pao row: If we learn anything from the virus outbreak, it should be the importance of free speech. *Hong Kong Free Press*. https://hongkongfp.com/2020/03/25/ming-pao-row-learn-anything-virus-outbreak-importance-free-speech/
Chan, T. C. (2019). 'Asianizing the field': Questioning critical tourism studies in Asia. *Tourism Geographies*, 1–18. https://doi.org/10.1080/14616688.2019.1674370.

Coole, D. and Frost, S. (Eds.). (2010). *New materialisms: Ontology, agency, and politics*. Duke University Press.

Easterling, K. (2014). *Extrastatecraft: The power of infrastructure space*. Verso.

Gibson, C. (2019). Critical tourism studies: New directions for volatile times. *Tourism Geographies*, 1–19. https://doi.org/10.1080/14616688.2019.1647453.

Laurance, W. F., Clements, G. R., Sloan, S., O'Connell, C. S., Mueller, N. D., Goosem, M., Venter, O., Edwards, D. P., Phalan, B., Balmford, A., Van Der Ree, R., & Arrea, I. B. (2014). A global strategy for road building. *Nature, 513*(7517), 229–232.https://doi.org/10.1038/nature13717.

Loong, L. H. (2020, June 4). The endangered Asian century: America, China, and the perils of confrontation. *Foreign Affairs*. https://www.foreignaffairs.com/articles/asia/2020-06-04/lee-hsien-loong-endangered-asian-century

Mionel, V., Negut, S., & Mionel, O. (2020). Pandemopolitics: How a public health problem became a geopolitical and geoeconomic issue. *Eurasian Geography and Economics*.

Mostafanezhad, M. (2020). Socialities of air pollution: An urban political ecology of tourism in Thailand. *Tourism Geographies*.

Mostafanezhad, M., Cheer, J., & Sin, H. L. (2020). Geopolitical anxieties of tourism: (Im)mobilities of the COVID-19 pandemic. *Dialogues in Human Geography, 10*(2), 182–186.

O'Callaghan-Gordo, C., & Anto, J. (2020). COVID-19: The disease of the anthropocene. *Environmental Research, 187*, 109683. https://doi.org/10.1016/j.envres.2020.109683.

Pan, C. (2020, May 21). Octopus and koala? Toward an anti-racist cultural politics against Yellow Perilism. Society and Space. https://www.societyandspace.org/articles/octopus-and-koala-anti-racist-cultural-politics

Rodenbiker, J. (2020, April 8). China's global reach: urban social lives of the more-than-human. Society & Space. https://www.societyandspace.org/articles/chinas-global-reach-urban-social-lives-of-the-more-than-human

Schell, O. (2020). US-China relations: the birth, life, and death of engagement. *The Wire China*. https://www.thewirechina.com/2020/06/07/the-birth-life-and-death-of-engagement/

Whatmore, S. (2006). Materialist returns: practising cultural geography in and for a more-than-human world. *Cultural Geographies, 13*(4), 600–609.

Winter, T., Teo, P., & Chang, T. (Eds.). (2009). *Asia on tour: Exploring the rise of Asian tourism*. Routledge.

Yuen, K. Y. 袁國勇, Lung, D. 龍振邦. (2020, March 18). 大流行緣起武漢 十七年教訓盡忘 [The origin of the pandemic in Wuhan: 17 years of forgotten lessons]. *Ming Pao*. https://news.mingpao.com/ins/文摘/article/20200318/s00022/1584457829823大流行緣起武漢-十七年教訓盡忘 (文-龍振邦-袁國勇)

Index

For Product Safety Concerns and Information please contact our EU
representative GPSR@taylorandfrancis.com
Taylor & Francis Verlag GmbH, Kaufingerstraße 24, 80331 München, Germany

www.ingramcontent.com/pod-product-compliance
Lightning Source LLC
Chambersburg PA
CBHW080931220326
41598CB00034B/5746

* 9 7 8 1 0 3 2 2 0 8 2 9 9 *